Springer

Milano
Berlin
Heidelberg
New York
Barcelona
Hong Kong
London
Paris
Singapore
Tokyo

S. Govoni • C.L. Bolis • M. Trabucchi (Eds)

Dementias
Biological Bases and
Clinical Approach to Treatment

 Springer

STEFANO GOVONI
Institute of Pharmacology
Faculty of Pharmacy
University of Pavia, Italy

CARLA LIANA BOLIS
Consultant Office of Research
Policy and Strategy Coordination of WHO
University of Milan, Italy

MARCO TRABUCCHI
Department of Neurosciences
University of Rome "Tor Vergata", Italy

© Springer-Verlag Italia, Milano 1999

ISBN 88-470-0048-3

Library of Congress Cataloging-in-Publication Data: Dementias: biological bases an clinical approach to treatment / S. Govoni, C.L. Bolis, M.Trabucchi (eds.). p. cm. Includes bibliographical references and index. ISBN 8847000483 1. Dementia--Physiological aspects. 2. Dementia--Treatment I. Govoni, Stefano, 1950-. II. Bolis, C.L. III. Trabucchi, Marco. [DNLM: 1. Dementia. WM 220 D376389 1999] RC521. D4578 1999 616.8'3--dc21 DNLM/DLC for Library of Congress 98-54980 CIP

Cover design: Simona Colombo (Milan)
Typesetting: Photo Life (Milan)
Printing and binding: Staroffset (Cernusco sul Naviglio, Milan)

Printed in Italy

SPIN: 10700076

Preface

To a certain extent the dementias have been forgotten diseases until just recently when they were brought to the attention of the general public and health authorities as a result of the increasing number of cases in the aging population, especially among famous people, and because of the efforts of private foundations.

The goals of the present volume are to present the dementias to health practitioners, to provide some basic information on their epidemiology and biological basis and to discuss the diagnostic and clinical problems that physicians and institutions face when caring for demented patients. This book explores the various types of dementias and is not limited to Alzheimer's disease although, as expected, more information is available and presented on this pathology. On the other hand, a few fundamental questions on dementia can only be answered through a comparison of the various forms. Examples of such questions are the following: Is the loss of cerebral tissue sufficient to cause dementia? Are there thresholds or is there a continuous progression toward the irreversible development of dementia? Are there common pathways in the dementing process? Are there common risk factors? Comparative analysis allows the common and distinctive patterns of the various dementias to be defined, ultimately leading to more focused therapeutic interventions.

This book is not a mere collection of reviews, but is based on the personal experiences of the authors who are all directly involved in dementia research or clinical care of the patient. Each chapter, however, can be read independently. In a few cases information on the epidemiology or biological basis may be redundant but, as editors, we considered that the perspective of a clinician taking care of a patient is different from that of a basic researcher, and therefore it was important to report both points of view. Finally, we included three "peculiar" chapters devoted to the care of the patient. These chapters illustrate those interventions providing care to the patient at home or in an institution, the economic burden of the illness for the family and the healthcare system, and finally, the ethical aspects. Notably, this last chapter is written by a philosopher.

We would like to close this preface with a quotation from Mother Theresa reported also in the chapter by Moyra Jones: "The biggest disease today is not leprosy or tuberculosis, but rather the feeling of being unwanted." The mission of the present book is exactly this: to help us remember through the experience of scientists, clinicians and care-givers that science and technology can cope with the needs of the demented by working together with all the people willing to recognize, care for and help the person inside the demented patient.

Finally, we wish to mention that this book would not have been possible without the financial support of several institutions and Italian divisions of pharmaceutical companies, including Bayer, Bracco, Fondazione Giovanni Lorenzini, Novartis, Pfizer, and the University of Pavia.

S. Govoni
C.L. Bolis
M. Trabucchi

Table of Contents

List of Contributors

GEORGE J. AGICH
F. J. O'Neill Chair in Clinical Bioethics, Department of Bioethics, Cleveland Clinic Foundation, Cleveland, Ohio 44122, USA

LUIGI AMADUCCI†
Italian National Research Council, Targeted Project on Aging, Via Leone Pancaldo 21, 50127 Florence, Italy

MARZIA BALDERESCHI
Italian National Research Council, Targeted Project on Aging, Via Leone Pancaldo 21, 50127 Florence, Italy

CLIVE BALLARD
MRC Neurochemical Pathology Unit, Newcastle General Hospital, Westgate Road, Newcastle upon Tyne NE4 6BE, UK

ANGELO BIANCHETTI
Geriatric Research Group, Via Romanino 1, 25122 Brescia, Italy
Department of Medicine, "Ancelle della Carità" Hospital, Cremona, Italy

FRANÇOIS BOLLER
INSERM U 324, 2ter rue d'Alésia, 75014 Paris, France

ARNE BRUN
Department of Pathology and Cytology, Lund University Hospital, 221 85 Lund, Sweden

STEPHEN T. CHEN
Department of Psychiatry and Biobehavioral Sciences, Neuropsychiatric Institute and Hospital, University of California, Los Angeles, 760 Westwood Plaza C8-849, Los Angeles, CA 90024, USA

JEFFREY L. CUMMINGS
Departments of Psychiatry and Biobehavioral Sciences and Neurology, Alzheimer's Disease Center, Reed Neurological Research Center, University of California, Los Angeles, 710 Westwood Plaza, Los Angeles, CA 90095-1769, USA

DIEGO DE LEO
Psychogeriatric Service, Department of Neurology and Psychiatry, University of Padova, Via Vendramini 7, 35137 Padova, Italy
Australian Institute for Suicide Research and Prevention, Griffith University, Nathan, Queensland 4111, Australia
IRCCS "S. Giovanni di Dio", "S. Cuore-Fatebenefratelli" Institute, Via Pilastroni 4, 25125 Brescia, Italy

ANTONIO DI CARLO
Italian National Research Council, Targeted Project on Aging, Via Leone Pancaldo 21,
50127 Florence, Italy

CRISTINA GEROLDI
Geriatric Research Group, Via Romanino 1, 25122 Brescia, Italy

EZIO GIACOBINI
Institutions Universitaires de Gériatrie de Genève, Route de Mon-Idée, 1226
Thônex-Geneva, Switzerland

STEFANO GOVONI
Institute of Pharmacology, University of Pavia, Viale Taramelli 14, 27100 Pavia, Italy
IRCCS "S. Giovanni di Dio", "S. Cuore-Fatebenefratelli" Institute, Via Pilastroni 4,
25125 Brescia, Italy

LARS GUSTAFSON
Department of Psychogeriatrics, Lund University Hospital, 221 85 Lund, Sweden

VLADIMIR HACHINSKI
Department of Clinical Neurological Sciences, London Health Sciences Centre,
University of Western Ontario, London, Ontario, Canada

DOMENICO INZITARI
Italian National Research Council, Targeted Project on Aging, Via Leone Pancaldo 21,
50127 Florence, Italy

MOYRA J.D. JONES
Moyra Jones Resources Ltd., 8264 Burnlake Dive, Burnaby, BC, Canada, V5A 3K9

PABLO MARTINEZ-LAGE
Department of Clinical Neurological Sciences, London Health Sciences Centre,
University of Western Ontario, London, Ontario, Canada
Centro Terapeutico Alzheimer, Centro Psicogeriátrico Landazabal, Burlada,
Navarra, Spain

WAYNE C. MCCORMICK
Department of Medicine, Division of Geriatrics, Box 359755, University of Washington,
Seattle, WA 98195, USA

CRISTOPHER MORRIS
MRC Neurochemical Pathology Unit, Newcastle General Hospital, Westgate Road,
Newcastle upon Tyne NE4 6BE, UK

WALTER PADOANI
Psychogeriatric Service, Department of Neurology and Psychiatry,
University of Padova, Via Vendramini 7, 35137 Padova, Italy

MARGARET PIGGOTT
MRC Neurochemical Pathology Unit, Newcastle General Hospital, Westgate Road,
Newcastle upon Tyne NE4 6BE, UK

MARCO RACCHI
Institute of Pharmacology, University of Pavia, Viale Taramelli 14, 27100 Pavia, Italy

MARCO TRABUCCHI
Geriatric Research Group, Via Romanino 1, 25122 Brescia, Italy
Department of Neurosciences, University of Rome "Tor Vergata", Rome, Italy
Alzheimer's Unit, IRCCS "S. Giovanni di Dio", "S. Cuore-Fatebenefratelli" Institute,
Via Pilastroni 4, 25125 Brescia, Italy

LATCHEZAR TRAYKOV
INSERM U 324, 2ter rue d'Alésia, 75014 Paris, France
State University Hospital of Neurology and Psychiatry, Medical University, blvd.
Tzarigradsko chosse IV km, 1113 Sofia, Bulgaria

ORAZIO ZANETTI
Alzheimer's Unit, IRCCS "S. Giovanni di Dio", "S. Cuore-Fatebenefratelli" Institute,
Via Pilastroni 4, 25125 Brescia, Italy

Dementias, the Dimension of the Problem: Epidemiology Notes

A. Di Carlo, M. Baldereschi, D. Inzitari, and L. Amaducci†

Introduction

The population of both the developed and developing world is aging due to the longer life expectancy and declining fertility. Owing to the consequent rapid growth of the elderly population, dementia has become a major problem for the health care systems and public health planning. The world population aged 60 and over was 488 million in 1990 and will be about 1,363 million in 2030, with an increase of 180%. In developed countries the number of elderly is projected to increase from 203 to 358 million during the period 1990–2030, with a percentage increase of 76.3%. Developing countries are facing an even more dramatic demographic transition: the number of elderly is estimated to grow from 286 million in 1990 to 1,005 million in 2030, with an increase of 251%. In 1990 developing countries contained 58% of the world's elderly, and in 2030 about two thirds of the total elderly population will be dwelling in these countries [1]. Given that age is the most substantiated risk factor for dementia, the aging of the population implies a growing number of persons at risk for dementia.

In 1993, the World Bank and World Health Organization analyzed the global burden of each disease in terms of Disability-Adjusted Life Years (DALYs), a unit measuring the future years of disability-free life that are lost as a result of premature deaths or disability. This standard unit represents an accurate measure of the burden of diseases and is essential for adequate planning of health care services and for the most effective allocation of resources [2].

The burden of dementia disorders has been estimated at 14 millions of DALYs for the entire world population (Table 1). Sixty-two percent of that burden is already being borne by the developing countries that are even less prepared than the industrialized world to face such a social emergency. Alzheimer's disease and the other dementias represent 2.4% of the 575.4 millions of DALYs lost in the world because of noncommunicable disorders and, in particular, 8.7% of the 161.2 millions DALYs lost as a result of noncommunicable disorders in people aged 60 years and over [1].

Epidemiological studies provide data on the incidence and age- and sex-specific prevalence of different types of dementia; they analyze the severity of dementia and the level of related disability and are thus essential for cost-effective planning of health care services in terms of assistance provided by the community and in institutions. Population-based studies, moreover, may give clues

Table 1. Burden of dementia in DALYs by sex and geographic region (hundreds of thousands of DALYs)

	India	Sub--Saharan Africa	China	Middle Eastern crescent	Other, Asia	Latin America	Formerly socialist economies	Established market economies	Demographically developing group	World
Males	9.8	4.1	12.6	4.6	6.8	4.0	5.1	16.7	41.9	63.7
Females	9.7	4.3	13.6	4.9	7.3	4.7	8.7	23.2	44.5	76.4

Source: The World Bank, World Development Report [1]

to help us understand the pathogenesis, and clinical research may help in the diagnosis and treatment of this group of diseases.

Prevalence

A large number of studies on dementia prevalence have been carried out in industrialized countries, which report roughly similar rates. The average prevalence for subjects aged 65 years and over is about 5%, ranging from 3.4% in the Lundby Study to 6.7% in the Hisayama Study (Table 2). The prevalence doubles approximately every 5 years of age, which is an almost exponential increase, at least between the ages of 65 and 84. The age-specific prevalence rates range from about 1% in subjects 65–69 years of age to 40% in the age group 85–89 years, according to the various surveys (Fig. 1).

The approximately exponential increase in the prevalence of dementia with age could lead to the conclusion that the disease may be inevitable in those who live long enough. Furthermore, some papers have claimed that dementia lies on a continuum with normal aging-related cognitive decline [19]. Data for subjects 90 years of age and over are still scanty and conflicting, showing in some studies a further increase, and in the others a plateau or a decrease in the prevalence rates

Table 2. Prevalence of dementia in people over 65 years of age

Authors	Geographical location	Prevalence
Rorsman et al. [3]	Lund (Sweden)	3.4%
Bachman et al. [4]	Framingham[a] (USA)	4.1%
Copeland et al. [5]	Liverpool (UK)	4.2%
Folstein et al. [6]	Baltimore (USA)	4.5%
Zhang et al. [7]	Shangai (China)	4.6%
Lobo et al. [8]	Zaragoza (Spain)	5.5%
Beard et al. [9]	Rochester (USA)	5.7%
O'Connor et al. [10]	Cambridge (UK)	6.0%
Rocca et al. [11]	Appignano[a] (Italy)	6.2%
Ueda et al. [12]	Hisayama (Japan)	6.7%

[a] > 60 years

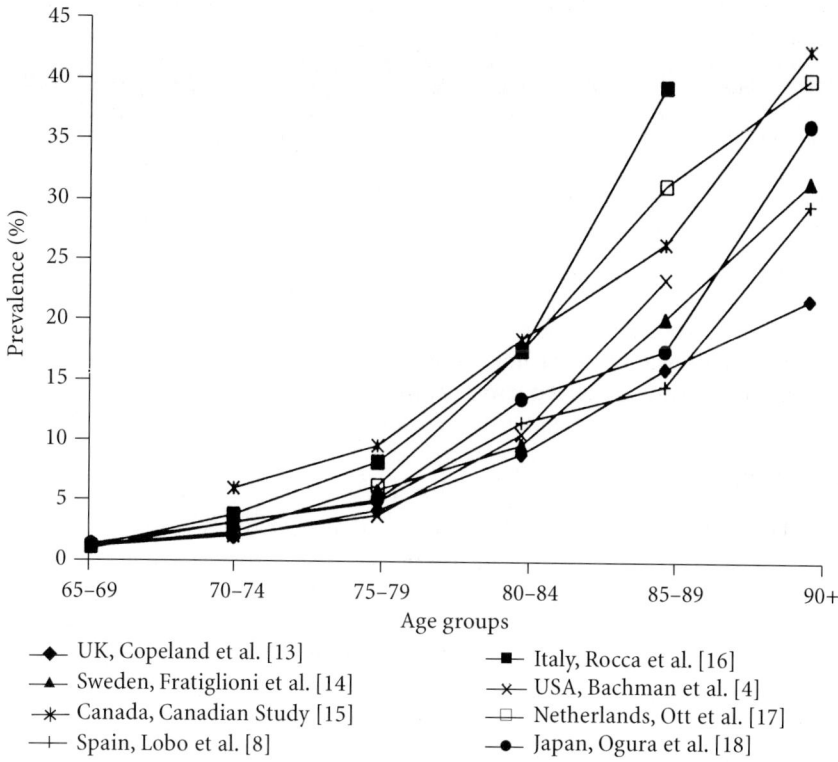

Fig. 1. Prevalence of dementia by age groups across various industrialized countries

[20]. The scarce number of subjects 90 years of age and over in the majority of the prevalence surveys may account for such discrepancies. In a recent meta-analysis the use of a logistic function was preferred to the exponential model to describe the prevalence of dementia at the oldest ages. The prevalence of dementia turned out not to continue to increase exponentially with age, but the rate of increase fell from the age 80–84 and flattened out to zero at age 95 [21]. The final model could be an "S" shaped curve, with a point of flexion at 95 years, at a 40% prevalence rate, suggesting that dementia is an age-related rather than an aging-related disorder: very elderly survivors may be at diminishing risk for dementia. Further studies are warranted to provide more information on this topic of great impact in a rapidly aging world.

The majority of the epidemiological studies focused on the estimation of *Alzheimer's disease* and vascular dementia rates, the most common types of dementia in industrialized countries, but there is conflicting evidence about the relative proportion of these diseases. Alzheimer's disease is reported to be the most frequent dementia disorder in the United States, Canada and Europe, accounting for 50%–80% of dementia cases [4, 8, 11, 14, 15, 17, 22, 23].

Vascular dementia is the second cause of dementia, with relative proportions ranging from 11% to 24% in the different surveys [8, 14, 15, 17]. In Italy, the Appignano Study [16] showed comparable rates for Alzheimer's disease and vascular dementia, and the prevalence was higher for vascular dementia in a Swedish study on people older than 85 years [24]. Population-based epidemiological surveys in Japan ranked vascular dementia as the first cause of dementia, followed by Alzheimer's disease [12, 25]. A reported higher frequency of cerebrovascular disease in Japan in comparison to other countries may partially explain these differences [26, 27].

Dementia syndrome has often been reported in *Parkinson's disease,* but its frequency varies from 35% to 90% in the different studies reviewed by Brown and Marsden [28]. Once applying the DSM-III-R criteria for dementia syndrome, these authors obtained a much lower estimate of about 15%, which is remarkable evidence of the diagnostic inconsistencies in the different surveys. Subsequent community-based studies have reported prevalences of dementia in Parkinson's disease ranging from 11% to 28% [29–31]. Dementia in Parkinson's disease is clearly age-dependent, as found also by Mayeux et al. [32]. The prevalences range from 12% in Parkinson's disease patients aged 35–64 years to 59% in patients aged 75 and over.

Over 60 different causes of dementia have been documented [33]. The most common dementing diseases and their distribution, pooled from 32 studies, are listed in Table 3. Among other causes of dementia, Creutzfeldt-Jakob's disease and HIV infection have to be mentioned. This latter disease has become a major cause of dementia among adult subjects. The prevalence of HIV dementia is 0.4% during the asymptomatic phase of HIV infection [35], raises to 7.3% in patients with

Table 3. Frequency of causes of dementia pooled from 32 studies. (Adapted from [34])

Cause	Occurrence (%)
Alzheimer's disease (AD)	57
Vascular dementia (VD)	13
Depression	4.5
Alcohol	4.2
Normal pressure hydrocephalus	1.6
Metabolic	1.5
Medications	1.5
Neoplasm	1.5
Parkinson's disease	1.2
Huntington's disease	0.9
Mixed AD and VD	0.8
Infection	0.6
Subdural hematoma	0.4
Post-trauma	0.4
Other	7.1
Not demented	3.7

AIDS [36] and may affect up to 60% of individuals in the late stages of the disease [37, 38]. The median survival of HIV patients was only 6 months after the diagnosis of dementia [39].

Cross-national differences in prevalence rates may provide intriguing clues about possible risk factors. Although many differences have already been reported, few studies have employed the same research design, so methodological problems may account for the different results. For instance, Schoenberg et al. [23] found that dementia was more frequent among blacks than among whites in the Copiah County Study, while other authors [40] claimed a relative absence of Alzheimer's disease in Nigeria both in clinical and in post-mortem series. A recent comparative cross-cultural study [41] showed that in the Nigerian town of Ibadan the prevalence of dementia was 2.3%, and the prevalence of Alzheimer's disease was 1.4%, lower than the values (4.8% and 3.7%, respectively) found in an Afro-American community of Indianapolis (USA).

Prevalence findings on dementia and Alzheimer's disease provide an important tool for understanding and "measuring" this age-related public health problem, but the search for risk factors requires incidence studies.

Incidence

In spite of the large number of studies on prevalence, there are few and often dissimilar data regarding the incidence of dementia. Incidence studies are expensive and require large samples, prolonged observation time, two or more examinations, a relatively stable population, and an accurate case ascertainment strategy, with a limited amount of subjects lost during follow-up. The definition of early cases, as well as the inclusion of mild and moderate forms of the disease, the problem of comorbidity, the use of sensitive screening instruments and of standardized diagnostic criteria and the representativeness of the study population are all other important questions to be addressed in incidence studies. These difficulties may in part explain some discrepancies found in the different surveys.

The result of an increase with age is consistent across all studies, and the annual incidence rate of dementia is estimated to be about 1% in people aged over 65 [5, 42–45], ranging from 0.2% to 0.8% in subjects aged 65–69 years, to 3% and more in subjects aged 80 and over [5, 42–48]. However, the age-specific rates may vary considerably in the different studies (Fig. 2). Differences are more striking in the oldest age groups, with rates that in subjects aged 80–84 years may range from 1.6% reported in the Framingham Study [42] to 4.6% in the Cambridge Project for Later Life [48].

Only few studies have analyzed the specific incidence of Alzheimer's disease. The overall annual incidence rate in people older than 65 years is reported to vary from 0.63% to 1.14% according to the different surveys [5, 43, 44, 47, 49], and age-specific figures range from 0.07% in the age group 65–69 to 3.3% in the age group 80–84 years [42, 44, 50–54], but with relevant differences across the studies (Table 4). As for vascular dementia, the annual incidence rate in the Lundby Study was

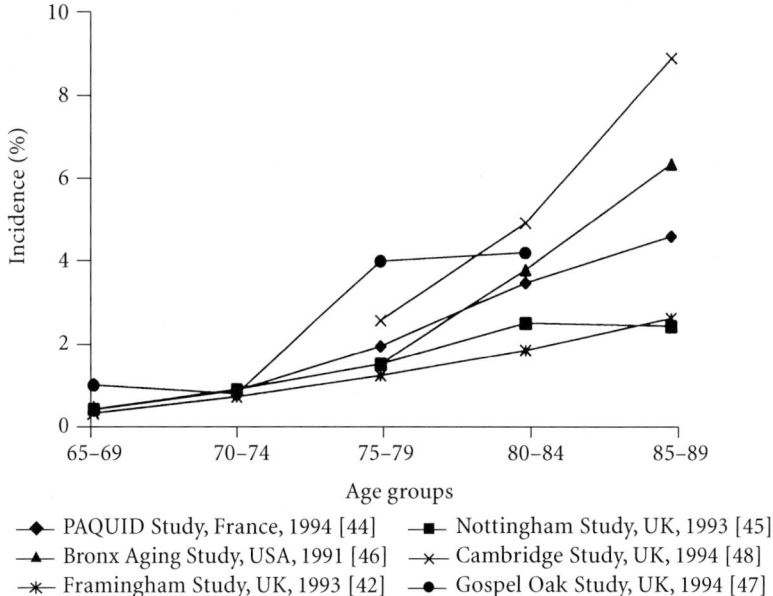

Fig. 2. Annual incidence rates of dementia by age groups in different surveys

estimated to be 0.13% for men and 0.10% for women in the total population [55]. Rates in individuals aged more than 65 years ranged from 0.19% in the Liverpool Study [5], to 0.44% in the Mannheim Study [43]. Higher values were reported in the Lundby Study in subjects aged more than 60 years, with annual incidence rates of 0.91% for men and 0.82% for women [55].

The age structure of the study sample, the sensitivity of the screening instru-

Table 4. Age-specific annual incidence rates (%) of Alzheimer's disease in different surveys

Survey	Age groups					
	60–64	65–69	70–74	75–79	80–84	85–89
Framingham Study [42]	–	0.07	0.31	0.6	1.07	1.46
PAQUID Study [44]	–	0.07	0.25	1.09	2.04	4.29
East Boston Study [51]	–	0.6	1.0	2.0	3.3	8.4
						(85 + years)
Rochester [53] – 1960–1964	0.096		0.53		1.43	
– 1970–1974	0.051		0.46		1.68	
	(60–69 years)		(70–79 years)		(80 + years)	
Lundby Study [49]	0.16		0.66		2.6	
	(60–69 years)		(70–79 years)		(80–89 years)	

ments, the number of dropouts, the duration of follow-up and the differential survival of patients with dementia may all in part explain these differences. The small sample size at the oldest ages is a limiting factor in most of the surveys. Similarly to prevalence, incidence rises markedly after the age of 65, but it is still unclear if the rates continue to increase in extreme old age, reach a plateau, or even decrease. Considering the different subtypes of dementia illnesses, further studies on the incidence of dementia, including larger numbers of subjects in the older age groups, are highly desirable. Incidence studies with longitudinal evaluations of the study sample may give a better estimate of the cognitive decline and may be less biased by confounding factors such as the education level. Accurate determination of age at onset of dementia may provide clues as to the etiology and pathogenesis, with a better description of the natural history, and identification of risk factors.

Demonstration of a decline in cognitive performances requires repeated evaluation, rather than a single demonstration of a low performance such as in prevalence studies. Moreover, the different survival rates of patients with dementia are influenced both by age and by the presence of the disease and may thus flaw the results of prevalence studies. Age-specific incidence rates are necessary to estimate the burden of demented patients in the future and the number of patients to be expected in the oldest group that will constitute a significant part of the population in the next few decades.

Risk Factors

Age is the single most-relevant risk factor for all the dementing disorders, including Alzheimer's disease. As for this risk factor, the epidemiological evidence from prevalence as well as incidence studies has already been reviewed above. One pressing issue still needs to be addressed: data on individuals over the age of 85 are insufficient to indicate whether the exponential relationship holds. Moreover, in the absence of a reliable biological marker, the diagnosis of dementia depends on cognitive and behavioral testing, whose standard errors increase markedly with age: data on the relationship between age and dementia at higher ages would theoretically require larger samples than for younger ages. The prevalence of dementia in the very old is an important issue, given that individuals aged 85 and over represent the fastest growing segment of the population in developed countries. Reliable projections of future cases need to clarify this point.

To date, analytic epidemiology for dementing diseases has used mainly case-control designs based on prevalent cases. Case-control studies are often hampered by low statistical power. To overcome this problem, the Concerted Action EURODEM Case-Control Studies Reanalysis pooled and re-analyzed data from 11 case-control studies of Alzheimer's disease [56]. Stronger evidence was related to a positive family history of dementia, a head injury with loss of consciousness, Down's syndrome in relatives and history of depression in late-onset cases.

The risk of developing Alzheimer's disease is increased three- to fourfold if one has a first-degree relative with the disease. A *family history* of dementia may indicate the presence of genetic mutations responsible for Alzheimer's disease (see below) and/or common exposure to some environmental factors. As for *head injury*, also a review by Graves et al. [57] confirmed that there is a striking consistency in regard to the strength of an association between head trauma and Alzheimer's disease. Head trauma may lead to Alzheimer's disease by accelerating the production of β-amyloid [58]. A previous myocardial infarction, reported as a possible risk factor for Alzheimer's disease in a longitudinal study [59], might act through the same mechanism [60]. The β-amyloid precursor protein contains "heat shock" promoter elements that could be activated by insults such as trauma or hypoxia [61].

The relationship between *depression* and the risk for dementia has been elucidated by a longitudinal study [62]. Depression at baseline moderately increases the risk of developing Alzheimer's disease (relative risk = 1.91; 95% confidence interval, 1.07–3.42). Whether depression is a very early manifestation of Alzheimer's disease, or increases susceptibility through another mechanism, remains to be determined.

Maternal age over 40 years at index delivery, family history of Parkinson's disease, and thyroid diseases showed only a borderline association with Alzheimer's disease [56].

Smoking habits turned out to have a negative correlation with Alzheimer's disease in various studies [63–65]. This inverse relationship might be due to an actual protective effect of nicotine which may up-regulate acetylcholine nicotinic receptors. Some authors suggest that compounds that act to stimulate nicotinic receptors may improve learning and memory in a variety of models of cognitive impairment in animals, and clinical studies have suggested positive effects on cognition of nicotine in human beings with and without Alzheimer's disease [61]. Finally, nicotine compounds might slow the progression of Alzheimer's disease, as suggested by preclinical models of cell death [66–68]. On the other hand, the association between smoking and Alzheimer's disease has been found to be absent in other recent studies [69, 70] and even positive in one study [71]. It has to be underlined that an effect of smoking may be to increase the risk of stroke in Alzheimer's disease patients, so that misdiagnosis of vascular dementia would occur. Furthermore, cigarette smoking is the single most important cause of premature death in developed countries [72], so that the differential survival between smokers and nonsmokers might bias the results of epidemiological studies.

The role of *educational attainment* is still a matter of discussion. Epidemiological studies provide conflicting evidence. The protective effect of education has been reported in various prevalence studies on dementia [7, 17, 69, 73–79], as well as in incidence studies [80, 81]. These findings provoked a discussion of possible biological mechanisms: the hypothesis is that education may increase synaptic density in the neocortex, the so-called brain reserve [82]. A reduction in hippocampal synapses [83] and in presynaptic terminals [84–86] has been observed in Alzheimer's disease patients. The onset of dementia might be delayed because education had improved neuronal networking, so that when neurons died, others could carry out similar function tasks, thus minimizing signs of functional and cognitive impairment. The capability of the brain to respond to different insults might be better pre-served in subjects with a higher education level. Not only previous education but also continued mental activities may be important for the elderly [87–89].

On the other hand, various studies have shown no association between low educational attainment and Alzheimer's disease, while the risk for developing other dementing disorders, mainly vascular dementia, has been reported to be significantly higher in the least-educated individuals [14, 80, 90–94]. This could be due to the fact that major risk factors for stroke, such as smoking and hyperten-sion, are more common in poorly educated people [95] and may contribute to the development of stroke and thus of stroke-related dementia.

In light of the above-mentioned conflicting evidence, incidence studies are rec-ommended to investigate the role of education accurately: high education level may delay the diagnosis of dementia, thus biasing the prevalence estimates.

As for *occupation*, there is one case-control study [93], and there are two preva-lence studies [74, 76] that report a significant increase in the prevalence of dementia in manual laborers. Furthermore, a longitudinal study [79] has shown that low occupation is an independent risk factor for cognitive impairment, tak-ing into account age, gender, education and history of stroke.

Findings on occupational exposure are controversial. The EURODEM meta-analysis provided negative results, while recent case-control studies have report-ed a significant increase in the risk of Alzheimer's disease for subjects exposed to glues, pesticides and several solvents. The effect of the latter type of chemical exposure could be modified by heavy alcohol consumption [96].

The potential benefit of post-menopausal *estrogen replacement therapy* in the prevention of Alzheimer's disease has been confirmed in several epidemiological studies [97–101], while only one paper gave negative results [102]. There are many mechanisms whereby estrogens might improve cognitive function: they may increase the activity of choline-acetyltransferase [103]; they may stimulate neu-ronal regeneration and modulate long- and short-term synaptic function [104, 105]. According to several pieces of scientific evidences, estrogens might also reduce ApoE production [106].

The use of *anti-inflammatory drugs* may decrease the risk of developing Alzheimer's disease. According to many studies, more than 40 proteins of the inflammatory response are accumulated in Alzheimer's disease brain lesions: cytokines, acute phase reactans, proteases and protease inhibitors. The protective role of nonsteroidal anti-inflammatory drugs is supported by several case-control studies that have recently been pooled and meta-analyzed by McGeer and McGeer [107]. The odds ratio of 0.496 (95% confidence interval, 0.343–0.716) for non-steroidal anti-inflammatory drugs, and the odds ratio of 0.656 (95% confidence interval, 0.431 to 0.999) for steroids indicated a significant negative association between Alzheimer's disease and the use of anti-inflammatory agents. These results are rather striking, but preliminary: the beneficial effect of such medica-tions must be confirmed through longitudinal observations and clinical trials.

The susceptibility to Alzheimer's disease is in part genetically determined. Our current knowledge can be summarized as follows:

- On chromosome 21 is the autosomal dominant gene for the amyloid precursor protein: so far, four mutations have been identified (20 families known to date).
- On chromosome 14 is the autosomal dominant gene for presenilin-1: 28 mutations identified so far (about 100 families known to date).
- On chromosome 1 is the autosomal dominant gene with variable penetrance for presenilin-2: three mutations identified (3 families known to date).
- On chromosome 19 is the autosomal dominant gene for apolipoprotein E (ApoE).

The first two genes are associated *only* with early-onset Alzheimer's disease.

The association between ApoE-ε4 and the risk for Alzheimer's disease was first reported in 1993 [108]. Many subsequent papers have established that ApoE genotype is the single most important genetic determinant of susceptibility to sporadic and late-onset Alzheimer's disease. ApoE is found in plasma, where it has an important role in the transport of cholesterol and the modulation of atherogenic lipoprotein metabolism. It is also produced by astrocytes in the brain, where its physiological role is less certain. The gene for human ApoE is on the long arm of chromosome 19 and exists in three common allelic forms (ApoE-ε2, ApoE-ε3, ApoE-ε4). These three alleles encode forms of the apolipoprotein that differ by amino acid substitutions at one or both of two sites, determining distinctive physical and biochemical properties for each isoform. The ApoE-ε4 allele has a frequency of about 15% in populations of European ancestry. ApoE-ε4 allele is found about three times more frequently among Alzheimer's disease patients than among age-matched controls, while ApoE-ε2 is slightly underrepresented in Alzheimer's disease. As most studies were based on prevalent cases, ApoE-ε2 carriers may have been selectively removed from the patient series over time. This may have resulted in an apparent decrease in the ApoE-ε2 allele frequency. ApoE genotype also affects the age at onset of Alzheimer's disease, with earlier onset associated with ApoE-ε4/4 and the latest with ApoE-ε2/3 and ApoE-ε2/2. ApoE-ε4 has also been shown to increase the risk of atherosclerosis, which may explain its association with vascular dementia as well. To date, no other form of dementia has been found to be associated with such a genotype [96].

The biological basis of this association is still unknown. It has been suggested that the different ApoE isoforms may differently affect amyloid deposition, tangle formation, neuronal plasticity, cholinergic functions and other biological aspects of Alzheimer's disease [109]. Subjects with an ApoE-ε4 allele can be considered to be at an increased risk for Alzheimer's disease compared to those without this allele, while the absence of ApoE-ε4 does not preclude development of the disease. One copy of ApoE-ε4 is associated with a moderate increase of Alzheimer's disease (reported odds ratios range from 2.2 to 4.4), while two copies convey a high risk (reported odds ratios range from 5.1 to 17.9). Nevertheless, some ApoE-ε4 carriers survive to old age and remain cognitively intact.

The current position on ApoE genotyping has been recently summarized in a consensus statement [110]: ApoE genotype testing should not be used in clinical routine either to diagnose Alzheimer's disease in subjects with dementia, or to screen and predict Alzheimer's disease in asymptomatic individuals. Based upon currently available data, ApoE testing alone is less accurate than standardized clinical diagnostic criteria. Furthermore, the presence of ApoE-ε4 genotype in asymptomatic individuals does not sufficiently predict the development of Alzheimer's disease, nor does absence of ApoE-ε4 genotype sufficiently eliminate the chance of developing the disease.

The discovery of autosomal dominant gene mutations in familial Alzheimer's disease has stimulated researchers not because they explain a high proportion of dementia, but because once the functions of the normal gene products are known, light may be thrown on the biological mechanisms of Alzheimer's disease, therapeutic targets being the ultimate objective. New susceptibility genes will probably be found.

The current epidemiological and biological evidence suggests that Alzheimer's disease is a chronic disease with a long preclinical period in which interventions to prevent the development of the disease might be possible. According to this proposed model, the diffuse plaques and tangles are laid down as the result of genetic, traumatic, anoxic and perhaps other events. At a certain point, the changes in the brain with Alzeimer's disease continue to progress on their own. Clinical symptoms begin to appear when the number of synapses falls below a threshold level or when acute stressors overcome the brain's capacity to respond effectively. Lack of education and later life cognitive inactivity may play a role by decreasing synaptic reserve.

There are still uncertainties regarding many putative factors. For several reasons, the use of prevalent cases limits the extent to which study findings can be used to answer the specific question of who is at risk for developing the disease and why. In such studies it is difficult to disentangle the extent to which a risk factor is related to the incidence of the disease or to survival time after the disease is diagnosed. Furthermore, conclusions drawn from case-control studies are necessarily limited by the influence of recall and selection bias. Longitudinal studies on dementia are warranted to allow a better estimation of risk factors, assess exposure prior to development of the disease, limit the effect of potentially confounding factors, allow analysis of the rate of cognitive and functional decline, evaluate comorbidity and to avoid survival bias.

Conclusions

There are still many uncertainties in regard to the evidence at hand. The assumed central role of amyloid is in dispute. Putative factors need to be confirmed prospectively. Long-term follow-up studies that are ongoing will provide reliable estimations of the role of genetic and environmental risk factors, even though nonresponse, competing mortality and comorbidity may challenge the validity of these results.

Recent epidemiological studies have led to preliminary findings that estrogen therapy and anti-inflammatory drugs may have a protective effect, which may prove to be of clinical relevance. Progress in the understanding of the genetics of

Alzheimer's disease and other types of dementia has opened new possibilities for epidemiological studies on the risk associated with these genetic factors. The risk of Alzheimer's disease associated with the various genetic factors identified, including ApoE, remains to be quantified by follow-up studies of incident cases. Moreover, the gene-environmental interaction needs to be studied, as the strength of association between an environmental factor and the risk of the disease may depend on the presence of a genetic factor and, conversely, the effect of a genetic factor may be conditional on the presence of environmental risk factors. For instance, there is some evidence of synergistic effects of ApoE and head trauma, as well as of an antagonistic effect of ApoE and smoking [96].

Given that dementia in all its forms is clearly a major issue for public health, scientific contributions are urgently required to clarify the etiology and develop therapies. In fact, many scientific research efforts are currently directed at etiology and risk factors (molecular biology and environmental exposure) and at the development of neuropharmaceutical devices to slow or reverse cognitive decline. In conclusion, epidemiological research on dementia is far from finished. Its achievements will help identify individuals at risk for the disease, as well as improve diagnostic accuracy and allow deferral of the onset or slow the progress of the disease.

The cost of dementia is high because it includes the expenses of caring for disabled individuals for a long time, and lost earnings both of patients and of those who have to quit working to look after an affected relative. Health and social services need an early solution because dementia is already burdening communities.

Acknowledgements. The authors thank Ms. Maria Elena Della Santa for her support in preparing the manuscript.

References

1. The World Bank (1993) World Development Report 1993. Oxford University Press, New York
2. Murray JLC, Lopez AD (1996) Evidence-based health policy lessons from the Global Burden of Disease Study. Science 274: 740-743
3. Rorsman B, Hagnell O, Lanke J (1985) Prevalence of age psychosis and mortality among age psychotics in the Lundby Study. Neuropsychobiology 13: 167-172
4. Bachman DL, Wolf PA, Linn R, Knoefel JE, Cobb J, Belanger A, D'Agostino RB, White LR (1992) Prevalence of dementia and probable senile dementia of the Alzheimer type in the Framingham Study. Neurology 42: 115-119
5. Copeland JRM, Davidson IA, Dewey ME, Gilmor C, Larkin BA, McWilliam C, Saunders PA, Scott A, Sharma V, Sullivan C (1992) Alzheimer's disease, other dementias, depression and pseudodementia: prevalence, incidence and three-year outcome in Liverpool. Br J Psychiatry 161: 230-239
6. Folstein MF, Bassett SS, Anthony JC, Romanoski AJ, Nestadt GR (1991) Dementia: case ascertainment in a community survey. J Gerontol 46: M132-M138

7. Zhang M, Katzman R, Salmon D, Jin H, Cai G, Wang Z, Qu G, Grant I, Yu E, Levy P, Klauber MR, Liu WT (1990) The prevalence of dementia and Alzheimer's disease in Shangai, China: impact of age, gender and education. Ann Neurol 27: 428-437

8. Lobo A, Saz P, Marcos G, Dia JL, De-la-Camara C (1995) The prevalence of dementia and depression in the elderly community in a southern European population. The Zaragoza Study. Arch Gen Psychiatry 52: 497-506

9. Beard CM, Kokmen E, Offord K, Kurland LT (1991) Is the prevalence of dementia changing? Neurology 41: 1911-1914

10. O'Connor DW, Pollitt PA, Hyde JB, Fellows JL, Miller ND, Brook CP, Reiss BB, Roth M (1989) The prevalence of dementia as measured by the Cambridge Mental Disorders of the Elderly Examination. Acta Psychiatr Scand 79: 190-198

11. Rocca WA, Hofman A, Brayne C, Breteler MMB, Clarke M, Copeland JRM, Dartigues JF, Engedal K, Hagnell O, Heeren TJ, Jonker C, Lindesay J, Lobo A, Mann AH, Mölsä P, Morgan K, O'Connor DW, da Silva Droux A, Sulkava R, Kay DWK, Amaducci L, for the EURODEM Prevalence Research Group (1991) The prevalence of vascular dementia in Europe: facts and fragments from 1980-1990 studies. Ann Neurol 30: 817-824

12. Ueda K, Kawano H, Hasuo Y, Fujishima M (1992) Prevalence and etiology of dementia in a Japanese community. Stroke 23: 798-803

13. Copeland JRM, Dewey ME, Wood N, Searle R, Davidson IA, McWilliam C (1987) Range of mental illness among the elderly in the community: prevalence in Liverpool using the GMS-AGECAT Package. Br J Psychiatry 150: 815-823

14. Fratiglioni L, Grut M, Forsell Y, Viitanen M, Grafström M, Holmén K, Ericsson K, Bäckman L, Ahlbom A, Winblad B (1991) Prevalence of Alzheimer's disease and other dementias in an elderly urban population: relationship with age, sex and education. Neurology 41: 1886-1892

15. Canadian Study of Health and Aging Working Group (1994) Canadian Study of Health and Aging: study methods and prevalence of dementia. Can Med Assoc J 150: 899-913

16. Rocca WA, Bonaiuto S, Lippi A, Luciani P, Turtù F, Cavarzeran F, Amaducci L (1990) Prevalence of clinically diagnosed Alzheimer's disease and other dementing disorders: a door-to-door survey in Appignano, Macerata Province, Italy. Neurology 40: 626-631

17. Ott A, Breteler MMB, van Harskamp F, Claus JJ, van der Cammen TJM, Grobbee DE, Hofman A (1995) Prevalence of Alzheimer's disease and vascular dementia: association with education. The Rotterdam Study. BMJ 310: 970-973

18. Ogura C, Nakamoto H, Uema T, Yamamoto K, Yonemori T, Yoshimura T and the Cosepo Group (1995) Prevalence of senile dementia in Okinawa, Japan. Int J Epidemiol 24: 373-379

19. Huppert FA, Brayne C, O'Connor D (eds) (1994) Dementia and normal ageing. Cambridge University Press, Cambridge

20. Ritchie K, Kildea D, Robine JM (1992) The relationship between age and the prevalence of senile dementia: a meta-analysis of recent data. Int J Epidemiol 21: 763-769

21. Ritchie K, Kildea D (1995) Is senile dementia "age-related" or "ageing-related"? – evidence from meta-analysis of dementia prevalence in the oldest old. Lancet 346: 931-934

22. Beard CM, Kokmen E, O'Brien PC, Kurland LT (1995) The prevalence of dementia is changing over time in Rochester, Minnesota. Neurology 45: 75-79

23. Schoenberg BS, Anderson DW, Haerer ME (1985) Severe dementia prevalence and clinical features in a biracial US population. Arch Neurol 42: 740-743

24. Skoog I, Nilsson L, Palmertz B, Andreasson LA, Svanborg A (1993) A population-based study of dementia in 85-year-olds. N Engl J Med 328: 153-158

25. Homma A, Niina R (1988) International views: research on Alzheimer's disease in Japan. Alzheimer Dis Assoc Disord 2: 366-374

26. Takeya S (1966) Epidemiological studies on cerebrovascular diseases in Hisayama, Kyushu, Japan. Fukuoka Acta Med 57: 994-1019

27. Urakami K, Igo M, Takahashi K (1987) An epidemiologic study of cerebrovascular disease in western Japan: with special reference to transient ischemic attacks. Stroke 18: 396-401

28. Brown RG, Marsden CD (1984) How common is dementia in Parkinson's disease? Lancet 1: 1262-1265

29. Aarsland D, Tandberg E, Larsen JP, Cummings JL (1996) Frequency of dementia in Parkinson disease. Arch Neurol 53: 538-542

30. Sutcliffe RL, Prior R, Mawby B, McQuillan WJ (1985) Parkinson's disease in the district of the Northampton Health Authority, United Kingdom. A study of prevalence and disability. Acta Neurol Scand 72: 363-379

31. Tison F, Dartigues JF, Auriacombe S, Letenneur L, Boller F, Alpérovitch A (1995) Dementia in Parkinson's disease: a population-based study in ambulatory and institutionalized individuals. Neurology 45: 705-708

32. Mayeux R, Denaro J, Hemenegildo N, Marder K, Tang MX, Cote LJ, Stern Y (1992) A population-based investigation of Parkinson's disease with and without dementia. Arch Neurol 49: 492-497

33. Haase GR (1977) Disease presenting as dementia. In: Wells CE (ed) Dementia. Davis, Philadelphia, pp 26-27

34. Clarfield AM (1988) The reversible dementias: do they reverse? Ann Intern Med 109: 476-486

35. McArthur JC, Cohen BA, Selnes OA, Kumar AJ, Cooper K, McArthur JH, Soucy G, Cornblath DR, Chmiel JS, Wang MC (1989) Low prevalence of neurological and neuropsychological abnormalities in otherwise healthy HIV-1-infected individuals: results from the Multicenter AIDS Cohort Study. Ann Neurol 26: 601-611

36. Janssen RS, Nwanyanwu OC, Selik RM, Stehr-Green JK (1992) Epidemiology of human immunodeficiency virus encephalopathy in the United States. Neurology 42: 1472-1476

37. McArthur JC (1987) Neurologic manifestations of AIDS. Medicine 66: 407-437

38. Price RW, Brew BJ (1988) The AIDS dementia complex. J Infect Dis 158: 1079-1083

39. McArthur JC, Hoover DR, Bacellar H, Miller EN, Cohen BA, Becker JT, Graham NMH, McArthur JH, Selnes OA, Jacobson L, Visscher BR, Concha M, Saah A for the Multicenter AIDS Cohort Study (1993) Dementia in AIDS patients: incidence and risk factors. Neurology 43: 2245-2252

40. Osuntokun BO, Ogunniyi AO, Lekwauwa UG (1992) Alzheimer's disease in Nigeria. Afr J Med Sci 21: 71-77

41. Hendrie HC, Osuntokun BO, Hall KS, Ogunniyi AO, Hui SL, Unverzagt FW, Gureje O, Rodenberg CA, Baiyewu O, Musick, BS Adeyinka A, Farlow MR, Oluwole SO, Class CA, Komolafe O (1995) Prevalence of Alzheimer's disease and dementia in two communities: Nigerian Africans and African Americans. Am J Psychiatry 152: 1485-1492

42. Bachman DL, Wolf PA, Linn RT, Knoefel JE, Cobb JL, Belanger AJ, White LR, D'Agostino RB (1993) Incidence of dementia and probable Alzheimer's disease in a general population: the Framingham Study. Neurology 43: 515-519

43. Bickel H, Cooper B (1994) Incidence and relative risk of dementia in an urban elderly population: findings of a prospective field study. Psychol Med 24: 179-192

44. Letenneur L, Commenges D, Dartigues JF, Barberger-Gateau P (1994) Incidence of dementia and Alzheimer's disease in elderly community residents of south-western France. Int J Epidemiol 23: 1256-1261

82. Katzman R (1993) Views and reviews: education and the prevalence of Alzheimer's disease. Neurology 43: 13-20

83. Hamos JE, DeGennaro LJ, Drachman DA (1989) Synaptic loss in Alzheimer's disease and other dementias. Neurology 39: 355-361

84. DeKosky ST, Scheff SW (1990) Synapse loss in frontal cortex biopsies in Alzheimer's disease: correlation with cognitive severity. Ann Neurol 27: 457-464

85. Masliah E, Terry RD, Alford M, DeTeresa R, Hansen LA (1991) Cortical and subcortical patterns of synaptophysinlike immunoreactivity in Alzheimer's disease. Am J Pathol 138: 235-246

86. Terry RD, Masliah E, Salmon DP, Butters N, DeTeresa R, Hill R, Hansen LA, Katzman R (1991) Physical basis of cognitive alterations in Alzheimer's disease: synapse loss is the major correlate of cognitive impairment. Ann Neurol 30: 572-580

87. Baltes PB (1991) The many faces of human ageing: toward a psychological culture of old age. Psychol Med 21: 837-54

88. Koh K, Ray R, Lee J, Nair A, Ho T, Ang PC (1994) Dementia in elderly patients: can the 3R mental stimulation programme improve mental status? Age Ageing 23: 195-199

89. Yesavage JA (1985) Nonpharmacologic treatments for memory losses with normal aging. Am J Psychiatry 142: 600-605

90. Amaducci L, Fratiglioni L, Rocca WA, Fieschi C, Livrea P, Pedone D, Bracco L, Lippi A, Gandolfo C, Bino G, Prencipe M, Bonatti ML, Girotti F, Carella F, Tavolato B, Ferla S, Lenzi GL, Carolei A, Gambi A, Grigoletto F, Schoenberg BS (1986) Risk factors for clinically diagnosed Alzheimer's disease: a case-control study of an Italian population. Neurology 36: 922-931

91. Beard CM, Kokmen E, Offord KP, Kurland LT (1992) Lack of association between Alzheimer's disease and education, occupation, marital status, or living arrangement. Neurology 42: 2063-2068

92. Cobb JL, Wolf PA, Au R, White R, D'Agostino RB (1995) The effect of education on the incidence of dementia and Alzheimer's disease in the Framingham Study. Neurology 45: 1707-1712

93. Fratiglioni L, Ahlbom A, Vitanen M, Winblad B (1993) Risk factors for late-onset Alzheimer's disease: a population based, case-control study. Ann Neurol 33: 258-266

94. Yoshitake T, Kiyohara Y, Kato I, Ohmura T, Iwamoto H, Nakayama K, Ohmori S, Nomiyama K, Kawano H, Ueda K, Sueishi K, Tsuneyoshi M, Fujishima M (1995) Incidence and risk factors of vascular dementia and Alzheimer's disease in a defined elderly Japanese population: the Hisayama Study. Neurology 45: 1161-1168

95. Garrison RJ, Gold RS, Wilson PW, Kannel WB (1993) Educational attainment and coronary heart disease risk: the Framingham Offspring Study. Prev Med 22: 54-64

96. van Duijn CM (1996) Epidemiology of the dementias: recent developments and new approaches. Neuroepidemiology 60: 478-488

97. Baldereschi M, Di Carlo A, Maggi S, Grigoletto F, Livrea P, Motta L, Bonaiuto S, Inzitari D, Loeb C, Canal N, Rengo F, Enzi G, Scarlato G, Amaducci L (1996) Estrogen replacement therapy and the risk of dementia in the Italian Longitudinal Study on Aging. Eur J Neurol 3: 85-86

98. Morrison A, Resnick S, Corrada M, Zonderman A, Kawas C (1996) A prospective study of estrogen replacement therapy and the risk of developing Alzheimer's disease in the Baltimore Longitudinal Study on Aging. Neurology 46: A435-A436

99. Paganini-Hill A, Henderson VW (1994) Estrogen deficiency and risk of Alzheimer's disease in women. Am J Epidemiol 140: 256

100. Tang MX, Jacobs D, Staern Y, Marder K, Schofield P, Gurland B, Andrews H, Mayeux R

(1996) Effect of oestrogen during menopause on risk and age at onset of Alzheimer's disease. Lancet 348: 429-432

101. Waring SC, Rocca WA, Petersen RC, Kokmen E (1997) Postmenopausal estrogen replacement therapy and Alzheimer's disease: a population-based study in Rochester, Minnesota. Neurology 48: A79

102. Brenner DE, Kukull WA, Stergachis A, van Belle G, Bowen JD, McCormick WC, Teri L, Larson EB (1994) Postmenopausal estrogen replacement therapy and the risk of Alzheimer's disease: a population-based case-control study. Am J Epidemiol 140: 262-267

103. Luine VN (1985) Estradiol increases choline acetyltransferase activity in specific basal forebrain nuclei and projection areas of female rats. Exp Neurol 89: 484-490

104. Gould E, Woolley CS, Frankfurt M, McEwen BS (1990) Gonadal steroids regulate dendritic spine density in hippocampal pyramidal cells in adulthood. J Neurosci 10: 1286-1291

105. Woolley CS, McEwen BS (1992) Estradiol mediates fluctuation in hippocampal synapse density during the estrous cycle in the adult rat. J Neurosci 12: 2549-2554

106. Honio H, Tanaka K, Kashiwagi T, Urabe M, Okada H, Hayashi M, Hayashi K (1995) Senile dementia-Alzheimer's type and estrogen. Horm Metab Res 27: 204-207

107. McGeer PL, McGeer EG (1996) Anti-inflammatory drugs in the fight against Alzheimer's disease. Ann N Y Acad Sci 777: 213-220

108. Strittmatter WJ, Saunders AM, Schmechel D, Pericak-Vance M, Enghild J, Salvesen GS, Roses AD (1993) Apolipoprotein E: high-avidity binding to beta-amyloid and increased frequency of type 4 allele in late-onset familial Alzheimer disease. Proc Natl Acad Sci USA 90: 1977-1981

109. Writing Committee, Lancet Conference 1996 (1996) The challenge of the dementias. Lancet 347: 1303-1307

110. National Institute on Aging/Alzheimer's Association Working Group (1996) Apolipoprotein E genotyping in Alzheimer's disease. Lancet 347: 1091-1095

The Biological Basis of Dementias

M. Racchi and S. Govoni

The Rationale for Searching the Biological Basis of Dementing Illnesses

In this chapter we review briefly the state of the art of research on the molecular mechanisms that underlie neurodegenerative dementing illnesses. There is still no cure for any of the dementing neurodegenerative diseases that lead to some form of dementia. Most of the treatments proposed for Alzheimer's disease (AD) and other dementias are, by design, effective only in ameliorating symptoms of the disease and, at best, slowing the pace of progression. The time of intervention, due to the complexity of early diagnosis, is often late with respect to the biological onset of the disease and therefore limits the final efficacy of the treatment. In neurodegenerative diseases where the progression of degeneration can be slowed, possibly stopped but theoretically not reversed, the time of intervention is an extremely important issue. Searching for the molecular and biological basis of dementing disorders provides an opportunity to identify targets of pharmacological intervention. Early biochemical markers of the onset of the disease could allow treatment at early stages; identification of genetic risk factors can direct new efforts at studying molecular targets for prevention. Clinical variability among populations of affected patients is certainly due to differences in genetic and biochemical backgrounds, and these can also account for the differences in effectiveness of therapeutic intervention. Only molecular dissection of the pathogenetic processes can provide the information needed to increase the likelihood that a specific treatment will be effective.

Although in this review we intended to examine the biological substrate of the various dementia illnesses, in the following section more emphasis has been placed on AD because of its greater prevalence and the greater abundance of studies on it that render AD the kind of model that may allow the establishment of guidelines useful for the study of all other forms of dementia.

Alzheimer's Disease

Neuropathology

The diagnostic criteria for AD (see chapter by Boller and Traykov, this volume) only allow a definite diagnosis to be made following a neuropathological examination of the brain of affected individuals. The macroscopic lesions that can be found at autopsy include marked atrophy of the cerebral hemispheres with severe involvement of associated areas of the cerebral cortex, while the motor, somatosensory, and visual cortex is relatively spared. Concerning the microanatomy, the type of neurons mostly affected by the pathology are large cortical neurons with sections > 90 μm^2, while no differences in the number of glial cells are observed. Conversely, there is a relative increase in the number of small neurons [1]. Loss of large neurons has also been demonstrated in the hippocampus and amygdala [2], in the entorhinal and visual cortex [3], in the locus ceruleus and in the nucleus basalis of Meynert [4] with a parallel relative increase in the number of small neurons and in some cases glial proliferation. The microscopic lesions that characterize the brain of AD patients are intracellular deposits, defined as neurofibrillary tangles (NFT) and extracellular agglomerates, defined as senile plaques. Neurofibrillary lesions have an intracellular localization: they are found in the neuronal soma and in the apical dendrites as NFT. Ultrastructurally, these lesions are mainly constituted by paired helical filaments (PHF) and, to a lesser extent, by straight filaments (SF). The major component of PHF is a multiple isoform phosphoprotein called "tau" protein. The function of tau is to participate in the formation of microtubules as an aid in the polymerization of tubulin. In the NFT tau protein is in a hyperphosphorylated form, and it is known that the phosphorylation of tau reduces its ability to bind and stabilize the microtubules.

The appearance and development of NFT lesions are not casual, but follow a stereotyped pattern that allows the definition of six stages for the neuropathology of AD [5, 6]. Stages I and II involve the transentorhinal region: this phase of the disease is virtually free of symptoms. The first cognitive deficits appear during stages III and IV when the neurofibrillary lesions in the transentorhinal region are more severe and the lesions begin to appear in the hippocampus, Ammon's horn and in some subcortical areas. Stages V and VI are characterized by a massive increase in the density of these lesions in the areas already affected with involvement also of the higher cortical associative areas.

Neuritic plaques are compact spheroidal extracellular deposits constituted mainly of filaments of a protein named β-amyloid or Aβ, surrounded by a variable number of dystrophic neurites. Activated microglial cells and reactive astrocytes are also found around the amyloid core of these plaques, also defined as mature plaques [7]. A number of proteins has been found in tight association with senile plaques, among which are α1 antichimotrypsin, complement cascade proteins, proteoglycans, cytokines and apolipoproteins. Besides mature plaques the brain of Alzheimer's patients also shows the presence of numerous diffuse plaques that are constituted by deposit of β-amyloid protein in amorphous form and are not associated with dystrophic neurites or activated glial cells [8]. Diffuse plaques are not detected with conventional amyloid stains such as Congo Red and thioflavin S; they are more abundant than mature plaques and involve larger areas of the cortex [9] and are present even in the cerebellum [10].

In contrast to NFT, β-amyloid deposits do not follow a definite pattern of distribution and show large individual variability. Areas where plaques are most abundant are the cerebral cortex and subcortical areas such as the caudate nucleus, putamen and thalamus. On the other hand, the areas with a higher density of NFT (olfactory nucleus, entorhinal cortex and CA1 region of the hippocampus) show the lower density of plaques. In the early stages of the disease primitive plaques are most abundant, while in the case of severe pathology the mature plaques are prevalent. A further type of amyloid deposit is observed in the intracerebral and meningeal microvessel wall: in particular, arterioles and capillaries show fibrillary deposit of β-amyloid. The vascular amyloid is prevalently located on the lumenal side of the basal lamina of the vessel. Deposits of β-amyloid immunoreactive material have been also described in the microvessel wall of extraneuronal tissues such as skin, spleen and gut [11] in AD patients, Down's syndrome and normal aging. Both NFT and senile plaques can be found in the brain of cognitively intact aged subjects. The major difference is therefore quantitative in defining their role in the development of dementia [12].

A novel type of lesion has recently been described and has so far been called "AMY" plaque since it was initially mistaken for amyloid plaques; however, the new plaque resembles the amyloid lesions in size and shape, but does not contain β-amyloid. It was discovered rather casually by the research group led by Trojanowsky and Lee [13], and the purification of the major proteinaceous component of these plaques may open new perspectives in AD research.

Neurochemistry

Neurotransmitters

Alterations of several neurotransmitter systems have been described, with the most important impairment being that of the cholinergic system. A consistent feature of AD is the degeneration of cholinergic neurons originating in the nucleus basalis of Meynert, the vertical diagonal band and septal nucleus and projecting to the cerebral cortex and the hippocampus. Acetylcholine concentration is reduced in AD brain, accompanied by a reduction of its synthesis and a deficiency of the high-affinity uptake of choline. A consistent and important alteration of the cholinergic system in AD brain is the dramatic deficit of choline acetyltransferase (ChAT) activity, which is reduced by 50%–85% in the cortex and hippocampus of the brains of AD patients. ChAT is considered the best marker of cholinergic terminals, and its reduction in AD is secondary to selective cholinergic neurodegeneration. The consistent alterations of the cholinergic system are, however, paralleled by controversial findings concerning muscarinic receptors. Some investigators have found no alterations in the number and pharmacological properties of muscarinic receptors in AD brain [14], while other authors have reported a reduced number of muscarinic receptors in the neocortex [15]. Concerning the subtypes of muscarinic receptors, it has been demonstrated that m2 receptors are selectively decreased, while m1 receptors are unaltered or in some cases reportedly upregulated, possibly as a function of compensation for the cholinergic denervation [16]. The alteration of the cholinergic system seems to be influenced also by the genetic background, in particular by the ApoE genotype (see below). In a study it has been observed that the activity of ChAT and other cholinergic parameters is reduced as a function of the number of ε4 alleles present. As a consequence of the increased cholinergic damage, the ε4/ε4 patients are more resistant to cholinergic drugs and suggest that such genetic susceptibility may divide patients into subgroups where the effectiveness of cholinergic treatment can be predicted [17].

Multiple neurochemical deficits have been reported aside from the consistent findings on the cholinergic system. Other transmitter systems that appear to be affected by the disease involve the noradrenergic, serotoninergic, GABAergic and glutamatergic systems in the cerebral cortex and the hippocampus of AD brain. In addition, alteration of peptidergic transmitters such as somatostatin, cholecystokinin, neuropeptide Y and corticotropin-releasing factor have been described in AD brain. In parallel to the severe neurochemical imbalance, various investigators have reported alterations in receptors such as 5HT, GABA, and α- and β-adrenergic receptors, without however being able to demonstrate a consistent alteration of one particular receptor type. It is clear that the selective neuronal degeneration that occurs in AD cannot be explained by a single specific deficit in neurotransmitters systems; it appears instead that there is a hierarchy of neurochemical alterations and the cholinergic system is most consistently altered in AD [16].

Transduction Mechanisms

Among the neurochemical alterations described in AD brain, one of the most complex aspects is represented by the alterations in signal-transduction mechanisms (Fig. 1). The alterations in neurotransmission in the AD brain seem to involve multiple levels, including modified receptor responses because of altered G-protein coupling, second messenger, synthesis, and protein kinases activation [18].

Guanine nucleotide regulatory binding proteins (G-proteins) are the first link between the activation of many receptors and the modulation of intracellular effector systems. The coupling of muscarinic receptors to G-proteins has been reported to be disrupted in AD cerebral cortex. More generally, a widespread, defective G-protein-stimulated adenylate cyclase activity has been reported. These alterations appear to be specifically correlated with defective G-protein coupling with their receptors [19, 20].

Receptor activation triggers the hydrolysis of phosphoinositides (PI), generating diacylglycerol and inositol trisphosphate (IP_3). These molecules serve as second messengers for the mobilization of calcium from internal stores and for the

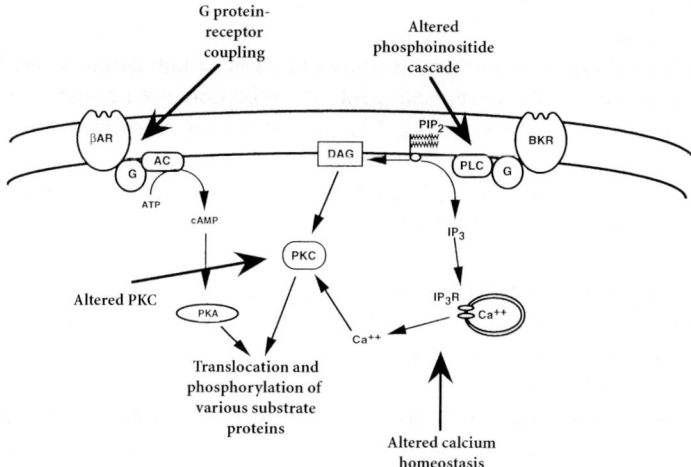

Fig. 1. Examples of altered signal transduction pathways in AD brain and peripheral tissues. Alteration of coupling of G-protein-associated receptors with their effectors is one of the most important alterations described concerning the current therapeutic strategies sustaining defective cholinergic neurotransmission. These alterations may pose an obstacle to effective treatment and compromise receptor response to activation. Nevertheless, the investigation of these processes may shed light on the mechanism of neuronal degeneration since the alteration of key effector enzymes, such as protein kinases, may influence cellular homeostasis negatively. *AC*, Adenylate cyclase; *βAR*, beta adrenergic receptor; *BKR*, bradykinin receptor; *DAG*, diacylglycerol; *G*, G-protein; *IP$_3$*, inositol trisphosphate; *IP$_3$R*, IP$_3$ receptor; *PIP$_2$*, phosphatidylinositol bisphosphate; *PKA*, protein kinase A; *PKC*, protein kinase C; *PLC*, phospholipase C

activation of protein kinase C (PKC). Also, these systems have been found altered in AD brain, with the observation of a reduction of PI in the brain of AD patients, suggestive of enhanced PI breakdown. These alterations, together with altered calcium homeostasis, have been implicated in the triggering of calcium-dependent cytoskeleton-disrupting mechanisms, ultimately leading to a disruption of neuronal structure.

PKC is an important kinase, particularly in the brain where it regulates many aspects of neuronal plasticity, including transmitter release, receptor sensitivity and, at more integrated levels, long-term synaptic potentiation, learning and memory. One of the most consistent findings in AD brains is the alteration of this kinase. The levels, activity and subcellular distribution of PKC have been found to be altered by several investigators. Reduction of PKC activity appears to be fairly specific for AD since it has not been observed in the brain of patients affected by other neurodegenerative diseases such as Pick disease or vascular dementia (VD). Moreover, PKC reduced activity is also consistently found in peripheral tissues from AD patients (see below), suggesting that these alterations are not secondary to neuronal loss and may be directly involved in the pathogenesis of AD. These observations assume importance in the definition of therapeutic targets for the disease since the pharmacological strategy of increasing defective neurotransmission may fail to be effective because of a defect in the transduction systems beyond receptor activation.

The studies on transduction systems in peripheral tissues from AD patients have supported the evidence that multiple pathways are affected by the disease and that altered responses can be elicited similarly in the brain and in peripheral tissues. A lower PKC activity has been demonstrated in cultured skin fibroblasts [21–23] of sporadic AD patients with respect to cells of control subjects. More detailed studies have demonstrated a specific change in the characteristics of the enzyme [24] perhaps due to a selective loss of one of the several [25] PKC isoforms present. Studies on fibroblasts from the Swedish kindred show that PKC is not altered in these cells [26], suggesting that the role of PKC in the pathogenesis of AD could be different in familial AD cases. Analysis of other transduction systems has demonstrated a reduction in β-adrenergic-stimulated cAMP increase [27] that after extensive studies using pharmacological probes was suggested to be the result of an abnormality in the coupling of G-proteins to the β-adrenergic receptor. Recent studies have shown that the alterations of transduction systems in AD fibroblasts are greatly affected by genetic background. Particularly because fibroblasts from sporadic AD patients show reduced β-adrenergic stimulated cAMP increase, fibroblasts from carriers of presenilin mutations show instead increased cAMP production. On the other hand, fibroblasts from carriers of the "Swedish" amyloid precursor protein (APP) mutation do not show alteration of these systems [28]. Finally, the phosphoinositide cascade has been found to be altered in cultured skin fibroblasts from AD subjects. In the presence of basal IP_3 levels similar in AD and control cells, the bradykinin (BK)-stimulated increase in IP_3 levels was much greater in AD fibroblasts than in controls, which has a positive correlation with an increase in the BK receptor number [29].

Genetic Factors

AD is now recognized as a leading cause for dementia in the elderly population over 65 years of age. The majority of late-onset cases are sporadic without any characteristics of familial inheritance; however, genetic traits play an important role in the disease both in terms of the presence or absence of susceptibility genes, and in a limited number of cases by the evidence of autosomal transmission of the disease in families characterized by an early onset of symptoms (Table 1).

Autosomal Transmission

Chromosome 21
The major component of senile plaques is β-amyloid, a 39-43 aminoacid peptide that was first isolated from blood vessels and meninges of affected brains in 1984

Table 1. Genes involved in Alzheimer's disease

Genetic influence	Chromosome	Gene	Characteristics
Autosomal dominant	21	APP	Age at onset 43–59 years; increased absolute levels of Aβ or increase of Aβ 1-42
Autosomal dominant	14	Presenilin 1	Age at onset 33–52 years; increased levels of Aβ 1-42
Autosomal dominant	1	Presenilin 2	Age at onset 50–65 years; increased levels of Aβ 1-42
Susceptibility gene	19	ApoE	ε4 allele anticipates age onset in late-onset familial AD and sporadic cases
Susceptibility gene	6	HLA-A2	Cofactor in anticipating age of onset in the presence of ε4 ApoE allele
Susceptibility gene	12	Unknown	Unknown

by Glenner [30]. Subsequent studies have demonstrated that the amyloid peptide is produced by post-translational processing of a large precursor called the "amyloid precursor protein" (APP) encoded by a gene on chromosome 21 that produces by alternative splicing three major APP isoforms of 695, 751 and 770 aminoacids [31–34]. The identification of the gene for APP on chromosome 21 [35] was an important finding that suggested a major role for the amyloid protein in the development of AD neuropathology. In fact, Down's syndrome patients characterized by trisomy of chromosome 21 invariably develop early in life the characteristic lesions (neuritic plaques) of AD. At the same time, the first studies of the genetic linkage with markers on chromosome 21 in families with early-onset AD were published. Six pathogenetic mutations have been identified in the APP gene [36] (Fig. 2).

It is important to notice that all these mutations are flanking or within the β-amyloid peptide sequence and appear to modify the normal metabolism of APP toward an increase in the production of amyloidogenic fragments. In some cases the effect seems to be a net increase in the production of β-amyloid, as was demonstrated for the "Swedish" mutation; in other cases such as the mutations at aminoacid 717, the effect of the mutation seems to produce a shift in the normal processing of the precursor and leads to the formation of longer forms of the peptide (42/43 aminoacids) that are more susceptible to aggregation and deposition of amyloid fibrils (reviewed in [37]).

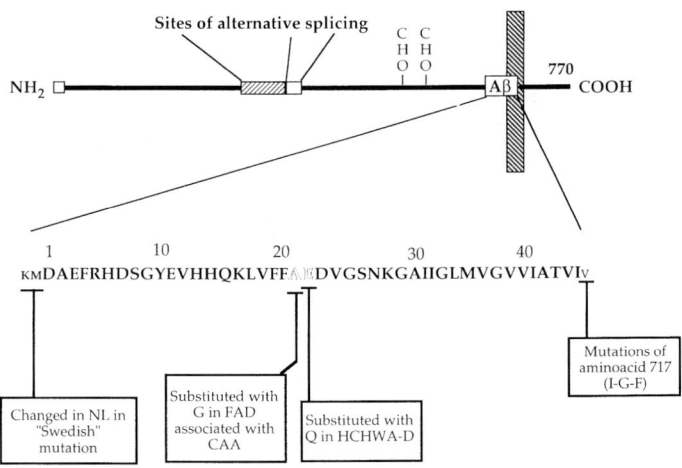

Fig. 2. Scheme representing the structure of the APP in its 770 aminoacid form. The sequence of the β-amyloid peptide is expanded to show the position of the known mutations associated with early-onset familial AD (FAD). In some early-onset families mutations at codon 717 (numbering according to the 770 aminoacid form of APP) that normally encodes a Val have been identified: the aminoacid is substituted with either Ile, Phe or Gly. These mutations are located close to the C terminus of the β-amyloid peptide. Another mutation identified on APP is a double mutant located at the N terminal side of β-amyloid where the wild-type Lys-Met is substituted with Asn-Leu. This mutation is characteristic of Swedish kindreds. Other APP mutations have been characterized in cases of hereditary cerebral hemorrhages with amyloidosis, Dutch type (HCHWA-D) and in a Dutch family with both presenile dementia and cerebral hemorrhagic disease. Both mutations lie within the β-amyloid peptide; in the first case a Gln at position 693 substitutes the wild-type Glu, while in the latter case the mutation is at codon 692 and results in the substitution of a Gly in place of the normal Ala. *CAA*, Cerebral amyloid angiopathy

Chromosome 14

The identification of a linkage between chromosome 14 and AD came from a purely genetic approach, studying families with early-onset disease where the presence of a mutation on APP was already excluded. The search led to the identification of a gene whose transcript has been called "presenilin 1" (PS1), coding for a protein of 467 aminoacids with the predicted structure of a membrane protein with seven/nine transmembrane domains [38] (Fig. 3). The original mutation identified were 5 in 11 non-related families. To date, more than 30 different mutations have been described in at least 60 independent pedigrees. The characteristics of PS1-linked familial AD are the very early onset, occurring as early as 30 years of age, and the particularly shorter duration of the disease compared to late-onset AD. A clinical feature characteristic of PS1-linked familial AD is the seizures, myoclonus and paratonia, observed in many subjects from different families. The inheritance is autosomal dominant, and penetrance is complete by the fifth/sixth decade of life.

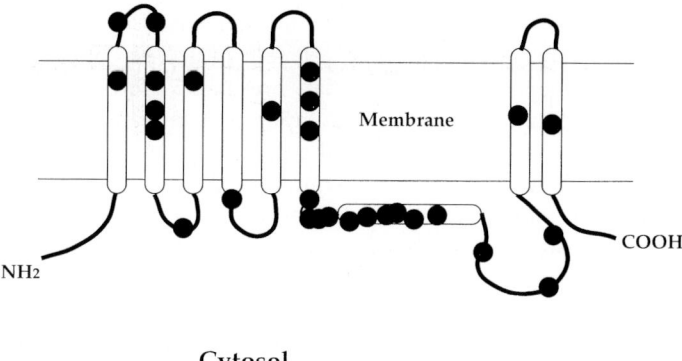

Fig. 3. Schematic drawing of the presenilin 1 gene product. The approximate topology according to a nine-transmembrane domain model is shown. The N and C terminal sides of the protein are oriented toward the cytoplasm of the cells, and the most accredited hypothesis about the intracellular localization of presenilins suggests that they are resident proteins of the endoplasmic reticulum. Presenilin 2 has a very similar predicted structure. The *black dots* show the approximate location of mutations so far identified on presenilin 1; their distribution is suggestive of a clustering of mutation in sites of the protein, which are possibly important for its physiological functions

Chromosome 1

Mutations on PS1 account for another proportion of early-onset familial AD (FAD). The evidence for the existence of a further gene different from those described above came from the study of a number of Volga German (VG) families. These familes originated from a group of German ancestors who in the late 1700s emigrated to the Volga region of Russia, remaining culturally distinct from the local population. Following a second emigration to the USA around the turn of the twentieth century, in recent years these families came to the attention of Alzheimer researchers. Bird et al. [39] identified seven VG kindreds with evidence of autosomal dominant transmission of early onset (mean age ranging from 51 to 56 years). Exclusion of linkage with mutations either on APP or PS1 and also exclusion of genetic linkage with the susceptibility gene on chromosome 19 (see below) has led to an intensive search for a new FAD gene that eventually yielded the identification of a gene on chromosome 1 initially called STM2 [40, 41]. The protein encoded by this gene is homologous to PS1 with a high degree of similarity (67%) in amino acid composition and a similar predicted structure of seven/nine membrane-spanning domains. Two mutations have been so far identified in this protein called "presenilin 2" (PS2), one present in the VG kindreds and a second identified in two Italian families.

The functions and physiological role of PS proteins have not yet been elucidated. Some clues come from the identification of homologous proteins in lower organisms such as the nematode *Caenorhabditis elegans*. In *C. elegans* the homolog of PS1, Sel-12 is involved in the mechanism of receptor signal transduction and in the determination of cell fate [42]. This latter function also appears to be important for mammalian presenilin proteins, as it has been shown that the PS mutant can facilitate apoptosis in transfected cells [43]. A major clue to the pathogenetic effect of the PS mutations came from the analysis of β-amyloid levels in the plasma and conditioned media of fibroblasts derived from carriers of mutant presenilin genes (see below). These studies have observed an increase in the levels of β-amyloid, ending at aminoacid 42 with apparently no effect on the levels of Aβ 1-40. As previously described with the APP717 mutation, the presence of PS mutations may affect APP metabolism (possibly by affecting its intracellular trafficking) toward the increased production of the more fibrillogenic Aβ 1-42 peptide.

Susceptibility Genes

Chromosome 19

The analysis of genetic factors involved in late-onset forms of AD is complex; however, intensive studies in this direction have yielded the identification of a major susceptibility gene on chromosome 19. Initially, a locus on chromosome 19 was identified and included a group of genes encoding for a series of apolipoproteins such as ApoCI, ApoCII, ApoCI pseudogene and ApoE [44]. Following the initial report identifying ApoE as the main Aβ binding factor in human CSF, a definitive identification of ApoE as a major genetic factor in late-onset AD was achieved by genotyping patients for ApoE polymorphism. The gene of ApoE consists of three different alleles ε2/ε3/ε4 whose frequency in the Caucasian population are 8%, 78%, and 14%, respectively. In AD, a significant disequilibrium in the frequencies of the three alleles has been reported. A higher frequency of the ε4 allele has been described in late-onset FAD (51%–52%) [45] but it is also elevated in sporadic forms of the disease (24%–40%). The ε4 allele appears to act as a dose-dependent modifier of the age of onset. In our own series each copy of the allele reduced the age of onset by approximately 2–3 years and also increased the disease duration in ε4 carriers [46].

The exact physiological role of ApoE in the brain is not completely understood; in AD brains ApoE is found deposited in tight association with amyloid plaques and, in the pathogenetic hypothesis suggested by Allen Roses, the role of ApoE4 would be that of favoring the deposition of β-amyloid into plaques. Other investigators favor the theory that the three isoforms all have a role in the neuroprotective mechanism and that it is interpreted differently by the three isoforms, with ApoE4 being the least efficient neuroprotective agent. It is important to remember that the main function of apolipoproteins is that of lipid transport, a function that at brain level could have very important implications in cellular repair mechanisms where lipid exchange and membrane activity are crucial.

Chromosome 6

A recent report suggests the existence of a correlation of the HLA-A2 allele with the age of onset of AD. Payami et al. [47] found a consistent pattern of reduction of age at onset as a function of the allele A2 in different sets of patients. Their research also suggested that the presence of the ε4 allele was additive with the HLA-A2 allele in an early shift of the age of onset.

Chromosome 12

Researchers have found evidence that a gene on chromosome 12 may be linked to late-onset forms of AD, and may play a role as a susceptibility gene, raising the risk of developing the disease in 15%–30% of all late-onset cases [48].

In Vitro Studies: Understanding the Pathogenesis of Alzheimer's Disease

From the previous paragraphs it is evident that no consensus has yet been reached on the causal mechanisms of AD – in part because of the practical limitations given by the lack of definite diagnostic criteria ante mortem. In view of these facts, there has been an intensive search for the identification of accessible tissues or body fluids suitable for exploring pathophysiological hypotheses and for providing a biological marker apt to confirm the diagnosis. The latter goal is still eluding the efforts of the investigators; however, several relevant data have been obtained, using peripheral cells and body fluids that are contributing to working hypotheses on the pathophysiological basis of the disease (see Fig. 4).

Alterations in Amyloid Precursor Protein Metabolism

The complex metabolism of APP is a process involving proteolytic cleavage of the precursor by a protease called "α secretase" releasing the ectodomain (sAPP or soluble APP) in the extracellular space. This proteolytic cleavage occurs within the Aβ sequence, therefore preventing the formation of amyloidogenic fragments [49]. On the other hand, there is evidence that Aβ can be formed and secreted as a physiological product of cell metabolism [50]. Several reports in the literature demonstrate that PKC activation with phorbol esters, while increasing sAPP secretion via the non-amyloidogenic pathway, also decreases β-amyloid secretion (Fig. 5) (reviewed in [51]). Using skin fibroblasts from sporadic AD patients and control subjects, we have demonstrated that the defective PKC activity (see above and [23]) is correlated with a reduced basal sAPP secretion from AD cells and reduced cell sensitivity to physiological levels of activation of PKC in terms of stimulated sAPP release [24]. The data described were the first report of an alteration in APP secretion in fibroblasts derived from sporadic AD patients and suggested the hypothesis that altered APP metabolism possibly underlying the pathology can be observed in peripheral cells, as well as in the brain from sporadic AD patients. This concept was more strongly supported by data obtained using cells derived from patients with FAD where a threefold increase in Aβ secretion from skin

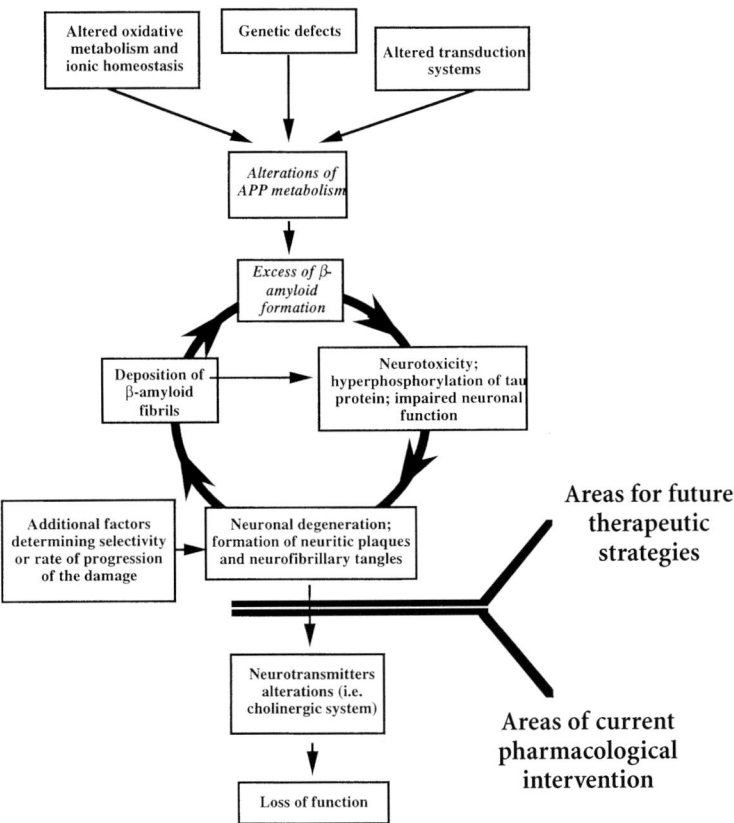

Fig. 4. A working hypothesis on AD pathogenetic mechanisms. This hypothesis views the accumulation of β-amyloid as a central event of the pathogenesis. Most of the data provided by the study of biological material obtained from patients support this hypothesis. It is established or indirectly suggested by the studies outlined in the text that either altered oxidative metabolism, altered transduction systems or genetic defects can result in an alteration of APP metabolism toward the production of amyloidogenic fragments. A vicious circle may then be caused by the unbalance of cellular homeostasis resulting from the initial insult. Other additional factors (in most cases still under investigation) contribute to the enhancement of neurodegenerative pathways. The figure shows a net distinction between the current levels of intervention of therapeutic strategies and the potential for new targets of intervention, stressing the importance of a broad investigation strategy

fibroblasts derived from FAD patients of the Swedish kindred [52] was demonstrated. Elevated β-amyloid levels were found in conditioned media of fibroblasts from both patients with clinical AD as well as normal subjects presumed to have presymptomatic AD. Increased Aβ secretion was also demonstrated in fibroblasts with FAD-linked PS1 and PS2 mutations [53]. Increased β-amyloid secretion can begin many years prior to the onset of symptoms, even in peripheral tissues, indicating that it does not require preexisting neural abnormalities.

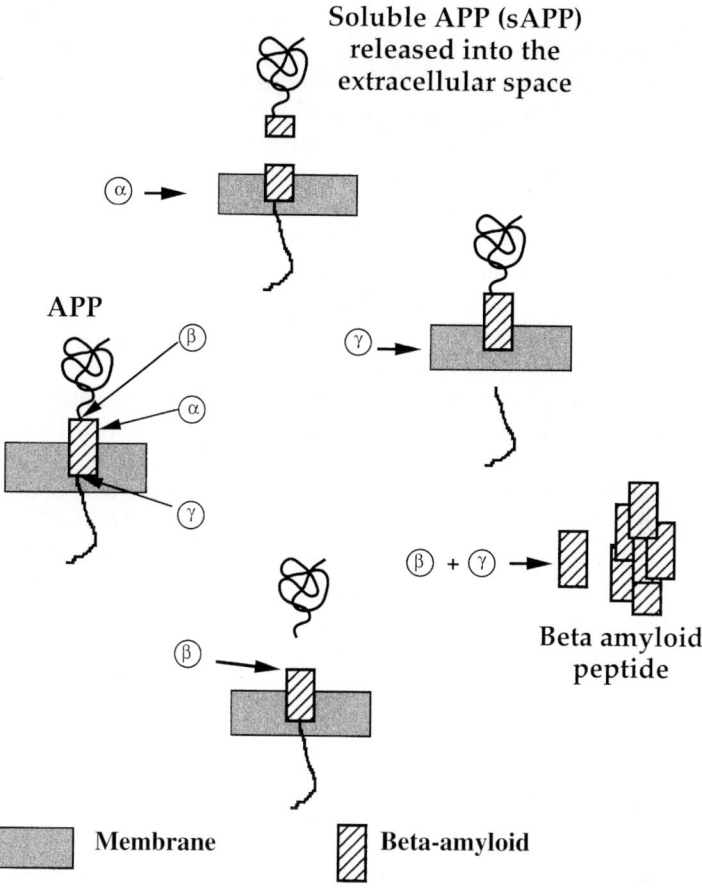

Fig. 5. Scheme of the APP metabolism. The figure summarizes the proteolytic pathways that lead to amyloidogenic and non-amyloidogenic processing of APP. Alpha secretase cleavage occurs within the sequence of β-amyloid, thereby preventing its formation. This pathway is accelerated by activation of PKC, dependent mechanisms and also by the activation of transduction pathways (both PKC dependent and independent) linked to G-protein-coupled receptors. The proteolytic pathways leading to the formation of Aβ from the action of β and γ secretase are also of physiological importance. These pathways seem to be alternatives to the non-amyloidogenic pathway and can be reduced by the activation of α-secretase cleavage of APP. Although our insight into the pharmacology of these processes is still limited, the proteases leading to Aβ formation are a theoretically important pharmacological target

Calcium Homeostasis and Oxidative Metabolism

Several findings (for review see [54]) support the hypothesis that altered processing of β-amyloid precursor protein contributes to loss of neuronal calcium homeostasis and hence neurofibrillary degeneration. These data suggest that a shift in

APP processing in favor of increased liberation of β-amyloid and reduced release of sAPP destabilizes intracellular calcium concentration ($[Ca^{++}]_i$) and endangers neurons in two ways: increased levels of β-amyloid would make neurons more vulnerable to excitotoxicity [55] and reduced levels of sAPP would deprive neurons of a neuroprotective substance that can stabilize $[Ca^{++}]_i$. Peripheral cells have been used to demonstrate alterations in calcium homeostasis involving both intracellular levels and compartmentalization, as well as influx following receptor challenges. Cellular calcium and oxidative metabolism are closely linked in fibroblasts, as in other tissues, and various parameters of altered energy metabolism have been reported in peripheral cells, and in particular in fibroblasts from Alzheimer patients (reviewed in [56, 57]). Increased lactate production and altered glucose utilization occur in AD fibroblasts. The activities of key mitchondrial enzymes such as α-ketoglutarate dehydrogenase are variably reduced in Alzheimer fibroblasts, possibly because of an inherent defect of the α-ketoglutarate dehydrogenase complex. Furthermore, reduced cytochrome c oxidase (CO) activity was demonstrated in platelets derived from AD patients [58] and also in AD fibroblasts from our bank [59]. Recently published data [60] describe genetic defects in the mitochondrial DNA (mDNA) coding for CO subunits CO_1 and CO_2. However, these results await independent confirmation on different series of patients. These results are consistent with the hypothesis that abnormal oxidation in AD brain is not just secondary to neurodegeneration, but this hypothesis needs to be further investigated.

Diagnostic Aids

Data Generated from In Vitro Studies

The successful identification of a peripheral marker of the disease may prove important either to predict or at least confirm the diagnosis. None of the alterations described in peripheral cells can yet be used as a diagnostic tool if considered alone. A possible novel approach to provide a profile for AD cells that is diagnostically distinguishable from that of controls was studied by Hirashima et al. [61]. Previous findings by the same authors have demonstrated the absence of a specific potassium channel and of tetraethylammonium (TEA)-induced calcium increase in AD fibroblasts [62]. The release from calcium stores elicited by bombesin or low BK concentrations was greatly enhanced in AD fibroblasts compared with cells from control donors [63]. Based on these results, Hirashima introduced a scoring system integrating the three responses in AD vs control fibroblasts, taking also into account the degree of responsiveness or unresponsiveness of each particular cell line. The index generated distinguishes AD patients from controls with both high specificity and sensitivity. These results indicate that by taking into account the overall profile of responses, it is possible to enhance the diagnostic value of these cellular alterations. More work is needed, however, to validate the results proposed by Hirashima, mostly to include variables such as patients bearing different kinds of neurological diseases, patients with different degrees of severity and length of the illness. Studies on platelets also suggest the possibility of detecting alterations of APP synthesis/metabolism of possible diagnostic significance in sporadic AD patients [64].

Body Fluids

The APP, Aβ and other Aβ-bearing fragments and also tau protein are detectable in cerebospinal fluid. Following the description by Seubert et al. [65] of the isolation of Aβ from human biological fluids (i.e. CSF and plasma), alterations of CSF Aβ levels have been reported in AD patients. Results in the measurement of absolute amounts of Aβ in AD and control CSF have varied. In particular, it appears that no disease specificity or direct correlation with disease severity can be found following these measurements, indicating that the total CSF Aβ level is not a useful marker for current diagnosis of AD (reviewed in [66]). The levels of Aβ 1-42 in CSF seem to be a more specific marker of AD pathology than total Aβ or Aβ 1-40. In fact, CSF Aβ 1-42 levels were significantly lower in AD patients relative to a non-AD group. Since Aβ 1-42 seems to participate in early deposits in the brain tissue of patients with AD [67] diminished clearance may account for its reduction in AD CSF. However, some overlapping of the control Aβ 1-42 levels with AD values was present (reviewed in [66]).

Aβ concentration seems to be altered as well in plasma from AD patients. Plasmatic Aβ concentrations are increased in carriers of familial AD linked to chromosome 14 and in carriers of the 670-671 mutation from the Swedish kindred [68]. The levels of Aβ 1-42 seem once again more reliable or at least more consistent among different laboratories. Younkin et al. [69] demonstrated that an elevated plasma concentration of Aβ 1-42 may play a role also in some patients with sporadic AD cases and that, in at least some of these patients, Aβ 1-42 may be elevated before the onset of symptoms.

Tau protein, the main component of NFT, can also be detected in CSF. CSF tau levels in AD cases are significantly more elevated [70] than in healthy control individuals, VD patients and neurological control subjects. While elevated tau levels are also found in other kinds of neurological diseases where neuronal death or damage occurs (i.e., hereditary cerebellar atrophy, olivopontocerebellar atrophy, Parkinson's disease), the simultaneous analysis of tau and Aβ 1-42 in the same CSF sample appears to be of particular interest. In fact, the presence of an elevated tau level and reduced Aβ 1-42 seems to be highly predictive of AD (specificity 96%). Conversely, high Aβ 1-42 and low tau has been observed only in control patients (specificity 100%). However, the combined CSF tau and Aβ 1-42 measurement were not informative in patients who fell into the low Aβ 1-42 /low tau group. Thus, the diagnostic potential of these tests needs to be further evaluated (see [66]).

Among other proteins unrelated to the major Alzheimer's markers such as Aβ or tau, the iron binding protein p97 has recently been proposed as a possible diagnostic marker for the disease [71]. Its level has been found significantly increased in freshly prepared plasma samples from AD patients. The promising data are represented by the virtual absence of overlaps with the levels measured in the plasma of control subjects, including individuals suffering from other neuropathologies. As for many other proposed diagnostic markers for AD, the plasma level of p97 awaits complete validation on a large number of patients in carefully designed multicenter studies.

Lessons Learned from the In Vitro Cellular Models

A considerable amount of data has been obtained, particularly from working on tissues from AD patients. These intense efforts of the scientific community have contributed to the definition of a working hypothesis for the pathophysiology of AD (see Fig. 4). Although most of these hypotheses need to be investigated further, our knowledge of the mechanisms underlying AD has been substantially improved. Moreover, the search for a biological marker of the disease has not led to a clear identification of a diagnostic marker, but it has helped to build a biological profile characteristic of AD cells. The association of different parameters clearly identifies a biological pattern of AD cells and lays the groundwork for more clinically oriented experimental studies that are carefully designed to detect the diagnostic potential of the observed biological differences.

Vascular Dementia

Neuropathology

Since the 1970s the concept of vascular dementia (VAD or VD) has had its ups and downs. Certainly in the late 1960s to early 1970s, vascular factors, i.e., "cerebral arteriosclerosis," were overestimated as a leading cause in determining loss of cognitive function in the aging brain. However, anatomopathology could not provide evidence that vascular damage played such an overwhelming role in dementing illnesses, pointing instead to the importance of primary degenerative forms. Later on, starting in the mid-1980s, the great effort made by neuroscientists to elucidate the biological basis of AD has somewhat limited the search in the field of VD in spite of the fact that recent data and imaging studies suggest that cerebrovascular disease may be a more common cause of dementia than usually recognized, in particular in the case of older patients. In addition, VD is ill defined and depends on variables such as size and position of the infarction, collateral supply, vascular density, size and symmetry of the lesion and in combination with other dementing diseases, including AD [72]. Hypoperfusion leading to incomplete infarction appears also to be an important component predisposing, when limited in size, for dementia, in particular in the case of additional infarcts, ultimately reaching a critical threshold for the development of dementia. While it is obvious that brain infarction does not mean dementia per se, the critical lesions leading to dementia (size, strategic location) have yet to be fully defined. The volume of infarction may be an important variable; however, volumes as low as 10 ml have been associated with dementia, and the number of infarcts also does not seem to be correlated with dementia. Small bilateral lesions in regions critical for cognitive functions and memory or lesions deep in the white matter may produce the maximal effect acting by disconnection [72].

Brun [73] analyzed pathoanatomically 175 consecutive cases autopsied during 1987 to 1992 to classify lesions in VD according to vascular size/level in combination with circulatory factors. Of the 175 cases examined, 63 were of the mixed type (AD and cerebrovascular dementia) and 59 pure cerebrovascular cases. In this series pure AD was relatively rare (28 cases), but the AD group including AD combined with vascular lesions (91 cases) was the largest one. Large-vessel disease was usually due to atheromatosis and thromboembolism, producing large infarcts, fitting, in the case of multipli-infarcts, the definition of multi-infarct dementia. Small infarcts, down to 20 mm in diameter, caused dementia if located in critical positions such as bilateral thalamic infarcts, a condition defined by Brun as strategic-infarct dementia. Small-vessel disease was usually due to hypertensive angiopathy, microembolism or collagen disorder, resulting in infarcts below 20 mm in diameter. The infarcts were both complete and incomplete. Binswanger's disease was the most frequent form with cortical plus subcortical hypertensive angiopathy with small infarcts, frequently incomplete. Several cases presented both small- and large-vessel disease. Brun defined also a group of patients in whom infarct, frequently incomplete, was due to episodes of hypoperfusion caused by a blood pressure fall associated with vascular stenosis. One important role in the pathogenesis of cerebrovascular diseases is also played by the structural and functional changes in the intracerebral blood vessels. Arterioles and capillaries can be altered, causing either obliterative lesions or impairing nutrition of the parenchyma and edema. These alterations may result from the degeneration of smooth muscle cells with deposition of collagens and other extracellular matrix components [74].

An unsolved puzzle is the forms of dementia such as mixed forms, where the borderline between AD and VD is not clearly defined. Both signs of vascular insults and plaques and/or NFT are observed on neuropathological examination. It is not understood whether the comorbidity is due to independent phenomena or interrelated events. According to the latter hypothesis, amyloid angiopathy may favor vessel fragility, leading to VD or hypoperfusion, and ischemia may promote neurofibrillary and plaque pathology. It is noteworthy that a significant proportion of stroke patients develop dementia within 4 years after the cerebrovascular accident [75]. Within this context it is suggested that brain ischemia may lead to amyloid deposition.

Neurochemistry

The levels and distribution of neurotransmitters and their metabolites show changes in the brain of VD patients. In a postmortem analysis Gottfries has shown [74] a peculiar pattern of changes in the brain neurochemistry. In particular, the brains of VD patients show severely reduced serotonin metabolism, as well as reduced activity of choline acetyltransferase, suggesting also depressed cholinergic activity. The possibility that the changes in cholinergic activity may be due to some extent to the presence of mixed dementias cannot be excluded. Loss of serotoninergic activity has been suggested as the cause of increased activity of the hypothalamic-pituitary-adrenal axis shown in patients with VD. The loss of serotonin may also be linked to depression or the depressive traits frequently observed in VD patients.

Genetic Factors

Families affected by cerebrovascular disease have been reported in several countries. Hereditary multi-infarct disease appears with clinical signs in young and middle-aged adults of either sex and culminates in progressive dementia following episodes of neurological disturbances. The disease has an autosomal dominant inheritance, and its locus has been identified on chromosome 19q12 [77]. Later the gene responsible for this phenotype is suggested to be Notch3 [78], which is part of a family of receptors triggering signal-transduction cascades involved in neuritogenesis. On neuropathological examination, this disorder is characterized by damage to smooth cells of the media of arterioles with deposition of collagens and extracellular matrix molecules, and therefore represents an example of an inherited disease, preferentially affecting cerebral arteries and arterioles [79].

   Other cases of inherited cerebrovascular disorders include the hereditary cerebral hemorrhages with amyloidosis Dutch type (HCHWA-D) and cerebral amyloid angiopathy associated with presenile dementia, both characterized by deposits of amyloid protein within the wall of cerebral vessels. In the case of HCHWA-D, the clinical course is characterized by recurrent strokes accompanied by vascular deposition of amyloid protein. In both cases a pathogenetic mutation segregating the disease has been found on the gene for the APP of AD within the domain coding for the β-amyloid peptide (see Fig. 2).

   The risk factors for VD in all other cases are supposed to be the same as those for stroke, including hypertension, diabetes mellitus, age, gender (males), smoking and cardiac conditions. Dementia involving white-matter lesions has been related to hypertension. ApoE has been suggested to be among the genetic risk factors associated with VD. In a selected series of clinically diagnosed VD patients, Frisoni et al. [80] have demonstrated an increase in the frequency of the ApoE4 allele similar to that observed in AD patients. However, the finding has been challenged by other investigators and still remains open to debate.

Non-Alzheimer Degenerative Dementias

Frontotemporal Dementia

Dementias that do not present the typical neuropathological lesions of AD are often ill defined, starting with the terminology. As a consequence, the description of such cases varies and many names, including non-specific dementia, dementia lacking distinctive neuropathology, Pick disease without Pick bodies, and frontotemporal dementia, have all been used. The clinical variability renders the study of these cases a difficult task and often a misdiagnosis of AD masks the disease, even though some distinctive features are present [81].

On neuropathological examination, atrophy of the fronto-temporal-parietal regions is a consistent feature. Microscopic evaluation reveals neuronal loss, neuropil vacuolation and gliosis, both affecting gray and white matter. In most cases the brain shows argyrophilic and tau-positive neuronal inclusion that, however, does not have the characteristics of typical Pick bodies. In all cases ballooned neurons and argyrophilic glial inclusion are consistent features.

Familial forms of frontotemporal dementia have been recently recognized and discussed as an independent entity in a consensus conference [82]. One of the dominant clinical features of the hereditary disorders is the presence of parkinsonian symptoms with a neuropathological correlate that shows severe neuronal loss, gliosis, status spongiosus in the frontal, temporal and cingulate cortices and marked neuronal loss in the substantia nigra. Moreover, the geneticists have identified a locus on chromosome 17 that segregates the disease. Because of these characteristics the disease has been given the unifying name of frontotemporal dementia and parkinsonism linked to chromosome 17 with the acronym FTDP-17. It is clear that more studies are needed for characterization of the gene or the genes responsible for the disease. Among others, tau protein has been a favorite candidate since its gene lies in the chromosomal locus involved.

Pick Disease

Pick disease is part of the diverse group of rare non-Alzheimer neurodegenerative diseases characterized on gross examination of the brain by fronto-temporal atrophy and on microscopical examination by remarkable neuronal loss, gliosis and spongiform changes. As in many other cases, there is not yet a clear consensus on the diagnostic criteria for Pick disease since not all investigators agree that the presence of Pick bodies is an absolute prerequisite in the presence of the other above-mentioned gross neuropathological features (reviewed in [83]). Pick bodies were described initially by Alzheimer as "argyrophilic, sometimes structureless balls near the nucleus, whose size varies from half to twice that of the nucleus." The inclusions, observed usually in small pyramidal cells, are composed of straight filaments 15–18 nm in diameter. These inclusions are stained with antibodies to ubiquitin, chromogranin A, phosphorylated neurofilaments epitopes, tau protein and β-amyloid precursor protein fragments. Tau protein is abnormally phosphorylated in Pick brains, but with a different pattern than that observed in paired helical filaments of Alzheimer brain. Ballooned neurons are cells that appear swollen with convex contours and present an eosinophil cytoplasm with eccentric nuclei. These cells share some of the immunocytochemical characteristics with Pick bodies and sometimes are filled with straight filaments similar to those found in Pick bodies. As stated before, in spite of extensive studies both on the morphology level and with new, improved immunocytochemical tools, a clear definition of Pick disease has not been obtained, emphasizing the need for a multidisciplinary strategy to uncover the molecular basis of this peculiar disease.

Lewy Body Disease

The term "dementia with Lewy bodies" generically describes various types of disorders such as diffuse Lewy body disease (DLBD), senile dementia of Lewy body type and Lewy body variant of AD [84]. The variable clinical features and the often concomitant presence of AD type neuropathology in many cases of DLBD have led to often confounding classification of dementias with Lewy bodies and also to misdiagnosis. Lewy body disorders are now recognized as the second cause of dementia following AD, but elucidation of the possible pathogenetic mechanisms is still eluding the efforts of investigators [85].

The characteristic neuropathologic finding that defines Lewy bodies diseases is the presence of rounded eosinophilic and intracitoplasmic bodies (Lewy bodies indeed) that are disseminated throughout the cerebral cortex, areas of the limbic system and in the brainstem. Identification of Lewy bodies upon neuropathological examination is achieved by means of newly developed immunocytochemistry tools such as monoclonal antibodies capable of distinguishing between Lewy bodies and NFT [86].

The terminology used above to define variants of Lewy body disorders reflects the quantitative differences in the density and anatomical distribution of Lewy bodies. In fact, in cases of Parkinson's disease, the localization of Lewy bodies is confined to the substantia nigra and is limited quantitatively to a small number of such inclusions. On the opposite side of the range diffuse localization of Lewy bodies in the cerebral cortex eventually overlaps with the classical neuropathological features of AD. The neuropathology provides a set of features that help differentiate Lewy body dementia from AD more clearly. Besides the presence or absence of Lewy bodies, one of the most peculiar characteristics is the rare frequency of NFT in the neocortex and the hippocampus in cases of Lewy body dementia, and also the levels of hyperphosphorylation of tau are less relevant in Lewy body disease than in AD (reviewed in [87]). Moreover, the neuronal damage in the nucleus basalis of Meynert in Lewy body cases is greater than that seen in Alzheimer brains and correlates with an extensive deficit in the cholinergic input to the frontal parietal and temporal cortices, suggesting a more severe cholinergic deficit in Lewy body patients [87].

Given the importance of genetics in AD and in particular of the ApoE alleles as risk factors for AD, research has concentrated on the influence of ApoE genotype also in correlation with the presence of disorders related to Lewy bodies. The results are controversial since some authors have shown that the ApoE ε4 allele is overrepresented in cases presenting AD neuropathology and also Lewy bodies, while no increased prevalence of ApoE4 can be found in cases with Lewy bodies but no AD neuropathology. Other authors, however, have reported that the ApoE ε4 allele increased in Lewy body cases without concomitant AD neuropathology [87, 88].

The differences observed in the neuropathological correlates of the cholinergic functions between AD patients and Lewy body patients are important in terms of responses to pharmacological treatments, in particular using anticholinesterase agents. It has been speculated that the response to tacrine of some AD patients was due to the concomitant presence of Lewy body pathology, but this hypothesis awaits more comprehensive neuropathological examination.

Prion Diseases

The prion diseases are a group of neurodegenerative diseases of both animals and humans. In the latter case they are distinguished in the more common sporadic forms such as Creutzfeldt-Jakob disease (CJD), transmissible forms such as kuru and possibly some cases of CJD, and more rare familial forms as Gerstmann-Sträussler-Scheinker syndrome (GSS) and fatal familial insomnia (FFI) [89].

Neuropathology

The most distinctive neuropathological finding of prion diseases is a widespread spongiform degeneration of the central nervous system, affecting mostly the cerebral cortex, striatum, thalamus (in particular in FFI) and the cerebellum. Neuronal loss and gliosis may vary according to the duration of the disease that can be as short as 3 months in particularly aggressive familial forms and up to more than 20 years in certain forms of GSS [90].

Another particular feature, not always present, is the formation of numerous amyloid deposits that are, however, different from the senile plaques in AD as they are formed mainly by a protease-resistant form of a protein called "prion protein." Cloning of the gene coding for the prion protein (PRNP) and the biochemical characterization of the normal and disease-related protein have led to the current advancement in knowledge about prion diseases [91]. The normal product of the PRNP gene is the prion protein, abbreviated to PrP^c, of 33-35 kDa, a protein normally expressed throughout the brain, particularly in neurons. The normal functions of PrP^c remains elusive, although some indirect evidence suggests a role of PrP^c in the synaptic transmission. Infective prions are mainly composed of a pathogenic prion protein abbreviated PrP^{Sc} (from scrapie, the ovine form of spongiform encephalopathy). According to the prion hypothesis, the PrPc converts to PrP^{Sc} through a conformational change that constitutes the central event in all prion diseases. The event would be a simple matter of chance in the sporadic forms, induced by the exogenous PrP^{Sc} in the transmissible or iatrogenic forms and facilitated in the familial forms as a consequence of the instability of the mutated PrP^c [91]. This hypothesis is still largely debated, but has received strong support in recent years (see also section animal models).

Etiology and Genetics

Transmissible Forms

Kuru is the prototype of human spongiform encephalopathy and is a disease restricted to an area in the eastern highlands of New Guinea. Its origin as an epidemic comes from the ritual cannibalism (now disappeared) practiced by this population. A modern form of human-transmissible prion has been the use of human CNS-derived tissues and proteins (see, for example, corneal grafting, and the use of purified human growth hormone). These procedures have led to a recent outbreak of iatrogenic forms of CJD. However, the use of recombinant growth hormone has reduced the possibility for this kind of transmission [89]. A new aspect of transmissible prion diseases has come to the forefront because of the recent outbreak of bovine spongiform ecephalopathy (BSE) in the UK. It seems possible that BSE can be transmitted to humans through the consumption of tainted beef products even though definitive proof of such transmission has not yet been obtained.

Autosomal Transmission

Familial forms of prion diseases are autosomal dominant disorders associated with the presence of mutations in the PRNP gene. Mutiple pathogenetic alleles are associated with phenotypes that may present different clinical characteristics. Two mutations in the PRNP gene are linked to familial forms of CJD. These forms are distinguished from the sporadic cases by a longer duration of the disease, but otherwise similar clinical signs. In the GSS syndrome, the clincal phenotype is distinguished from CJD by a slower progression, by the presence of cerebellar dysfunction and severe spastic paraparesis and by the invariable presence of multicentric amyloid plaques in all phenotypes of the disease.

One mutation is linked to a form of prion disease called FFI. FFI is linked to a mutation that also characterizes a form of familial CJD. The phenotypic difference between CJD and FFI is determined by a polimorphism in the PRNP gene at codon 129, a neutral polimorphism with the frequency of alleles of 0.62 for the aminoacid Met and of 0.38 for the aminoacid Val in the Caucasian population. The genotype segregating with CJD is 129 Val and the genotype segregating with FFI is 129 Met. In the presence of the same mutation at codon 178 substituting Asp with Asn, the polymorphic site determines a phenotype that in one case is characterized by a dementing illness with severe and diffuse spongiform degeneration, and in the second case a disease expressed as untreatable fatal insomnia with selective atrophy of the thalamic nuclei (reviewed in [90]).

Genetic Control and Susceptibility Factors

The arguments discussed above concerning the natural polymorphism of the PRNP gene apply also for the susceptibility to sporadic and transmissible forms of the disease. In the majority of sporadic CJD cases in fact, the homozygosity for either amino acid at position 129 has been proposed as a predisposing factor. In familial cases homozygosity is considered an aggravating factor.

Diagnostic Markers

As with AD, in the diagnosis of prion-related disorders, the definition of diagnostic certainty is also limited to examination of autopsy tissues. The only criteria ante mortem are limited to the identification of a rapidly progressive dementia and EEG abnormalities in the presence also of myoclonus, cerebellar symptoms, pyramidal and extrapyramidal signs or akinetic mutism. These criteria have only a limited sensitivity and specificity. A proposed additional marker of the disease is the measurement of CSF levels of neuronal specific enolase, a marker protein that increases in CJD patients. Since false-positives from patients with ischemia, subdural hematoma or tumors can occur using this marker, Zerr and coworkers [92] have proposed a novel test on CSF proteins that measures the presence of a doublet of proteins of molecular weight of 130/131 kDa by two-dimensional gel electrophoresis. The results of this study claim a test specificity of 100%. More recently, the 130/131 proteins have been identified as members of the 14-3-3 protein family, leading to a simple antibody-based immunoassay that yields a CJD diagnosis sensitivity of 98% and a specificity of 99%.

Animal Models of Dementia

The theoretical and practical problems related to the design of animal models are similar for AD and most of the neurodegenerative disorders, in particular those leading to the development of dementia. The most important difficulties arise when trying to interpret the development of symptoms involving cognition and behavior. The ideal animal model should reproduce simultaneously the neuropathological lesions characteristic of the disease and the neurochemical alterations and most importantly, the specific alterations in cognition. The importance of animal models in the study of neurodegenerative diseases is reflected by the need for appropriate models for the screening of potential therapeutic treatments. The two diseases where the majority of the work has been focused are AD and prion disease [93].

Animal Models of Alzheimer's Disease

Concerning AD, aged non-human primates have been shown to develop with aging cognitive disturbances, deposition of amyloid and neurochemical changes related to those characterizing the human disease. Dogs, bears and also sheep have shown the appearance with aging of amyloid deposits, but these findings have not been correlated with other neuropathological findings and neurological impairment. Other approaches toward an animal model of AD have been pursued by direct injection of β-amyloid peptide in the brain of rodents; however, findings in these studies have been difficult to reproduce between different laboratories [94]. The advancements in genetics and the use of transgenic technology have intensified efforts to try to reproduce the disease in an animal model. The first approaches toward the production of a transgenic model of AD were guided by the evidence coming from familial AD and Down's syndrome, suggesting that the level of β-amyloid was strictly related to the deveopment of the pathogenesis. Therefore, numerous groups have attempted to generate transgenic mice that overexpress full-length APP or APP derivatives either as wild-type genes or carrying one of the mutations linked to FAD. Various strategies during the course of the early 1990s all led to limited results. However, more recently new transgenic strategies have finally led to two mice modes that express many pathological features characteristic of AD and in one case also memory deficits. The transgenic mouse developed at Athena Neurosciences was based on the overexpression of APP carrying the 717 mutation (Val-Phe) in a system such that brain-specific (tenfold over endogenous APP) overexpression of the transgene was achieved and all three isoforms of APP were expressed [95]. The results of these manipulation are mice that produce detectable levels of Aβ 1-40 and 1-42 in brain exctracts and CSF and develop by 6–9 months amyloid deposits in the brain whose numbers increase with aging and are associated with activated astrocytes and dystrophic neurites. Also, synaptic density and dendrites were found to be reduced in the dentate gyrus of transgenic animals. However, no neurofibrillary pathology was observed in mice up to 13 months of age.

A second transgenic model has been created overexpressing the 695 amino acid isoform of APP carrying the double mutant characteristic of Swedish FAD. These mice demonstrated increased levels of Aβ 1-40 and 1-42, respectively, 5-fold and 14-fold over the endogenous levels and formation of brain amyloid plaques. In direct correlation with these neuropathological findings, these mice also showed the appearance with aging (9–10 months) of cognitive disturbances such as in learning and memory [96]. New transgenic animals are now being developed with the expression of mutant presenilins, and future development of transgenic technology will be the study of the interaction of different AD related genes by cross-breeding of appropriate transgenic animals. One example will be the cross-breeding of animals in which the ApoE genotype has been manipulated with animals expressing AD pathology due to altered APP or PS genes [94].

Animal Models of Prion Disease

The longest dated animal models of prion disease are those in which the spongiform encephalopathy was transmitted experimentally by injecting brain extracts of animals or humans affected by the disease in the brain of rodents and primates. Genetics once again suggested that in human familial prion disease such as GSS and FFI, mutations present on the prion gene (PRNP), which is a normal constituent of the genome, lead to increased susceptibility to the conformational changes that produce the infectious and pathogenetic form of the prion protein called PrP^{Sc}. The initial transgenic research on the production of an animal model of prion disease was conducted by introducing human-disease-linked mutations into the mouse *PrP* gene [97]. Mice expressing the mutant protein develop a neurological disease with neuropathological alterations similar to that in the human disease. The nervous tissue obtained from these animals can induce prion infections in animals carrying the same mutation, but at levels too low to develop the disease spontaneously. The information obtained from transgenic animals also provided evidence that a host-encoded protein plays a role in the pathogenesis of prion disease [98]. Transgenic mice expressing only the human *PrP* carrying a pathogenetic mutation did not develop neurological abnormalities; however, when the human mutant protein was expressed as a chimeric protein with elements of the mouse *PrP*, the animal developed the neurological syndrome. These results suggest that the mutant protein must interact with an endogenous protein to be converted to PrP^{Sc} and only the chimeric protein has the features necessary for the interaction. It is noteworthy that mice that express wild-type human *PrP* in a normal mouse environment are not sensitive to human prions, but the expression of human *PrP* in mice lacking the endogenous *PrP* create animals that are extremely sensitive to human prions.

Distinctive and Common Elements among the Various Dementias

The general and most obvious common element in all dementing disorders is the selective degeneration of populations of neurons, affecting specific regions of the brain and often suggesting a functional correlation with the clinical expression of the disease. What distinguishes the different types of dementia is not only the nature and localization of degenerative changes, but also the type of triggering neurotoxic insults that lead to neurodegeneration.

This is not an academic discussion, since based on this information the design of an intervention protocol could evolve that might be distinctive for each type of dementia or might have a more broad spectrum (see Fig. 6).

Signals Inducing Neurodegeneration as Distinctive Elements

Taking as an example the most studied types of dementia, i.e., AD, VD and prion diseases, it is clear that the origin of neurodegeneration comes from insults of different origin. Ischemia and subsequently reduced oxygen supply are the leading causes of neurodegeneration in the VDs. The discussion in such cases converges on where and to what extent the hypoxic status is maintained.

The situation becomes more complicated when dealing with AD or prion diseases. In both cases is clear that a toxic event occurs within the brain and if for prion diseases the pathogenetic prion protein can be assumed as the culprit of the neurodegenerative changes, in AD the putative causal role of Aβ is still largely debated. An essential corollary to the role of Aβ or the prion protein in neurodegeneration is that these polypeptides need to exert a neurotoxic action on neurons. The toxicity of β-amyloid and peptide fragments of the prion protein has been demonstrated in vitro by using primary neuronal cell cultures and neuronal cell lines [99]. These mechanisms have been largely discussed concerning the role of β-amyloid in AD neurodegeneration, and many aspects of β-amyloid neurotoxicity have been reviewed by Iversen et al. [37].

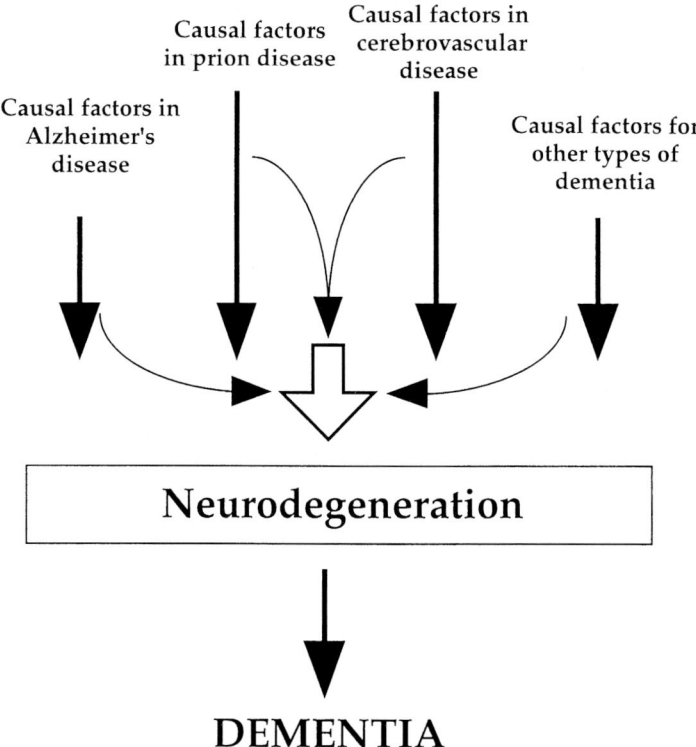

Fig. 6. Schematic drawing of possible convergence of various neurodegeneration triggering events. The causal factors of all neurodegenerative diseases (whether known or putative) may directly activate pathways of cell death and consequently neurodegeneration (*black arrows*). However, in some cases these pathways are common to other neurotoxic insults and can all merge (*thin arrows*) into a common final pathway that leads to dementia (*open arrow*). The clear recognition of these common events and of their level of interaction may prove extremely useful for the development of a common and possibly early intervention in the biological onset of neurodegenerative diseases

Fundamental Mechanisms of Neurodegeneration as Common Elements

In recent years intense research efforts have concentrated on the mechanism of neuronal degeneration, in particular concerning the aspects of apoptosis. Although there are both morphological and biochemical criteria to distinguish apoptosis from necrosis, it is now recognized that most of the mechanisms leading to either type of cell death are similar. In most cases of neurodegeneration the mechanisms eventually leading to cell death include disruption of oxidative metabolism and/or calcium homeostasis. It should be stressed, however, that, in turn, oxidative stress may lead to excess Aβ formation, sustaining therefore a vicious circle. As examples, the metabolic alterations occurring after an ischemic event, the extent of which is the determinant for the clinical expression of the disease, result in severe disruption of ionic homeostasis. The same insult can kill the same population of neurons by either apoptosis or necrosis, depending on the severity and duration. In the case of β-amyloid the neurotoxicity caused by the peptide has been linked to increased oxidative stress to the cells and disruption of calcium homeostasis. The mechanism of apoptosis has also been inferred in studies in vitro, suggesting that β-amyloid or fragments of the prion protein [99, 100] may activate (through a free-radical-mediated mechanism) the typical pathways seen in apoptosis [101]. Depending on the concentration of the peptide, a mechanism of necrotic cell death has also been described [102].

In conclusion, examination of the common elements leading to neurodegeneration and eventually dementia derived by the analysis of the literature suggests that alteration of the delicate homeostatic balance of the cell is always the final effect of any putative pathogenetic insult. These events trigger a mechanism of cell death that is balanced on the foggy border between apoptosis or necrosis with a distinction that lies within the frame of "how much" and "how long."

References

1. Terry RD, Peck A, De Teresa R, et al (1981) Some morphometric aspects of the brain in senile dementia of the Alzheimer type. Ann Neurol 10: 184-192
2. Vereecken TH, Vogels OJ, Nieuwenhuys R (1994) Neuron loss and shrinkage in the amygdala in Alzheimer's disease. Neurobiol Aging 15: 45-54
3. Leuba G, Kraftisik R (1994) Visual cortex in Alzheimer's disease: occurrence of neuronal death and glial proliferation, and correlation with pathological hallmarks. Neurobiol Aging 15: 29-43
4. Vogels OJM, Broere CAJ, ter Laak HJ, et al (1990) Cell loss and shrinkage in the nucleus basalis Meynert complex in Alzheimer's disease. Neurobiol Aging 11: 3-13
5. Braak H, Braak E (1991) Neuropathological stageing of Alzheimer-related changes. Acta Neuropathol 82: 239-259
6. Goedert M (1993) Tau protein and the neurofibrillary pathology of Alzheimer's disease. Trends Neurosci 16: 460-465
7. Selkoe DJ (1994) Normal and abnormal biology of the β-amyloid precursor protein. Ann Rev Neurosci 17: 489-517
8. Selkoe DJ (1989) Aging, amyloid, and Alzheimer's disease. N Engl J Med 320: 1484-1487
9. Tagliavini F, Giaccone G, Frangione B, Bugiani O (1988) Preamyloid deposits in the cerebral cortex of patients with Alzheimer's disease and non demented individuals. Neurosci Lett 93: 191-196
10. Li Y-T, Woodruff-Pack DS, Trojanowski JQ (1994) Amyloid plaques in cerebellar cortex and the integrity of Purkinje cell dendrites. Neurobiol Aging 15: 1-9
11. Joachim CJ, Mori H, Selkoe DJ (1989) Amyloid β-protein deposition in tissues other than brain in Alzheimer's disease. Nature 341: 226-230
12. Mann DMA (1994) Pathological correlates of dementia in Alzheimer's disease. Neurobiol Aging 15: 357-360
13. Schmidt ML, Lee VM, Forman M, Chiu TS, Trojanowski JQ (1997) Monoclonal antibodies to a 100-kd protein reveal abundant Aβ-negative plaques throughout gray matter of Alzheimer's disease brains. Am J Pathol 151: 69-80

14. Davies P, Verth AH (1978) Regional distribution of muscarinic acetylcholine receptors in normal and Alzheimer's type dementia brains. Brain Res 138: 385-392

15. Shimohama S, Taniguchi T, Fujiwara M, Kameyama M (1986) Changes in nicotinic and muscarinic cholinergic receptors in Alzheimer-type dementia. J Neurochem 46: 288-293

16. Horsburg K, Saitoh T (1994) Altered signal transduction in Alzheimer disease. In: Terry RD, Katzman R, Bick KL (eds) Alzheimer disease. Raven, New York, pp 387-404

17. Poirier J, Delisle MC, Quirion R, Aubert I, Farlow M, Lahiri D, Hui S, Bertrand P, Nalbantoglu J, Gilfix BM, et al (1995) Apolipoprotein E4 allele as a predictor of cholinergic deficits and treatment outcome in Alzheimer disease. Proc Natl Acad Sci USA 92: 12260-12264

18. Cowburn RF, Fowler CJ, O'Neill C (1996) Neurotransmitters, signal transduction and second-messengers in Alzheimer's disease. Acta Neurol Scand Suppl 165: 25-32

19. Cowburn RF, Fowler CJ, O'Neill C (1996) Neurotransmitter receptor/G-protein mediated signal transduction in Alzheimer's disease brain. Neurodegeneration 5: 483-488

20. Fowler CJ, Garlind A, O'Neill C, Cowburn RF (1996) Receptor-effector coupling dysfunctions in Alzheimer's disease. Ann N Y Acad Sci 786: 294-304

21. Van Huyn T, Cole G, Katzman R, Huang K-P, Saitoh T (1989) Reduced protein kinase C immunoreactivity and altered protein phosphorylation in Alzheimer's disease fibroblasts. Arch Neurol 46: 1195-1199

22. Bruel A, Cherqui G, Columelli S, Margelin D, Roudier M, Sinet PM, Prieur M, Perignon JL, Delabar J (1991) Reduced protein kinase C activity in sporadic Alzheimer's disease fibroblasts. Neurosci Lett 133: 89-92

23. Govoni S, Bergamaschi S, Racchi M, Battaini F, Binetti G, Bianchetti A, Trabucchi M (1993) Cytosol protein kinase C downregulation in fibroblasts from Alzheimer's disease patients. Neurology 43: 2581-2586

24. Bergamaschi S, Binetti G, Govoni S, Wetsel WC, Battaini F, Trabucchi M, Bianchetti A, Racchi M (1995) Defective phorbolester-stimulated secretion of β-amyloid precursor protein from Alzheimer's disease fibroblasts. Neurosci Lett 201: 1-4

25. Racchi M, Bergamaschi S, Govoni S, Wetsel WC, Bianchetti A, Binetti G, Battaini F, Trabucchi M (1994) Characterization and distribution of protein kinase C isoforms in human skin fibroblasts. Arch Biochem Biophys 314: 107-111

26. Vestling M, Adem M, Lannfelt L, Cowburn RF (1995) Protein kinase C levels and activity in cultured skin fibroblasts from affected and non-affected members of the Swedish family with the amyloid precursor protein 670/671 mutation. Soc Neurosci Abs 21: 774.14

27. Huang HM, Gibson GE (1993) Altered β-adrenergic receptor stimulated cAMP formation in cultured skin fibroblasts from Alzheimer donors. J Biol Chem 268: 14616-14621

28. Vestling M, Adem A, Racchi M, Gibson GE, Lannfelt L, Cowburn RF (1997) Differential regulation of adenylyl cyclase in fibroblasts from sporadic and familial Alzheimer's disease cases with PS1 and APP mutations. Neuroreport 8: 2031-2035

29. Huang HM, Lin TA, Sun GY, Gibson GE (1995) Increased inositol 1,4,5-trisphosphate accumulation correlates with an up-regulation of bradykinin receptors in Alzheimer's disease. J Neurochem 64: 761-766

30. Glenner GC, Wong CW (1984) Alzheimer's disease: initial report of the purification and characterization of a novel cerebrovascular amyloid protein. Biochem Biophys Res Commun 120: 885-890

31. Kang J, Lemaire HG, Unterbeck A, et al (1987) The precursor of Alzheimer's disease amyloid A4 protein resembles a cell-surface receptor. Nature 325: 733-736

32. Ponte P, Gonzalez-De Whitt P, Schilling J, et al (1988) A new A4 amyloid mRNA contains a domain homologous to serine proteinase inhibitors. Nature 331: 525-527

33. Tanzi RE, McClatchey AI, Lamperti ED, et al (1988) Protease inhibitor domain encoding by an amyloid protein precursor mRNA associated with Alzheimer disease. Nature 331: 528-530

34. Kitaguchi N, Takahashi Y, Tokushima Y, et al (1988) Novel precursor of Alzheimer's disease amyloid protein shows protease inhibitory activity. Nature 331: 530-532

35. Goldgaber D, Lerman MI, McBride OW, Saffiotti U, Gajdusek C (1987) Characterization and chromosomal localization of a cDNA encoding brain amyloid of Alzheimer disease. Science 235: 877-880

36. Schellenberg GD (1995) Genetic dissection of Alzheimer disease, a heterogeneous disorder. Proc Natl Acad Sci USA 92: 8552-8559

37. Iversen LL, Mortishire-Smith RJ, Pollack SJ, Shearman MS (1995) The toxicity in vitro of beta-amyloid protein. Biochem J 311: 1-16

38. Sherrington R, Rogaev EI, Liang Y, et al (1995) Cloning of a gene bearing missense mutations in early-onset familial Alzheimer's disease. Nature 375: 754-760

39. Bird TD, Lampe TH, Nemens EJ, et al (1988) Familial Alzheimer's disease in American descendants of the Volga-Germans: probable founder effect. Ann Neurol 23: 25-31

40. Rogaev EI, Sherrignton R, Rogaeva EA, et al (1995) Familial Alzheimer's disease in kindreds with missense mutations in a gene on chromosome 1 related to the Alzheimer's disease type 3 gene. Nature 376: 775-778

41. Levy-Lahad E, Wasco W, Poorkaj P, et al (1995) Candidate gene for the chromosome 1 familial Alzheimer's disease locus. Science 269: 973-977

42. Levitan D, Greenwald I (1995) Facilitation of lin-12-mediated signalling by sel-12, a *Caenorhabditis elegans* S182 Alzheimer's disease gene. Nature 377: 351-354

43. Wolozin B, Iwasaki K, Vito P, Ganjei JK, Lacana E, Sunderland T, Zhao B, Kusiak JW, Wasco W, D'Adamio L (1996) Participation of presenilin 2 in apoptosis: enhanced basal activity conferred by an Alzheimer mutation. Science 274: 1710-1713

44. Pericak-Vance MA, Bebout JL, Gaskell PC Jr, et al (1991) Linkage studies in familial Alzheimer's disease - evidence for chromosome 19 linkage. Am J Hum Genet 48: 1034-1050

45. Saunders AM, Strittmatter WJ, Schmechel D, et al (1993) Association of apolipoprotein E allele epsilon 4 with late-onset familial and sporadic Alzheimer's disease. Neurology 43: 1467-1472

46. Frisoni GB, Govoni S, Geroldi C, Bianchetti A, Calabresi L, Franceschini G, Trabucchi M (1995) Gene dose of the e4 allele of Apolipoprotein E and disease progression in sporadic late onset Alzheimer's disease. Ann Neurol 37: 596-604

47. Payami H, Schelllenberg GD, Zareparsi Z, Kaye J, Sexton GJ, Head MA, Matsuyama SS, Jarvik LF, Miller B, Mc Manus DQ, Bird TD, Katzman R, Heston L, Nornan D, Small GW (1997) Evidence for association of HLA-A2 allele with onset age of Alzheimer's disease. Neurology 49: 512-518

48. Pericak-Vance MA, Bass MP, Yamaoka LH, Gaskell PC, Scott WK, Terwedow HA, Menold MM, Conneally PM, Small GW, Vance JM, Saunders AM, Roses AD, Haines JL (1997) Complete genomic screen in late-onset familial Alzheimer disease. Evidence for a new locus on chromosome 12. JAMA 278: 1237-1241

49. Sisodia SS (1992) Beta-amyloid precursor protein cleavage by a membrane bound protease. Proc Natl Acad Sci USA 89: 6075-6079

50. Shoji M, Golde TE, Ghiso J, Cheung TT, Estus S, Shaffer LM, Cai X-D, McKay DM, Tintner R, Frangione B, Younkin SG (1992) Production of the Alzheimer amyloid b protein by normal proteolytic processing. Science 258: 126-129

51. Checler F (1995) Processing of the β-amyloid precursor protein and its regulation in Alzheimer's disease. J Neurochem 65: 1431-1444

52. Citron M, Vigo-Pelfrey C, Teplow DB, Miller C, Schenk D, Johnston J, Winblad B, Venizelos N, Lanfelt L, Selkoe DJ (1994) Excessive production of amyloid b-protein by peripheral cells of symptomatic and presymptomatic patients carrying the Swedish familial Alzheimer disease mutation. Proc Natl Acad Sci USA 91: 11993-11997

53. Scheuner D, Eckman C, Jensen M, Song X, Citron M, Suzuki N, Bird TD, Hardy J, Hutton M, KukullW, Larson E, Levy-Lahad E, Vitanen M, Peskind E, Poorkaj P, Schellenberg G, Tanzi R, Wasco, W, Lannfelt L, Selkoe D, Younkin S (1996) Secreted amyloid b-protein similar to that in the senile plaques of Alzheimer's disease is increased in vivo by the presenilin 1 and 2 and APP mutations linked to familial Alzheimer's disease. Nat Med 2: 864-870

54. Mattson MP, Barger SW, Cheng B, Lieberburg I, Smith-Swintosky VL, Rydel RE (1993) β-amyloid precursor protein metabolites and loss of neuronal Ca^{2+} homeostasis in Alzheimer's disease. Trends Neurosci 16: 409-414

55. Mattson MP, Cheng B, Davis D, Bryant K, Lieberburg I, Rydell RE (1992) Beta amyloid peptides destabilize calcium homeostasis and render human cortical neurons vulnerable to excitotoxicity. J Neurosci 12: 376-389

56. Govoni S, Gasparini L, Racchi M, Trabucchi M (1996) Peripheral cells as an investigational tool for Alzheimer's disease. Life Sci 59: 461-468

57. Gibson G, Martins R, Blass J, Gandy S (1996) Altered oxidation and signal transduction systems in fibroblasts from Alzheimer patients. Life Sci 59: 477-489

58. Parker WD, Filley CM, Parks JK (1990) Cytochrome c oxidase deficiency in Alzheimer's disease. Neurology 40: 1302-1303

59. Curti D, Rognoni F, Gasparini L, Cattaneo A, Paolillo M, Racchi M, Zani L, Bianchetti A, Trabucchi M, Bergamaschi S, Govoni S (1997) Oxidative metabolism in cultured fibroblasts derived from sporadic Alzheimer's disease (AD) patients. Neurosci Lett 235: 1-4

60. Davis RE, Miller S, Herrnstadt C, Ghosh SS, Fahy E, Shinobu LA, Galasko D, Thal LJ, Beal MF, Howell N, Parker DW (1997) Mutations in mitochondrial cytochrome c oxidase genes segregate with late onset Alzheimer's disease. Proc Natl Acad Sci USA 94: 4526-4531

61. Hirashima N, Etcheberrigaray R, Bergamaschi S, Racchi M, Battaini F, Binetti G, Govoni S, Alkon DL (1996) Calcium responses in human fibroblasts: a diagnostic molecular profile for Alzheimer's disease. Neurobiol Aging 17: 549-555

62. Etcheberrigaray R, Ito E, Oka K, Tofel-Grehl B, Gibson GE, Alkon DL (1993) Potassium channel dysfunction in fibroblasts identifies patients with Alzheimer disease. Proc Natl Acad Sci USA 90: 8209-8213

63. Ito E, Oka K, Etcheberrigaray R, Nelson TJ, McPhie DL, Tofel-Grehl B, Gibson GE, Alkon DL (1994) Internal Ca^{2+} mobilization is altered in fibroblasts from patients with Alzheimer disease. Proc Natl Acad Sci USA 91: 534-538

64. Di Luca M, Pastorino L, Cattabeni F, Zanardi R, Racagni G, Smeraldi E, Perez J (1996) Abnormal pattern of platelet APP isoforms in Alzheimer's disease and Down syndrome. Arch Neurol 53: 1162-1166

65. Seubert P, Vigo-Pelfrey C, Esch F, Lee M, Dovey H, Davis D, Sinha S, Schlossmacher M, Whaley J, Swindlehurst C, et al (1992) Isolation and quantification of soluble Alzheimer's beta-peptide from biological fluids. Nature 359: 325-327

66. Gasparini L, Racchi M, Binetti G, Trabucchi M, Solerte SB, Alkon D, Etcheberrigaray R, Gibson G, Blass J, Paoletti R, Govoni S (1998) Peripheral markers in testing pathophysiological hypotheses and diagnosing Alzheimer's disease. FASEB J 12: 17-34

67. Iwatsubo T, Odaka A, Suzuki N, Mizusawa H, Nukina N, Ihara Y (1994) Visualization of Aβ42(43) and Aβ40 in senile plaques with end-specific monoclonals: evidence that an initially deposited species is Aβ42(43). Neuron 13: 45-52

68. Jensen M, Song XH, Suzuki N, Lanfelt L, Younkin SG (1995) The βAPP670/671 Alzheimer mutation (Swedish) increases plasma amyloid b protein concentration. Soc Neurosci Abs 21: 1501

69. Younkin LH, Eckman CB, Yager D, Graff-Rafford N, Younkin SG (1996) Analysis of plasma Aβ concentration in sporadic AD patients and non-demented subjects of all ages. Neurobiol Aging 17[4S]: S167

70. Arai H, Terajima M, Miura M, Higuchi S, Muramatsu T, Machida N, Seiki H, Takase S, Clark CM, Lee VM-Y, Trojanowski JQ, Sasaki H (1995) Tau in cerebrospinal fluid: a potential diagnostic marker in Alzheimer's disease. Ann Neurol 38: 649-652

71. Kennard ML, Feldman H, Yamada T, Jefferies WA (1996) Serum levels of the iron binding protein p97 are elevated in Alzheimer's disease. Nat Med 2: 1230-1235

72. O'Brien MD (1994) How does cerebrovascular disease cause dementia? Dementia 5: 126-133

73. Brun A (1994) Pathology and pathophysiology of cerebrovascular dementia: pure subgroups of obstructive and hypoperfusive etiology. Dementia 5: 145-147

74. Zhang WW, Badonic T, Hooh A, Jiang MH, Ma KC, Nie JX, Olsson Y, Sourander P (1994) Structural and vasoactive factors influencing Intracerebral arterioles in cases of vascular dementia and other cerebroavscular disease: a review. Dementia 5: 153-162

75. Pasquier F, Leys D (1997) Why are stroke patients prone to develop dementia? J Neurol 244: 135-142

76. Gottfries CG, Blennow K, Karlsson I, Wallin A (1994) The neurochemistry of vascular dementia. Dementia 5: 163-167

77. Tournier-Lasserve E, Joutel A, Melki J, Weissenbach J, Lathrop GM, Chabriat H, Mas JL, Cabanis EA, Baudrimont M, Maciazek J, et al (1993) Cerebral autosomal dominant arteriopathy with subcortical infarcts and leukoencephalopathy maps to chromosome 19q12. Nat Genet 3: 256-259

78. Joutel A, Corpechot C, Ducros A, Vahedi K, Chabriat H, Mouton P, Alamowitch S, Domenga V, Cecillion M, Marechal E, Maciazek J, Vayssiere C, Cruaud C, Cabanis EA, Ruchoux MM, Weissenbach J, Bach JF, Bousser MG, Tournier-Lasserve E (1996) Notch3 mutations in CADASIL, a hereditary adult-onset condition causing stroke and dementia. Nature 383: 707-710

79. Baudrimont M, Dubas F, Joutel A, Tournier-Lasserve E, Bousser MG (1993) Autosomal dominant leukoencephalopathy and subcortical ischemic stroke. A clinicopathological study. Stroke 24: 122-125

80. Frisoni GB, Calabresi L, Geroldi C, Bianchetti A, D'Acquarica AL, Govoni S, Sirtori CR, Trabucchi M, Franceschini G (1994) Apolipoprotein E epsilon 4 allele in Alzheimer's disease and vascular dementia. Dementia 5: 240-242

81. The Lund and Manchester Groups (1994) Clinical and neuropathological criteria for frontotemporal dementia. J Neurol Neurosurg Psychiatry 57: 416-418

82. Foster NL, Wilhelmsen K, Sima AAF, Jones MZ, D'Amato CJ, Gilman S, and Conference Participants (1997) Frontotemporal dementia and Parkinsonism linked to chromosome 17: a consensus conference. Ann Neurol 41: 706-715

83. Von Sattel JP, Binetti G, Welley LM (1976) Pick disease from molecular mechanism of dementia. In: Wasco W, Tanzi R (eds) Molecular basis of dementia. Humana, Totowa NJ, pp 253-269

84. Kosaka K, Iseki E (1996) Dementia with Lewy bodies. Curr Opin Neurol 9: 971-975

85. Kalra S, Bergeron C, Lang AE (1996) Lewy body disease and dementia. A review. Arch Intern Med 156: 487-493

86. Dickson DW, Wu E, Crystal HA, Mattiace LA, Yen SHC, Davies P (1992) Alzheimer's disease and age related pathology in difuse Lewy body disease. In: Boller F, Forette F, Khachaturian Z, Poncet M, Christen Y (eds) Heterogeneity of Alzheimer's disease. Springer, Berlin Heidelberg New York, pp 168-186

87. Liberini P, Valerio A, Memo M, Spano P (1996) Lewy body dementia and responsiveness to cholinesterase inhibitors: a paradigm for heterogeneity of Alzheimer's disease? Trends Pharmacol Sci 17: 155-160

88. Katzman R, Galasko D, Saitoh T, Tal LJ, Hansen L (1995) Genetic evidence that the Lewy body variant is indeed a phenotypic variant of Alzheimer's disease. Brain Cogn 28: 259-265

89. Goldfarb LG, Brown P (1995) The transmissible spongiform encephalopathies Annu Rev Med 46: 57-65

90. Parchi P, Gambetti P (1995) Human prion diseases. Curr Opin Neurol 8: 286-293

91. Prusiner SB (1996) Molecular biology and pathogenesis of prion diseases. Trends in Biochem Sci 21: 482-487

92. Zerr I, Bodemer M, Otto M, Poser S, Windl O, Kretzschmar HA, Gefeller O, Weber T (1996) Diagnosis of Creutzfeldt-Jakob disease by two-dimensional gel electrophoresis of cerebrospinal fluid. Lancet 348: 846-849

93. Lee MK, Borchelt DR, Wong PC, Sisodia SS, Price DL (1996) Transgenic models of neurodegenerative diseases. Curr Opin Neurobiol 6: 651-660

94. Cole GM, Frautschy SA (1997) Animal models for Alzheimer's disease. Alzheimer's Dis Rev 2: 33-41

95. Games D, Adams D, Alessandrini R, Barbour R, Berthelette P, Blackwell C, Carr T, Clemens J, Donaldson T, Gillespie F, Guido T, Hagopian S, Johnson-Wood K, Khan K, Lee M, Leibowitz P, Lieberburg I, Little S, Masliah E, McConlogue L, Montoya-Zavala M, Mucke L, Paganini L, Penniman E (1995) Alzheimer-type neuropathology in transgenic mice overexpressing V717F beta-amyloid precursor protein. Nature 373: 523-527

96. Hsiao K, Chapman P, Nilsen D, Eckman C, Harigaya Y, Younkin S, Yang F, Cole G (1996) Correlative memory deficits, Aβ elevation and amyloid plaques in transgenic mice. Science 274: 99-102

97. Hsiao K, Scott M, Foster D, Groth DF, Dearmond FJ, Prusiner SB (1990) Spontaneous neurodegeneration in transgenic mice with mutant prion protein. Science 250: 1587-1590

98. Telling GC, Scott M, Mastrianni J, Gabizon R, Torchia M, Cohen FE, Dearmond SJ, Prusiner SB (1995) Prion propagation in mice expressing human and chimeric PrP transgenes implicates interaction of cellular PrP with another protein. Cell 83: 79-90

99. Forloni G (1996) Neurotoxicity of beta amyloid and prion peptides. Curr Opin Neurol 9: 492-500

100. Forloni G, Angeretti N, Chiesa R, Monzani E, Salmona M, Bugiani O, Tagliavini F (1993) Neurotoxicity of a prion protein fragment. Nature 362: 543-546

101. Yankner BA (1996) Mechanisms of neuronal degeneration in Alzheimer's disease. Neuron 16: 921-932

102. Behl C, Davis JB, Klier FG, Shubert D (1994) Amyloid beta peptide induces necrosis rather than apoptosis. Brain Res 645: 253-264

Classification and Diagnosis of Dementias

F. Boller and L. Traykov

Diagnosis of Dementia

Definition of Dementia

Dementia is a clinical state characterized by a decline in multiple cognitive functions that interferes with a subject's ability to function on a daily basis, with a preserved level of consciousness, at least up to its most advanced stage. In an effort to operationalize the definition of dementia, attempts have been made to specify the type and degree of the cognitive decline. Most criteria require that multiple domains of cognitive functions be affected and that the cognitive dysfunction be clinically significant. This approach helps distinguish the demented patient from patients with restricted brain lesions [e.g., with a cerebrovascular accident (CVA) producing aphasia] and from the cognitive changes that occur with normal aging.

The study of dementias has had important developments in the last 20 years. At first, research was mostly limited to the investigation of Alzheimer's disease (AD), but more recently the interest of neurologists, neuropsychologists and neuroscientists has included the study of vascular dementias (VaD), the non-Alzheimer degenerative dementias (NADD), and other disorders causing dementia.

Dementia is not a disease. It is a symptom-complex reflecting a pathological process that is not necessarily located in the brain. The final result is well known to the physician and it is not difficult to make the diagnosis of dementia, especially in its late phases. However, several problems remain in terms of diagnosis, particularly in its early stages. To provide a few examples that will be developed in this chapter, the clinical diagnosis of very early dementia and its differentiation from normal aging remain problematic. Recent studies [1, 2] indicate that performance which is below age norms but not low enough to meet conventional criteria for dementia identifies a population at high risk for developing a progressive dementing disorder. Another problem is the differential diagnosis between different forms of dementia, which is complicated by the fact there is a genuine overlap between the pathology of AD and the pathology observed in many other degenerative disorders.

Yet early detection is crucial for many reasons. Even an early clinical diagnosis or late preclinical diagnosis of dementia may still be too late for effective therapy. The neuropathological lesions may have progressed without the appearance of severe behavioral changes, leaving too little intact tissue to benefit from a pharmacological intervention. If the ultimate treatment or, even better, prevention of dementias is to be used effectively, clinicians must be equipped with methods for identifying the dementing syndrome at its earliest possible stage before the disability has reached significant proportions.

The present chapter will describe the diagnostic criteria currently in use and will point out their discrepancies concerning the early identification of the dementing syndrome. We will then discuss the differentiation between early dementia and normal aging. Finally, after a short description of the existing classifications of dementias, the diagnostic criteria of some of them will be presented and the differential diagnosis discussed.

Diagnostic Criteria for Dementia

The most widely used criteria for dementia are those developed by the World Health Organization (WHO) for the *International Classification of Diseases* (ICD), tenth revision, 1989 [3] and by the American Psychiatric Association (APA) for the *Diagnostic and Statistical Manual of Mental Disorders* (DSM), third and fourth editions [4, 5]. ICD-10 gives a simple definition of dementia as "a syndrome due to disease of the brain, usually of a chronic or progressive nature, in which there is an impairment of multiple higher cortical functions, including memory, thinking, orientation, comprehension, calculation, learning capacity, language and judgment." It states that dementia must occur in the presence of clear consciousness, must be of such a degree as to impair activities of daily living (ADL), and that symptoms should be present for at least 6 months.

DSM defines dementia in greater detail and suggests specific criteria essential for its diagnosis. The third edition published in 1980 (with a 1987 revision, DSM-III-R) by the APA [4] (Table 1), represented important progress compared to the term "organic brain syndrome" used in DSM-I and DSM-II, which on one the hand considered dementia as an invariably irreversible condition and on the other did not encourage investigations aimed at establishing a precise neuropathological etiology.

The concept of reversibility or irreversibility is absent from DSM-III, which has introduced instead the need for a clinical decision concerning whether the main disorders (memory loss, impaired judgment, aphasia and personality disorders) are of sufficient severity to impair everyday life.

Even though the ICD-10 criteria for dementia are not very different from those of DSM-III-R, a comparison between them shows some dissimilarities in item composition and diagnostic algorithms. ICD-10 does not include the assessment of "other disturbances of higher cortical functions such as aphasia, apraxia, agnosia and constructional difficulty" that are useful for interrater reproducibility. Moreover, only the DSM-III-R requires a clear effort aimed at establishing the

Table 1. DSM-III-R criteria for dementia. (Adapted from [4])

A. Demonstrable evidence of short-term and long-term memory impairment.

B. At least one of the following:
   (1) Impairment in abstract thinking.
   (2) Impaired judgment.
   (3) Other disturbances of higher cortical functions such as aphasia, apraxia, agnosia
       and constructional difficulty.
   (4) Alteration of personality, i.e., alteration or accentuation of premorbid traits.

C. A and B must be of sufficient severity to interfere with work or the usual social
   activities or relationships with others.

D. Does not occur exclusively in the context of altered consciousness or delirium.

E. Either (1) or (2):
   (1) Evidence from the history, physical examination or laboratory tests of a specific
       organic factor that is judged to be etiologically related to the disturbance.
   (2) In the absence of such evidence, an etiologic organic factor can be presumed if
       conditions other than organic mental disorders have been reasonably excluded.

presence of a central nervous system (CNS) lesion. On the other hand, the advantage in the diagnostic algorithm of ICD-10 is that it suggests a time period over which symptoms should be present (6 months).

The DSM-III criteria are highly reliable. A multicentric study led by Françoise Forette, which besides Paris included the University of Bordeaux (Dr. Orgogozo and Dr. Dartigues) and of Grenoble (Dr. Hugonot and Dr. Israel), has shown that the clinical diagnosis of dementia based on these criteria is quite robust [6]. The reliability of these criteria was estimated by comparing the diagnosis made at 1-year intervals on 55 consecutive subjects with suspected cognitive impairment. Fifty-two of the 55 subjects were given the same diagnosis a year later, indicating a reliability of 95% for the DSM-III criteria.

On the other hand, the use of both DSM-III-R and ICD-10 criteria for diagnosis of dementia in a cross national research study proved to be equally reliable and demonstrated high interrater reliability for the diagnosis of dementia syndrome [7]. This multicentric study with the participation of research centers in six countries suggests that clinicians from different cultures and medical traditions can use the two sets of criteria effectively and thus reliably identify dementia cases in cross national research.

A modification of DSM-III was published recently [5] (Table 2). The criteria shown in the table do not refer to dementia but to "dementia of the Alzheimer type." The table shows that reference to abstract thinking and judgment has been replaced by the term "executive functioning," which is more modern and perhaps easier to test objectively. The diagnosis puts even more emphasis on the definition of the concept of cognitive and functional loss ("significant decline"). On the

Table 2. DSM-IV criteria for dementia of the Alzheimer type. (Adapted from [5])

A. The development of multiple cognitive deficits as manifested by both:
   (1) Memory impairment (inability to learn new information and inability to recall previously learned knowledge).
   (2) Cognitive impairment manifested by at least one of the following:
       (a) Aphasia (language disturbance).
       (b) Apraxia (inability to carry out motor activities despite intact comprehension and motor function).
       (c) Agnosia (failure to recognize or identify objects despite intact sensory function).
       (d) Disturbance in executive functioning (i.e., planning, organizing, sequencing, abstracting).

B. The course is characterized by continuing and gradual cognitive and functional decline.

C. The cognitive deficits cause significant impairment in social or occupational functioning and represent a significant decline from a previously higher level of functioning.

D. Other specific causes of dementia have been considered and ruled out, as evidenced by ALL of following:
   (1) Absence of other nonpsychiatric medical conditions of the CNS that cause progressive deficits in memory and cognition (e.g., cerebrovascular disease, Parkinson's disease, Huntington's disease, subdural hematoma, normal pressure hydrocephalus).
   (2) Absence of any systemic nonpsychiatric medical conditions that are known to cause dementia (e.g., hypothyroidism, vitamin B_{12} or folic acid deficiency, niacin deficiency, hypercalcemia, neurosyphilis).
   (3) Absence of alcohol or other substance-induced dementia.

E. The deficits do not occur exclusively during the course of delirium.

F. Not better accounted for by another axis I disorder (e.g., major depressive disorder, schizophrenia)

---

other hand, the presence of personality disorders is no longer included among the diagnostic criteria.

In recent years, progress in our ability to diagnose dementia has been achieved thanks to the efforts of several publications originating from different consensus groups.

A considerable improvement in the approach to dementing illnesses and especially to AD was made possible by the criteria set by a consensus conference convened in Bethesda and known under the acronym of NINCDS/ADRDA [8] (see below). Prior to the development of the NINCDS/ADRDA criteria, the most commonly used definitions were those of the DSM-III-R. While compatible, there are several notable differences in the NINCDS/ADRDA criteria: (1) intellectual dysfunction must be demonstrable on formal mental status and neuropsychological testing; (2) at least two areas of cognition (unspecified) are required for diagnosis but, unlike the DSM-III-R, memory does not necessarily have to be one of them; (3) specific evidence of deterioration in social or occupational functioning is not required.

The requirements of the NINCDS/ADRDA criteria for diagnosing dementia, while useful, may delay the diagnosis of some individuals, especially in the early stages of the disease. Several problem areas can be outlined.

First, brief cognitive scales can be insensitive to mild degrees of cognitive impairment, particularly in well-educated subjects. Although cutoff scores for dementia have been recommended for most of these tests, the diagnosis of dementia can often be made before these levels have been reached. Second, the criteria established by the work group do not require evidence of functional decline, but this information can often be invaluable in making an early diagnosis. Finally, if the diagnosis of dementia requires the impairment of at least two areas of cognition, many patients who are initially affected in one area such as language, visuospatial ability or executive functions do not the meet criteria for diagnosis until later in the course of their illness.

The criteria currently recommended for the diagnosis of VaD are discussed in detail in a later section of this chapter. One important preliminary question is the differentiation of early dementia from normal aging.

Dementia and Normal Aging

Our understanding of neuropsychological changes in normal aging has evolved considerably in recent years [9, 10]. Most studies agree that age is accompanied by an overall decline in cognitive functions as measured by IQ tests. A more detailed analysis shows that, in general, verbal functions (corresponding to so-called crystallized intelligence [11]) are relatively preserved in aging subjects, while non-verbal functions ("fluid intelligence") tend to show a decline. This is reflected by the results of classical intelligence tests such as the Wechsler Adult Intelligence Scale (WAIS): up to age 75, there tends to be little or no changes in verbal subtests such as vocabulary or arithmetic. In contrast, digit-symbol substitution, object assembly and other performance subtests often show changes starting at age 60 [12]. More detailed neuropsychological tests show that in general, the most important cognitive changes involve attention (particularly divided attention), some visuospatial functions and some processes involved in learning and remembering. Some authors have found, however, that if assessment is restricted to elderly subjects who are completely free of any disease, the effect of aging on cognitive functions is mild.

A working party of the International Psychogeriatric Association (IPA) in collaboration with the WHO proposed diagnostic criteria for "aging-associated cognitive decline" (AACD) to categorize subjects with cognitive decline falling short of dementia [2]. The criteria include the presence for at least 6 months of subjective gradual cognitive and objective evidence of abnormal performance in any principal domain of cognition. The abnormality is defined as performance at least one standard deviation below the age and education norms in well-standardized neuropsychological tests. Furthermore, there must be no evidence of any medical condition known to cause cerebral dysfunction. Thus, the diagnosis of AACD identifies persons with subjective and objective evidence of cognitive decline, which does not evolve toward a dementia. AACD has to be differentiated from "mild cognitive disorder" (MCD), a classification included in the research criteria for ICD-10 by WHO [3]. This diagnosis is used only when there is an indication of a disease known to cause cerebral dysfunction. A concept akin to AACD is included in DSM-IV [5] as "age-related cognitive decline" (ARCD) and defined as "an objectively identified decline in cognitive functioning consequent to the aging process that is within normal limits given the person's age."

In everyday life, memory complaints are by far the most frequent "cognitive symptom" of elderly persons. How often does this correspond to impaired performance, and how often is it the prelude to an actual dementia? Having found that at least in certain cases partial amnesia does not evolve toward an actual dementia, Kral first proposed the existence of a particular entity called "benign senescent forgetfulness" [13]. A somehow comparable notion, but more closely tied to neuropsychological tests, has been introduced by Crook et al. under the name of "age-associated memory impairment" (AAMI) [14]. A recent paper [1] has proposed that the memory performance of elderly persons be looked at not only in light of the performance of young adults, but also in relation to the performance of other elderly subjects. On this basis, it is proposed that "age-consistent memory impairment" (ACMI) be identified for elderly subjects who perform within one standard deviation (SD) of the mean for their age. Subjects who perform less well (between 1 and 2 SD) would fall within the category of "late life forgetfulness" (LLF) [15]. Those who perform below two SD are suspected to have pathological impairment. On the other hand, the same paper [1] has proposed that we take a new look at the relationship of memory and aging by considering separately memory complaints and memory performance. By so doing, four groups of elderly subjects could theoretically be identified:

1. Subjects with no complaints and with normal performance. This group is considered normal or perhaps "supernormal." In other words, this group might represent persons who have had a high educational level and are completely disease-free (like the group studied by Rapoport et al. [16]).

2. Subjects with high-level complaint, but with performances corresponding to those of normal subjects of the same age. Psychoaffective disorders (such as depression) might be responsible in some of these cases, but this group may also include subjects with a high intellectual level, a mild impairment which may not be detected by common neuropsychological tests.

3. Subjects with a memory complaint associated with an abnormally low performance. This group represents the greatest challenge and is in fact probably heterogeneous. It may include some depressed subjects and some subjects in the early phase of AD (where anosognosia is frequent but is not always present). It seems reasonable to hypothesize that a certain percentage of these subjects may have a benign, non-evolving form of amnesia, perhaps similar to the deficit postulated by Crook et al. under the label of AAMI [14].

4. The last group includes subjects with no memory complaints, but clearly impaired performance. These are likely to be patients in the early phases of AD or related disorders.

The Consortium to Establish a Registry for Alzheimer's Disease (CERAD) is also in the process of undertaking a study of very early changes affecting memory. This has led to the development of a protocol entitled "possible dementia prodromes" (PDP). For entry into the PDP protocol, subjects should be at least 50 years old with mild memory loss of at least 6 months' duration, minor behavioral or language disorders, or episodes of confusion that do not justify a diagnosis of dementia, and clinical dementia rating (CDR) of 0.5.

All these criteria leave open the question of progression. Only a large-scale longitudinal study will be able to ascertain the reality of the cognitive decline in elderly subjects, its possible pathology and its evolution. In conclusion, it is important to stress that major loss of memory and a sizable intellectual decline are not part of normal aging and represent clear indications for further workup.

Classification of Dementia

Several classification systems for dementia are currently in use, taking into account: (1) the specific disease underlying the dementia; (2) the localization of the pathological process; (3) the brain structures that are involved; (4) the DSM and ICD multiaxial coding system in which each axis corresponds to a different class of information. Other classification criteria could take into account the incidence and prevalence of the different forms according to age, sex, and other factors that contribute to the development of dementia (see chapter by Di Carlo et al., this volume).

Etiologic Classification

Once a diagnosis of dementia has been confirmed, the diagnostic task becomes that of determining the likely underlying pathological cause. Dementia, defined as multiple domains of cognitive impairment, is a syndrome with different causes encompassing infections, intoxication, trauma, and vascular, genetic, metabolic and neoplastic disease. The causes most likely to occur in clinical practice can be classified as follows:

a. Degenerative: AD, NADDs [Pick's disease, frontal lobe degeneration (FLD), primary progressive aphasia (PPA), diffuse Lewy body disease (DLBD), Huntington's and Parkinson's disease (PD), progressive supranuclear palsy (PSP)]
b. Vascular: multi-infarct dementia, lacunar state, Binswanger's disease
c. Infections: neurosyphilis, AIDS-dementia complex (ADC) and possibly prion-related diseases such as Kuru, Creutzfeldt-Jakob disease, fatal familial insomnia and Gerstmann-Sträussler-Scheinker (GSS) syndrome
d. Trauma: sequelae of both open and closed head injury
e. Metabolic/endocrine: hypothyroidism, vitamin B_{12} deficiency, etc.

f. Toxic: alcohol-related syndromes, medication, etc.

g. Tumors: primary or metastatic brain tumor

This etiologic approach is not entirely consistent because of lack of a defined cause for many of the dementias such as AD. For classifying these diseases, the term "degenerative" is usually applied in spite of the absence of a uniformly accepted definition of degenerative disease. The search for the cause for a dementing state leads to a division of dementia: treatable and untreatable. The supporters of this division emphasize the need to exclude all treatable causes for dementia in the diagnostic algorithm of degenerative dementias that therefore run the risk of becoming a synonym for untreatable.

Classification According to the Localization of the Pathological Process

Conditions affecting practically every organ or organ system can produce a dementia or even present as a dementia. These conditions can be classified into three main categories according to the localization of the pathological process:

1. Those that affect other organs such as the liver, the kidneys, etc. without directly affecting the brain. This category also includes alcohol and other drugs. It has been stated that polypharmacy is the commonest cause of potentially treatable dementia.

2. Conditions affecting the brain and accompanied by neurological symptoms. These include tumors, traumas, CVAs, Huntington's, Creutzfeldt-Jakob and PD. Because of the characteristic clinical picture, these conditions do not usually raise any problems of differential diagnosis with AD except sometimes in the early stages. VaD will be discussed later.

3. Conditions characterized by a clinical picture essentially limited to dementia, sometimes referred to as primary dementias. They include AD as well as non-Alzheimer dementias such as Pick's disease and DLBD.

Cortical and Subcortical Dementia

The idea that lesions of the central gray nuclei and other subcortical structures may produce dementia is an old one, and the term "subcortical dementia" was introduced as far back as the 1930s [17], in reference to VaD. While discussing a case of PSP, Albert et al. [18] proposed the idea of a clinical picture characterized by memory loss, difficulties in manipulation of old knowledge and personality disorders (apathy with episodes of irritability) and slowing of the thinking process. At about the same time, McHugh and Folstein [19], while describing the clinical picture of Huntington's disease, proposed a picture consisting of a decline of "all cognitive functions," but without involvement of cortical functions (i.e, no aphasia, apraxia, etc.) and with marked personality changes (apathy and inertia). In their comments, they did not mention diseases other than PD proposed as a possible model of neurotransmitter deficit.

Several diseases (listed in Table 3) have been labeled subcortical dementia. Cummings [21] has proposed a series of clinical criteria that may differentiate cortical from subcortical dementia (Table 4).

Table 3. Diseases associated with the syndrome of subcortical dementia. (Adapted from [20])

Degenerative disorders
    Parkinson's disease
    Huntington's disease
    Progressive supranuclear palsy
    Idiopathic basal ganglia calcification ("Fahr's disease")
    Spinocerebellar degeneration
    Thalamic degeneration
    Progressive subcortical gliosis
Vascular disorders
    Lacunar state
    Thalamic vascular lesions
    Binswanger's disease
Metabolic disorders
    Wilson's disease
    Hypoparathyroidism
Multiple sclerosis and other demyelinating diseases
AIDS encephalopathy
Sarcoidosis
Normal pressure hydrocephalus
Dementia pugilistica
Neuro-Behcet disease
Dementia associated with depression ("pseudodementia")

Several researchers have expressed doubts concerning the validity of the concept of subcortical dementia. These doubts are based on different arguments ranging from neuroanatomical [22, 23], to conceptual [24] and empirical [25]. Several anatomical and neuroimaging studies have shown that lesions extend beyond the subcortical areas in PD [26], in Huntington's disease [27], in PSP [28], and in HIV encephalopathy [29]. It would therefore be much more rational to change the term "subcortical dementia" to "fronto-subcortical dementia" [30, 31].

Table 4. Contrasting characteristics of "cortical" and "subcortical" dementia. (Adapted from [21])

Characteristic	Subcortical dementia	Cortical dementia
Language	No aphasia	Aphasia
Memory	"Tendency of forgetfulness"	Amnesia
Cognition	"Dysexecutive" syndrome; bradyphrenia	Dementia (DSM-III-R)
Personality	Apathy	Unconcerned
Mood	Depression or other affective disturbances	Normal
Speech	Dysarthria	Normal
Posture	Impaired	Normal
Coordination	Impaired	Normal
Abnormal movements	Present frequently	Present rarely

In the conclusion of a review written a few years ago, Dubois et al. [32] concluded that the concept of subcortical dementia has a "heuristic" value in the sense that it underscores the anatomical and functional relationships between the cortex and subcortical structures, particularly the striatum. Like several other clinical concepts used in everyday practice, that of subcortical dementia remains probably useful, even though it does not always hold up to scientific analysis. Further discussion of this topic will be found in the sections dealing with Parkinson's and Huntington's disease, VaD and HIV dementia (ADC).

ICD and DSM Classifications of Dementia

The WHO's ICD is a systematic classification that is widely used to classify mortality and morbidity information for statistical purposes and to index hospital records by diseases. The main purpose of the tenth revision [3] is to provide more categories (there is a new alphanumeric format with 26 letters in the alphabet but only ten digits) and thus leave space for future expansion without the whole classification having to be changed.

The general format of Chapter V entitled "Mental, Behavioral and Developmental Disorders" (F00–F99) is very similar to that of the APA's recent classification because it incorporates many of the innovations introduced in DSM-III. Most categories are provided with operational criteria for each diagnosis, but the clinical descriptions and diagnostic guidelines in ICD-10 are less detailed and less restrictive. There is also provision for multiple axes, as in DSM. The multiaxial system of DSM-IV allows the systematic recording of different information sets: the dementing disorders are coded to axis I (clinical disorders), while the specific etiology, if known, is coded to axis III (general medical conditions). These conditions are classified outside the "Mental Disorders" chapter of ICD-9-CM (or outside Chapter V of ICD-10).

The classification of "organically" determined psychiatric states has been a source of confusion to clinicians for many years. In both systems of classification, the three main syndromes in this section are delirium, dementia and amnestic syndrome, but the term "organic" has been eliminated from DSM-IV "because it implies that the other disorders in the manual do not have an 'organic' component" [5]. The classification of dementias is broadly similar in both systems, including dementia of the Alzheimer's type (with early and with late onset), VaD, dementia due to Pick's, Creutzfeldt-Jakob's, Huntington's, Parkinson's and HIV diseases, and a specific listing of a variety of dementias due to general medical and neurological conditions. In DSM-IV are also added dementia due to head trauma, substance-induced persisting dementia, and dementia due to multiple etiologies. To indicate the predominant feature of the clinical presentation, only dementia of the Alzheimer's type and VaD have codable subtypes: with delirium, with delusions, with depressed mood, and uncomplicated. This organization is provided to assist in differential diagnosis. In addition, the categories correspond to current views on the frequency and importance of various dementias.

Alzheimer's Disease [33]

AD is a degenerative disease of the CNS that is characterized clinically by progressive dementia and, histologically, by senile plaques (SP) and neurofibrillary tangles (NFT). The disease usually starts after the age of 40, and its incidence increases with age. Despite a marked sharpening of diagnostic criteria and some new ancillary tests, the diagnosis is still based on the exclusion of other conditions and on probability.

NINCDS/ADRDA Criteria

We have already presented the DSM-IV criteria for the diagnosis of AD. A considerable improvement in the approach to AD was made possible by the NINCDS/ADRDA criteria [8] (Table 5). The most important contribution of these criteria, which have now been adopted by many research groups throughout the world, is a "probabilistic" approach to the diagnosis. According to the elements available and to the more or less typical symptoms and clinical history, the diagnosis of AD can be considered as either definite, probable or possible. Besides the criteria listed in Table 5, the NINCDS/ADRDA conference also listed a series of neuropsychological tests to be used in the evaluation of patients with suspected AD (see also [34] for a review). One may note that the NINCDS/ADRDA criteria pay little attention to the age of onset of the symptoms (as long as they occur after the age of 40) and reflect a current tendency to de-emphasize possible differences between presenile and senile forms of the disease.

The validity of the NINCDS/ADRDA criteria has been evaluated in relation to the clinical diagnosis of AD [35], as well as by the clinical-pathological correlations [36].

The Consortium to Establish a Registry for Alzheimer's Disease (CERAD) Criteria

Following the creation of the first ten Alzheimer Disease Research Centers (ADRC) by the National Institute of Aging in the United States, it became apparent that there was a need to establish clinical diagnostic criteria that would allow comparison of groups at different centers. Under the leadership of researchers from different centers (particularly Duke University in Durham and Washington University in St. Louis), the CERAD was established in the late 1980s [37]. Since its creation, CERAD has extended considerably and there are now many centers that use the battery, not only in the US, but also in Canada and in many others European countries.

It is important to stress that the CERAD criteria derive, with minor modifications, from those of the NINCDS/ADRDA Work Group [8]. The Clinical Assessment Battery contains semistructured interviews with the subject and with a person who knows him or her very closely (so-called "informant"). CERAD also includes general physical, neurological, and laboratory examinations (e.g., tests to exclude thy-

Table 5. NINCDS/ADRDA criteria for the diagnosis of AD. (Adapted from [8])

I.   Criteria for the clinical diagnosis of PROBABLE Alzheimer's disease include:
     – Dementia established by clinical examination and documented by the Mini-Mental
       Test, Blessed Dementia Scale, or some similar examination, and confirmed by
       neuro-psychological tests; deficits in two or more areas of cognition.
     – Progressive worsening of memory and other cognitive functions;
     – No disturbance of consciousness;
     – Onset between ages 40 and 90, most often after age 65; and
     – Absence of systemic disorders or other brain diseases that in and of themselves
       could account for the progressive deficits in memory and cognition.

II. The diagnosis of PROBABLE Alzheimer's disease is supported by:
     – Progressive deterioration of specific cognitive functions such as language (apha-
       sia), motor skills (apraxia), and perception (agnosia);
     – Impaired activities of daily living and altered patterns of behaviour;
     – Family history of similar disorders, particularly if confirmed neuropathologically; and
     – Laboratory results of: normal lumbar puncture as evaluated by standard tech-
       niques, normal pattern or nonspecific changes in EEG, such as increased slow-
       wave activity, and evidence of cerebral atrophy on CT with progression document-
       ed by serial observation.

III. Other clinical features consistent with the diagnosis of PROBABLE Alzheimer's
     disease, after exclusion of causes of dementia other than Alzheimer's disease, include:
     – Plateaus in the course of progression of the illness;
     – Associated symptoms of depression, insomnia, incontinence, delusions, illusions,
       hallucinations, catastrophic verbal, emotional, or physical outbursts, sexual disor-
       ders, and weight loss;
     – Other neurological abnormalities in some patients, especially with more advanced
       disease and including motor signs such as increased muscle tone, myoclonus, or
       gait disorder; seizures in advanced disease; and
     – CT normal for age.

IV. Features that make the diagnosis of PROBABLE Alzheimer's disease uncertain or
     unlikely include:
     – Sudden, apoplectic onset;
     – Focal neurologic findings such as hemiparesis, sensory loss, visual field deficits,
       and incoordination early in the course of the illness; and
     – Seizures or gait disturbances at the onset or very early in the course of the illness.

V.   Clinical diagnosis of POSSIBLE Alzheimer's disease:
     – May be made on the basis of the dementia syndrome, in the absence of other neu-
       rologic, psychiatric, or systemic disorders sufficient to cause dementia, and in the
       presence of variations in the onset, in the presentation, or in the clinical course;
     – May be made in the presence of a second systemic or brain disorder sufficient to
       produce dementia, which is not considered to be the cause of the dementia; and
     – Should be used in research studies when a single, gradually progressive severe cog-
       nitive deficit is identified in the absence of other identifiable cause.

VI. Criteria for diagnosis of DEFINITE Alzheimer's disease are: the clinical criteria for
     probable Alzheimer's disease and histopathologic evidence obtained from a biopsy
     or autopsy

VII. Classification of Alzheimer's disease for research purposes should specify features
     that may differentiate subtypes of the disorder, such as:
     – Familial occurrence;
     – Onset before age of 65;
     – Presence of trisomy-21; and
     – Coexistence of other relevant conditions such as Parkinson's disease.

roid disease or vitamin B$_{12}$ deficiency), a drug inventory, a depression scale, and a general medical history. The neuropsychological battery assesses language, memory, constructional praxis, and general intellectual status. Global dementia severity is staged in accordance with the Mini-Mental State Examination (MMSE), the dementia rating scale of Blessed for ADL, and the CDR scale [38]. Finally, each subject is given a diagnosis (no dementia, probable AD, or possible AD).

Thanks in part to the recently introduced criteria, the ability of clinicians to diagnose AD has increased considerably in recent years. If an autopsy is taken as a "gold standard," the percentage of correct diagnoses has raised from less than 50% [39] to about 80% 15 years later [40]. Two studies based on small series [41, 42] went as far as to state that diagnostic accuracy could reach 100%, but the most recent CERAD data [43] indicate a probably more realistic reliability of the order of 85%.

Vascular Dementia

Recent research on VaD has shown on the one hand a convergence of thoughts regarding the definition and relative ease of the diagnostic process, and on the other a number of problems, both theoretical and practical. These are related to the great varieties of etiologies involved and a great number of clinical syndromes and patterns of evolution [44, 45]. The causal relationship between the cerebrovascular events and the dementia continues to be debated.

The concept of VaD includes all dementing processes that follow vascular accidents hemorrhagic or ischemic [46, 47]. However, dementias that follow hemorrhagic accidents are relatively rare and have been much less studied than dementias of ischemic origin. This section will therefore deal mainly with ischemic vascular dementias (IVD).

The diagnostic approach to VaD generally the requires the presence of three elements: presence of a dementia, a cerebrovascular event and a causal relationship between the two. A number of international efforts have led to the elaboration of more reliable diagnostic criteria: the Alzheimer's disease diagnosis and treatment centers (ADDTC) criteria [48] and the NINDS-AIREN criteria [49]. These two sets of criteria allow specific studies of VaD. As stated above, the DSM-IV criteria [5] define dementia as a disorder of memory and at least one of the following disorders: aphasia, apraxia, agnosia and impairment of executive functions. The diagnosis of cerebrovascular disease requires either the observation of focal signs upon neurological examination, or the objective finding of vascular lesions by neuroimaging techniques. However, DSM-IV fails to propose operational criteria for establishing a causal relationship between the vascular events and the dementia.

Chui et al. [48] consider that a clinical judgment of global alterations of cognitive functions and their influence on ADL is more important than a detailed neuropsychological battery in making a precise diagnosis of dementia. Clinical examination and detection of cerebral lesions by neuroimaging techniques allow the determination of the presence of cerebrovascular lesions. Documentation of a temporal relationship between a vascular accident and dementia is necessary and sufficient to postulate a causal relationship between the two.

The NINDS-AIREN criteria [49] have an approach basically similar to DSM-IV, requiring the presence of deficits in two or more cognitive domains with impairment of ADL. Neuroimagery plays a determining role in diagnosing a CVA. Lack of proof of a CVA is indeed a decisive argument against the diagnosis of VaD. The causal relationship between CVA and dementia is established on the basis of their causal relationship and the evolution of the cognitive decline.

The typical symptoms of VaD are: sudden onset and stepwise evolution; presence of focal signs and symptoms and demonstration of cerebrovascular disease [4]. However, one of these elements is absent in about 15% of cases [50], leading to obvious diagnostic problems. The differential diagnosis becomes even more difficult to establish if one considers the frequent coexistence of vascular lesions with lesions typical of AD [40, 51].

It should be stressed that the use of "vascular scores" such as the one proposed by Hachinski et al. [52] and Loeb and Gandolfo [53] allows differentiation of degenerative dementias from VaD, but are of no use in cases of mixed dementias. In addition, these scores do not allow differentiation of subgroups of VaD.

Non-Alzheimer Degenerative Dementias

"Primary" degenerative diseases other than AD were considered to be mere curiosities until recently. Their importance has increased considerably thanks to the description of "new" entities. They are characterized clinically by focal signs, accompanied by neuroradiological and neuropathological evidence of focal atrophy, not always fitting the diagnostic criteria for Pick's disease. Increasing data suggest that NADD differ from AD not only from a clinical and neuropathological point of view, but also in their pattern of transmission and perhaps even in their response to therapy. This section will include a presentation of diagnostic criteria for the following entities:
1. Pick's disease
2. PPA
3. FLD
4. DLBD

Pick's disease

Pick's disease was long considered to be the main differential diagnosis to AD, but is actually quite uncommon. In his review, Knopman [54] suggests a prevalence rate between 0% and 5% among all dementias. Most clinical studies are based on series of less than six cases [55]. The onset of Pick's disease occurs earlier than that of AD (around 60) and men are affected more often than women. The diagnosis is based on the triad of focal neurological and neuropsychological signs in the context of clinically obvious dementia, cerebral atrophy with focal accentuation and, at autopsy, typical argyrophilic inclusions.

Following older contributions to a rational diagnosis [56, 57], Mendez et al. [55] have proposed elements allowing the differential diagnosis of Pick's disease from AD. Based on the retrospective review of 24 cases of autopsy-verified Pick, compared to 42 carefully matched cases of AD, each patient with Pick's disease had at least three of the following clinical signs, which were not found in any AD cases:
- "Presenile" onset of the disease
- Early personality disorders
- Hyperorality
- Disinhibition
- Roaming behavior

To these, one should add:
- Disorders of speech with relative preservation of auditory comprehension and absence of fluent aphasia
- Neuroradiological evidence of lobar atrophy

It must be emphasized that the neuroradiological signs are often found only in the later stages of the disease and that, in contrast, the neuropsychological profile is often "evanescent" [58]. As the disease progresses, it becomes harder and harder to distinguish it from other dementias.

Primary Progressive Aphasia

In 1982, Mesulam [59] described six patients affected by an aphasia that progressed quite slowly, without evidence of dementia. No neuropathological data were provided. Mesulam emphasized the difference with other known pathologies and stated that it was a new nosological entity. Other cases described previously [60, 61] seem to fit the same label. Since then, several other observations have been added and in a recent review, Weintraub and Mesulam [62] present 63 cases. The diagnostic criteria proposed by Mesulam and his co-workers [62] include mainly disorders of language affecting not only speech, with insidious onset and gradual progression. This deficit should remain practically isolated for at least 2 years. It may, however, be associated with disorders such as constructional apraxia and acalculia. Recent neuropathological data [63, 64] suggest that PPA might be considered a subgroup of Pick's disease (on this point see the contradictory point of view by Kertesz [63] and Duyckaerts et al. [66]). This conclusion does not contradict the existence of PPA as a separate clinical entity. PPA may have a familial occurrence [67].

Frontal Lobe Degeneration or Frontal Lobe Dementia

FLD was first described under that name by researchers from Lund [68]. Neary and his collaborators in Manchester [69] had previously shown that 25% of cases of a sample of 24 cases of dementia, followed clinically, and who had undergone a biopsy, did not fit the clinical and pathological criteria of AD. They concluded that cerebral atrophy and cognitive deterioration were not necessarily synonymous with AD. Since then, the two groups have presented numerous cases of FLD, showing that it occurs more often in males below 65 with half the cases having a positive family history for dementia.

The following diagnostic criteria have been proposed [70]:

A. Core diagnostic features

Behavioral disorder
- Insidious onset and slow progression
- Early loss of personal awareness
- Early loss of social awareness
- Early signs of disinhibition
- Mental rigidity and inflexibility
- Hyperorality
- Stereotyped and perseverative behavior
- Utilization behavior [71]
- Distractibility, impassivity, and impersistence
- Early loss of insight into the fact that the altered condition is due to a pathological change of own mental state.

Affective symptoms
- Depression, anxiety, excessive sentimentality, suicidal and fixed ideation, delusion
- Hypochondriasis, bizarre somatic preoccupation
- Emotional unconcern
- Amnesia

Speech disorder
- Progressive reduction of speech
- Verbal stereotypes
- Echolalia and perseveration
- Late mutism

Spatial orientation and praxis tend to be preserved

Physical signs
- Early primitive reflexes
- Early incontinence
- Late akinesia, rigidity, tremor
- Low and labile blood pressure

Investigations
- Normal EEG despite clinically evident dementia
- Brain imaging: predominant frontal or anterior temporal abnormality, or both
- Neuropsychology

B. Supportive diagnostic features
- Onset before 65
- Positive family history of similar disorder in a first-degree relative

– Bulbar palsy, muscular weakness and wasting, fasciculations
Finally, a set of exclusion features are proposed.

Diffuse Lewy Body Disease

DLBD is characterized by progressive dementia, often accompanied by psychiatric and extra-pyramidal manifestations. Histologically, one finds cytoplasmic inclusions in cortical neurons resembling the Lewy bodies found in subcortical structures in cases of PD [72]. Lesions identical to those of AD are very frequent. The prevalence of the disease seems to vary according to different centers. Hansen [73] found a histological picture compatible with DLBD in 36% of 36 demented patients. Other studies [74, 75] suggest that about 20% of demented patients may have DLBD, thus putting this conditioning second place, right after AD, among degenerative dementias. Other centers, however, have not found such a high prevalence [76]. Like in PD, males are more affected than females. The mean duration of the disease is around 6 years, much less than in AD or PD alone, but comparable to PD with dementia [77]. Table 6 summarizes the clinical features of the disease. Dementia shares some characteristics with that of AD, with important deficits of memory, language, visuo-spatial abilities and gestual behavior

Neuropsychological tests show greater visuo-spatial deficits than those of AD, while frontal signs are said to be less prominent [73]. It has been argued that a qualitatively differentiating feature between DLBD and AD is that the DLBD patients do worse in the copy than in the draw part of the clock-face test [78]. Dementia is often preceded by psychiatric symptoms such as confusion and visual hallucinations. A marked fluctuation of cognitive deficits is also said to be characteristic of DLBD. Among other neurological signs, one finds extrapyramidal signs and gait disorders [79], but these "parkinsonian" signs are not always found with the same frequency [80].

Table 6. Main clinical features in DLBD. (Adapted from [75])

Feature	Frequency	Note
Dementia	100	Present by definition
Fluctuation	>10	Noted in 80% of cases in some recent reports
Psychosis	33	Mainly visual hallucinations and paranoid ideation
Depression	15	
Parkinsonism	90	Includes rigidity (80), tremor (50), bradykinesia (40), gait disorders (50) and flexed posture (30)
Myoclonus	10	
Other abnormal movements	10	Include dystonia, chorea and drug-induced dyskinesias
Pyramidal signs	25	Usually occur late
Other brainstem signs	10	Include dysphagia and supranuclear gaze palsy
Autonomic failure	10	Hypotension orthostatic

Parkinson's Disease

PD affects mainly the motor system by producing tremor, rigidity, bradykinesia and gait disorders. Other disorders, cognitive and non-cognitive, contribute to the "ill-being" of the patients. It should be stated clearly, however, that many patients remain intellectually normal throughout the illness and that dementia is not a necessary consequence of PD.

How many PD patients become demented? The answer varies, according to different studies; the problem is complicated by the fact that the instruments customarily used to diagnose dementia (such as the DSM) or to quantify it (such as the MMS) are not adapted to PD. Mahieux et al. have developed a test that better suits that purpose [81]. Leaving aside these important methodological problems, most authors find that the prevalence of dementia is greater in PD than in age-matched controls. On average it is said to range from 30 to 40%. The features of dementia in PD are far from uniform. As stated above, several authors have included the dementia of PD among subcortical dementias. Current experience shows that especially in advanced cases "no consistent qualitative features are present to constitute a specific pattern of the dementia of PD" [82].

The differential diagnosis potentially includes all the condition where dementia is associated with parkinsonism. In practice, the differentiation rests mainly between AD and PD because cognitive impairment and extrapyramidal signs can occur in both. PD is the probable diagnosis when two of four cardinal motor features are present without evidence of dementia prior to the onset of the motor signs. Patients with AD and parkinsonism have milder extrapyramidal signs than

Table 7. CERAD criteria for the diagnosis of Parkinson's dementia

---

1. PROBABLE Parkinson's dementia
    a. Presence of dementia with memory loss and other cognitive impairment established by the CERAD clinical battery (CDR > 0.5);
    b. Absence within previous 6 months of medications capable of inducing extrapyramidal side effects;
    c. An established temporal relationship between Parkinson's disease and dementia, i.e., onset of Parkinson's disease occurred 1 year or more before dementia was recognized;
    d. At least two of the major extrapyramidal signs (tremor, rigidity, Parkinsonian gait, bradykinesia body) found on examination

2. POSSIBLE Parkinson's dementia
    a. Presence of dementia established by the CERAD clinical battery (CDR > 0.5); with or without memory loss;
    b. Absence within previous 6 months of medications capable of inducing extrapyramidal side effects;
    c. Only one of the major extrapyramidal signs found on examination;
    d. Two of more of the minor signs of extrapyramidal dysfunction (rising from a chair, posture, stability on displacement, turning, bradykinesia face) were found on examination, with a severity rating of 2 or greater.

---

seen in PD. Rigidity is the most common feature, whereas tremor and bradykinesia are relatively frequent. Taking in considerations these suggestions, CERAD has developed diagnostic criteria for the differentiation of AD patients with parkinsonism from those with dementia caused by PD (Table 7).

There must be a positive answer to all of the items in Part 1 or in Part 2 to warrant the diagnosis of, respectively, probable or possible Parkinson's dementia.

Progressive Supranuclear Palsy

PSP or Steele-Richardson-Olszewski disease [83] is a degenerative process involving several structures located mainly in the rostral portions of the brainstem. This is the condition for which the term "subcortical dementia" was proposed by Albert et al. [18].

Clinical features of PSP include, in addition to an extrapyramidal syndrome with marked axial rigidity, an ophthalmoplegia affecting mainly vertical (and especially downward) eye movements. There is also disturbance of sleep, while dementia is present in 20%–60% of cases. This "dementia" affects particularly memory, without being a genuine amnestic syndrome and produces also psychomotor slowing, personality changes and difficulty organizing visual perception [18, 84]. The personality changes consist mainly of apathy, depression and episodes of aggressivity.

The variability of symptoms, especially in the early phases of the disease, may lead to diagnostic difficulties in relation to other diseases with extrapyramidal syndromes. Many of the diseases shown in Table 8 are easily ruled out on the basis of the clinical history and ancillary procedures such as modern imaging techniques. Cerebrovascular diseases, communicating hydrocephalus and space-occupying lesions fall in that category. Other conditions listed in the table are exceedingly rare (Hallervorden-Spatz, neuroacanthocytosis), confined mainly to specific geographical areas (Parkinson-dementia complex of Guam) or seem to have disappeared (encephalitis lethargica). In practice, the differential diagnosis rests mainly with DLDB, PD and less commonly AD. On occasion, there can be confusion with Creutzfeldt-Jakob disease , but in that condition the evolution is generally much quicker. Pick's disease and the other non-Alzheimer's disease forms of cerebral atrophy [69], particularly dementia of frontal lobe type [85], as well as corticobasal degeneration, may also produce some confusion. Progressive subcortical gliosis remain mainly neuropathological diagnoses.

Lees [86] has proposed a series of diagnostic criteria requiring supranuclear ophthalmoplegia with abnormality of downward gaze plus at least two of the following disorders:
a. Dystonia and axial rigidity
b. Pseudo-bulbar syndrome
c. Bradykinesia and rigidity
d. Frontal signs
e. Postural instability with backward falls

Table 8. Some diseases where dementia and extrapyramidal signs may be associated

Alzheimer's disease
Parkinson's disease
Pick's disease
Frontal-lobe-type dementia
Cerebrovascular disease
Creutzfeldt-Jakob disease
Cerebral anoxia
Huntington's disease
Space-occupying lesions
Wilson's disease
Communicating hydrocephalus
Dementia pugilistica
Progressive supranuclear palsy
Encephalitis lethargica
Hallervorden-Spatz
Progressive subcortical gliosis
Multiple system atrophy
Corticobasal degeneration
Parkinsonism dementia of Guam
Neuroacanthocytosis

Huntington's Disease

Huntington's disease also known as Huntington's chorea, is a rather uncommon condition. Its prevalence varies between 30 and 100 per million, while its incidence is estimated between 3 and 5 per million [87, 88]. However, its interest is considerable for at least two reasons. It is one of the best described forms of "genetic dementia." In addition, it is accompanied by a characteristic series of neuropsychological deficits and is often considered a typical form of subcortical dementia.

The average age of onset of the disease is between 35 and 40 years. Its average duration spans about 14 years. In the initial phases of the illness, modifications of the personality, depression, mania, and a hallucinatory state of the schizophrenic type can precede by several years the appearance of abnormal choreic movements. Memory problems also appear early. They affect the anterograde memory but can also implicate the retrograde memory. Memories of past events, whatever the period of life during which they occurred, are equally affected, thus suggesting the existence of a problem of retrieval rather than of encoding [89].

One of the most tragic feature of Huntington's disease occurs from its genetic mode of transmission. The disease is autosomal dominant with very high penetrance. Several tests have been proposed in order to obtain an early diagnosis and to predict which subjects will develop the disease: motor, pharmacological, neuroimagery tests [90, 91]. This research, aside from its medical consequences, also involves considerable ethical implications.

AIDS Dementia Complex

Since the publication of the first case of HIV infection at the beginning of the 1980s, it has become apparent that a significant number of symptoms of the disease concern the CNS. One of the symptoms most frequently associated with this infection is a subacute encephalitis accompanied by progressive dementia [92]. Although this dementia was first attributed to superimposed infections, the majority of researchers now consider it to be a direct consequence of the action of the virus. This dementia was first defined as "progressive global deterioration." In order to differentiate it from other forms of dementia and to highlight its specificity, it is now defined as the "AIDS dementia complex" (ADC [93]).

The syndrome consists of three elements:

a. Motor problems affecting different kinds of movement: gestures become awkward; trembling affects the extremities of limbs. Movement becomes difficult and balance is unstable. Neurological examination confirms this lack of motor coordination and shows an increase in reflexes.

b. Behavioral troubles of a psychiatric nature: a loss of spontaneity induces a state of apathy in the patient who tends to become increasingly isolated. This behavior has been compared to the athymormia (severe abulic state) described in patients with vascular bilateral lesions of the striatum [94]. Maruff et al. [31] have found that the CES-D [95] allow better differentiation of ADC from HIV than from AIDS subjects without neurological signs.

c. Cognitive problems: a slowing down of intellectual activities, difficulty in concentrating, a tendency to forget. Neuropsychological examination confirms a psychomotor slow-down and indicates a problem with "executive" functions and a deficiency of memory and of divided attention, without these troubles being the sign of a general deficit. Higher cortical functions are generally unimpaired.

In light of this manner of disorder, it would be conceivable to propose a minimum number of tests for patients suspected of having an encephalopathic disorder consisting of the grooved pegboard, finger tapping, reaction times, finger-to-finger test, trails A and B, digit symbols from the WAIS, and verbal fluency. Obviously, the diagnosis of ADC cannot be reached on the basis of the neuropsychological tests alone. However, psychometric assessment can provide useful information that would clearly strengthen the diagnosis of ADC and quantify its severity.

Do HIV-seropositive subjects without symptoms also manifest cognitive troubles? The response to this question remains very controversial. The first studies [96] had revealed that nearly half of these subjects manifested a deterioration of their performance on neuropsychological tests, but research conducted on a very important sample of patients does not seem to confirm these results [97].

On the basis of longitudinal studies, we can conclude with certainty that patients do not manifest a gradual decline in their cognitive functions in the course of the asymptomatic phase [98], which suggests that they develop an encephalopathy and a full-blown dementia in a fairly brief interval of time.

References

1. Derouesne C, Kalafat M, Guez D, Malbezin M, Poitrenaud J (1994) The age-associated memory impairment construct revisited. Comments and recommendations of a French-speaking work group. Int J Geriatr Psychiatry 9: 577-587
2. Levy R (1994) Aging-associated cognitive decline. Int Psychogeriatr 6: 63-68
3. World Health Organisation (1993) The ICD-10 classification of mental and behavioural disorders. Diagnostic criteria for research. World Health Organisation, Geneva
4. American Psychiatric Association (1987) Diagnostic and statistical manual of mental disorders, 3rd edn, revised. American Psychiatric Association, Washington DC
5. American Psychiatric Association (1994) Diagnostic and statistical manual of mental disorders, 4th edn. American Psychiatric Association, Washington DC
6. Forette F, Henry JF, Orgogozo JM, Pere JF, Hugonot L, Israel L, Loria Y, Goulley F, Lallemand A, Boller F (1989) The reliability of clinical criteria for the diagnosis of dementia: a longitudinal multicentric study. Arch Neurol 46: 646-648
7. Baldereschi M, Amato MP, Nencini P, et al (1994) Cross-national interrater agreement on the clinical diagnostic criteria for dementia. Neurology 44: 239-242
8. McKhann G, Drachman D, Folstein M, Katzman R, Price D, Stadlan EM (1984) Clinical diagnosis of Alzheimer's disease: report of the NINCDS-ADRDA work group under the auspices of the Department of Health and Human Services Task Force on Alzheimer's Disease. Neurology 34: 939-944
9. Albert MS (1988) Cognitive function. In: Albert MS, Moss MB (eds) Geriatric neuropsychology. Guilford, New York, pp 33-53
10. Boller F, Marcie P, Traykov L (1996) Neuropsychologie du vieillissement normal. In: Botez MI (ed) Neuropsychologie clinique et neurologie du comportement. Masson/Presse Universitaire du Quebec, Paris, pp 527-548
11. Horn JK, Cattell RB (1967) Age differences in fluid and crystallized intelligence. Acta Psychol 26: 107-129
12. Hochanadel G, Kaplan E (1984) Neuropsychology of normal aging. In: Albert ML (ed) Clinical neurology of aging. Lexington Books, Lexington, pp 121-132
13. Kral VA (1962) Senescent forgetfulness: benign and malignant. J Can Med Assoc 86: 257-260
14. Crook T, Bartus RT, Ferris SH, Whitehouse P, Cohen GD, Gershon S (1986) Age-associated memory impairment: proposed diagnostic criteria and measures of clinical change. Report of a National Institute of Mental Health work group. Dev Neuropsychol 2: 261-276
15. Blackford RC, La Rue A (1989) Criteria for diagnosing AAMI: proposed improvement from the field. Dev Neuropsychol 5: 295-306
16. Creasey H, Schwartz M, Frederickson H, Haxby J, Rapoport S (1986) Quantitative computed tomography in dementia of the Alzheimer type. Neurology 36: 1563-1568
17. Von Stockert F (1932) Subkorticale Demenz. Archiv Psychiatrie 97: 77-100
18. Albert M, Feldman R, Willis A (1974) The "subcortical dementia" of progressive supranuclear palsy. J Neurol Neurosurg Psychiatry 37: 121-130

19. McHugh P, Folstein M (1975) Psychiatric syndromes of Huntington's chorea: a clinical and phenomenological study. In: Benson FD, Blumer D (eds) Psychiatric aspects of neurologic disease. Grune and Stratton, New York, pp 267-286

20. Mandell A, Albert M (1990) History of subcortical dementia. In: Cummings J (ed) Subcortical dementia. Oxford University Press, New York, pp 17-30

21. Cummings JL, Benson DF (1983) Dementia. A clinical approach. Butterworths, Stoneham, pp 89-103

22. Bruyn G, Bots G, Dom R (1979) Huntington's chorea: current neuropathological status. In: Chace T, Wexler N, Barbeau A (eds) Huntington's disease (Advances in neurology, vol 23). Raven, New York, pp 83-93

23. Bondareff W, Mountjoy C, Roth M (1982) Loss of neurons of origin of the adrenergic projection to cerebral cortex (nucleus locus ceruleus) in senile dementia. Neurology 32: 164-168

24. Whitehouse P (1986) The concept of subcortical dementia. Another look. Ann Neurol 19: 1-6

25. Mayeux R, Stern Y, Rosen J, Benson DF (1983) Is "subcortical dementia" a recognizable clinical entity? Ann Neurol 14: 278-283

26. Taylor AE, Saint-Cyr JA, Lang AE (1986) Frontal lobe dysfunction in Parkinson's disease. The cortical focus of neostriatal outflow. Brain 109: 845-883

27. Brandt J (1991) Cognitive impairments in Huntington's disease: insights into the neuropsychology of the striatum. In: Boller F, Grafman J (eds) Handbook of neuropsychology, vol 5. Elsevier, Amsterdam, pp 241-264

28. Blin J, Baron J, Dubois B, Pillon B, Cambon J, Agid Y (1990) Positron emission tomography study in progressive supranuclear palsy: brain hypometabolic pattern and clinico-metabolic correlations. Arch Neurol 47: 747-752

29. Everall I, Luthbert P, Lantos P (1991) Neuronal loss in the frontal cortex in HIV infection. Lancet 337: 1119-1121

30. Freedman M, Albert M (1985) Subcortical dementia. In: Frederiks JAM (ed) Handbook of clinical neurology. Elsevier, Amsterdam, pp 311-316

31. Maruff P, Currie J, Malone V, McArthur-Jackson C, Mulhall B, Benson E (1994) Neuropsychological characterization of the AIDS dementia complex and rationalization of a test battery. Arch Neurol 51: 689-695

32. Dubois B, Boller F, Pillon B, Agid Y (1991) Cognitive deficits in Parkinson's disease. In: Boller F, Grafman J (eds) Handbook of neuropsychology, vol 5. Elsevier, Amsterdam, pp 333-356

33. Boller F, Duyckaerts C (1996) Alzheimer's disease: clinical and anatomical aspects. In: Feinberg TE, Farah MJ (eds) Behavioral neurology and neuropsychology. McGraw Hill, New York, pp 521-544

34. Boller F (1991) Hallazgos neuropsicologicos cn la enfermidad de Alzheimer. In: Tolosa E, Bermejo F, Boller F (eds) Demencia senil. Springer-Verlag Iberica, Barcelona, pp 65-73

35. Huff FJ, Becker JT, Belle SH, Nebes R, Holland A, Boller F (1987) Cognitive deficits and diagnosis of Alzheimer's disease. Neurology 36: 1198-1214

36. Tierny MC, Fisher RH, Lewis AJ, Zorzitto ML, Snow WG, Reid DW, Nieuwstraten P (1988) The NINCDS-ADRDA work group criteria for the clinical diagnosis of probable Alzheimer's disease: a clinicopathological study of 57 cases. Neurology 38: 359-364

37. Morris JC, Heyman A, Mohs RC, Hughes JP, van Belle G, Fillenbaum G, Mellits ED, Clark C, and the CERAD investigators (1989) The Consortium to Establish a Registry for Alzheimer's Disease (CERAD). I. Clinical and neuropsychological assessment of Alzheimer's disease. Neurology 39: 1159-1165

38. Hughes CP, Berg L, Danziger WL, Cohen LA, Martin RL (1982) A new clinical scale for the staging of dementia. Br J Psychiatry 140: 566-572
39. Todorov A, Go R, Constantinidis J, Elston R (1975) Specificity of the clinical diagnosis of dementia. J Neurol Sci 26: 81-98
40. Boller F, Lopez OL, Moossy J (1989) Diagnosis of dementia: clinicopathologic correlations. Neurology 39: 76-79
41. Martin EM, Wilson RS, Penn RD, Fox JH, Clasen RA, Savoy SM (1987) Cortical biopsy results in Alzheimer's disease: correlation with cognitive deficits. Neurology 37: 1201-1204
42. Morris JC, Berg L, Fulling K, Torack AM, McKeel DW (1987) Validation of clinical diagnostic criteria in senile dementia of the Alzheimer type. Ann Neurol 22: 122
43. Heyman A, Fillenbaum G, Mirra S, and participating CERAD neuropathologists (1992) Clinical misdiagnosis of Alzheimer's disease. A review of CERAD autopsy findings. Ann Neurol 32: 270-271
44. Hachinski V (1992) Preventable senility – a call for action against the vascular dementias. Lancet 340: 645-648
45. Tatemichi TK, Sacktor N, Mayeux R (1994) Dementia associated with cerebrovascular disease, other degenerative disease, and metabolic disorders. In: Terry R, Katzman R, Bick KL (ed) Alzheimer disease. Raven, New York, pp 123-167
46. Erkinjuntti T (1994) Clinical criteria for vascular dementia: the NINDS-AIREN criteria. Dementia 5: 189-192
47. Caplan L (1996) From vascular lesions to dementia. In: Forette F, Christen Y, Boller F (eds) La demence: pourquoi? Fondation Nationale de Gerontologie, Paris, pp 157-174
48. Chui HC, Victoroff JI, Margolin D, Jagust W, Shankle R, Katzman R (1992) Criteria for the diagnosis of ischemic vascular dementia proposed by the State of California Alzheimer's Disease Diagnostic and Treatment Centers. Neurology 42: 473-480
49. Roman GC, Tatemichi TK, Erkinjuntti T, et al (1993) Vascular dementia: diagnostic criteria for research studies. Neurology 43: 250-260
50. Erkinjuntti T (1987) Differential diagnosis between Alzheimer's disease and vascular dementia: evaluation of common clinical methods. Acta Neurol Scand 76: 433-442
51. Mirra SS, Heyman A, McKeel D, Sumi SM, Crain BJ, Brownlee LM, Vogel FS, Hughes JP, van Belle G, Berg L, and participating CERAD neuropathologists (1991) The Consortium to Establish a Registry in Alzheimer's disease (CERAD). II. Standardization of the neuropathologic assessment of Alzheimer's disease. Neurology 41: 479-486
52. Hachinski VC, Iliff LD, Zilhka E, et al (1975) Cerebral blood flow in dementia. Arch Neurol 32: 732-737
53. Loeb C, Gandolfo C (1983) Diagnostic evaluation of degenerative and vascular dementia. Stroke 14: 399-401
54. Knopman D (1993) The non-Alzheimer degenerative dementias. In: Boller F, Grafman J (eds) Handbook of neuropsychology, vol 8. Elsevier, Amsterdam, pp 295-313
55. Mendez MF, Selwood A, Mastri AR, Frey WH (1993) Pick's disease versus Alzheimer's disease: a comparison of clinical characteristics. Neurology 43: 289-292
56. Escourolle R (1956) La maladie de Pick. Etude d'ensemble et synthèse anatomo-clinique. Masson, Paris
57. Tissot R, Constantinidis J, Richard J (1975) La maladie de Pick. Masson, Paris
58. Knopman DS, Christensen KJ, Schut LJ, et al (1989) The spectrum of imaging and neuropsychological findings in Pick's disease. Neurology 39: 362-368
59. Mesulam MM (1982) Slowly progressive aphasia without generalized dementia. Ann Neurol 11: 592-598
60. Cole M, Wright D, Banker BQ (1979) Familial aphasia due to Pick's disease. Ann Neurol 6: 158

61. Wechsler AF (1977) Presenile dementia presenting as aphasia. J Neurol Neurosurg Psychiatry 40: 303-305

62. Weintraub S, Mesulam MM (1993) Four neuropsychological profiles in dementia. In: Boller F, Grafman J (eds) Handbook of neuropsychology, vol 8. Elsevier, Amsterdam, pp 253-282

63. Kertesz A, Munoz DG (1994) The pathology and nosology of primary progressive aphasia. Neurology 44: A259

64. Fustinoni O, Mangone CA, Abiusi GRP, et al (1994) Primary progressive aphasia: clinical subtypes, with one postmortem study. Neurology 44: A387

65. Kertesz A (1996) Pick complex and Pick's disease, the nosology of frontal lobe dementia, primary progressive aphasia, and corticobasal ganglionic degeneration. Eur J Neurol 3: 280-282

66. Duyckaerts C, Dürr A, Uchihara T, Boller F, Hauw JJ (1996) Pick complex: too simple? Eur J Neurol 3: 283-286

67. Morris JC, Cole M, Banker BQ, Wright D (1984) Hereditary dysphasic dementia and the Pick-Alzheimer spectrum. Ann Neurol 16: 455-466

68. Brun A (1987) Frontal lobe degeneration of non-Alzheimer type. I. Neuropathology. Arch Gerontol Geriatr 6: 193-208

69. Neary D, Snowden JS, Bowen DM, et al (1986) Neuropsychological syndromes in presenile dementia due to cerebral atrophy. J Neurol Neurosurg Psychiatry 49: 163-174

70. The Lund and Manchester Groups (1994) Clinical and neuropathological criteria for frontotemporal dementia. J Neurol Neurosurg Psychiatry 57: 416-418

71. Lhermitte F, Pillon B, Serdaru M (1986) Human autonomy and the frontal lobes. Ann Neurol 19: 326-334

72. Kosaka K (1993) Dementia and neuropathology in Lewy body disease. In: Narabayashi H, Nagatsu T, Yanagisawa N, Mizuno Y (eds) Parkinson's disease. From basic research to treatment (Advances in neurology, vol 60). Raven, New York, pp 456-463

73. Hansen L, Salmon D, Galasko D, et al (1990) Lewy body variant of Alzheimer's disease: a clinical and pathological entity. Neurology 40: 1-8

74. Joachim CL, Morris JH, Selkoe DJ (1988) Cinically diagnosed Alzheimer's disease: autopsy results in 150 cases. Ann Neurol 24: 50-56

75. Lennox G (1992) Lewy body dementia. In: Rossor MN (ed) Baillière's clinical neurology. Baillière Tindall, London, pp 653-676

76. Boller F, Duyckaerts C (1994) Dementia in diffuse Lewy body disease. In: Alberca R, Boller F (eds) Early diagnosis of Alzheimer disease. Aula Medica, Madrid, pp 156-165

77. Boller F, Mizutani T, Roessmann U, Gambetti PL (1980) Parkinson disease, dementia and Alzheimer disease: clinico-pathological correlations. Ann Neurol 7: 329-335

78. Gnanalingham K, Byrne J, Thornton A (1996) Clock-face drawing to differentiate Lewy body and Alzheimer type dementia syndromes. Lancet 347: 696-697

79. Crystal HA, Dickson DW, Lizardi JE, Davies P, Wolfson LI (1990) Antemortem diagnosis of diffuse Lewy body disease. Neurology 40: 1523-1528

80. Burkhardt CR, Filley CM, Kleinschmidt-DeMasters BK, de la Monte S, Norenberg MD, Schneck SA (1988) Diffuse Lewy body disease and progressive dementia. Neurology 38: 1520-1528

81. Mahieux F, Michelet D, Manifacier MJ, Fenelon G, Boller F, Guillard A (1992) Mini Mental Parkinson: validation study of the revised form. J Neurol 239: S120

82. Brown R, Marsden C (1988) "Subcortical dementia": the neuropsychological evidence. Neurosciences 25: 363-387

83. Steele JC, Richardson JC, Olszewski J (1964) Progressive supranuclear palsy. A heterogeneous degeneration involving the brain stem, basal ganglia and cerebellum, with vertical gaze and pseudobulbar palsy, nuclear dystonia and dementia. Arch Neurol 10: 333-359
84. Pillon B, Dubois B, Ploska A, Agid Y (1991) Severity and specificity of cognitive impairment in Alzheimer's, Huntington's, and Parkinson's diseases and progressive supranuclear palsy. Neurology 41: 634-643
85. Mann DMA, South PW, Snowden JS, Neary D (1993) Dementia of frontal lobe type: neurophatology and immunohistochemistry. J Neurol Neurosurg Psichiatry 56: 605-614
86. Lees AJ (1987) The Steele-Richardson-Olszewski syndrome (progressive supranuclear palsy). In: Marsden CD, Fahn S (eds) Movement disorders. Butterworths, London, pp 272-287
87. Kurtzke J (1979) Huntington's disease: mortality and morbidity data from outside the United States. Adv Neurol 23: 13-25
88. Conneally P (1984) Huntington disease: genetics and epidemiology. Am J Hum Gen 36: 506-526
89. Albert M, Butters N, Brandt J (1981) Patterns of remote memory in amnesic and demented patients. Arch Neurol 38: 495-500
90. Klawans H, Goetz C, Perlik S (1980) Presymptomatic and early detection of Huntington's disease. Ann Neurol 8: 343-347
91. Mazziotta JC, Phelbs ME, Pahl JJ, et al (1987) Reduced cerebral glucose metabolism in asymptomatic subjects at risk for Huntington's disease. N Engl J Med 316: 357-362
92. Snider W, Simpson D, Nielsen S, Gold J, Metroka C, Posner J (1983) Neurological complications of acquired immune deficiency syndrome: analysis of 50 patients. Ann Neurol 14: 403-418
93. Navia B, Jordan B, Price R (1986) The AIDS dementia complex. I. Clinical features. Ann Neurol 19: 517-524
94. Habib M, Poncet M (1988) Loss of drive, interest and affect ("athymormia syndrome") with lacunar lesions of the striatum. Rev Neurol 144: 571-577
95. Fuhrer R, Rouillon F (1989) La version française de l'echelle CES-D (Center for Epidemiologic Studies-Depression Scale). Description et traduction de l'echelle d'auto/valuation. Psychiatr Psychobiol 4: 163-166
96. Grant I, Atkinson JH, Hesselink JR, Kennedy CJ, Richman DD, Spector SA, McCutchan JA (1987) Evidence for early central nervous system involvement in the acquired immunodeficiency syndrome (AIDS) and other human immunodeficiency virus (HIV) infections. Ann Intern Med 107: 828-836
97. McArthur J, Cohen B, Selnes O, et al (1989) Low prevalence of neurological an neuropsychological abnormalities in otherwise healthy HIV-1 infected individuals: results from the multicentral AIDS cohort study. Ann Neurol 26: 601-609
98. Selnes OA, Miller E, McArthur J, et al (1990) HIV-1 infection: no evidence of cognitive decline during the asymptomatic stages. Neurology 40: 204-208

Alzheimer's Disease

A. Bianchetti and M. Trabucchi

Alzheimer's disease (AD) is the most common of the dementing disorders. Once considered a rare disorder, AD is now recognized as a major cause of death and a growing public health problem. In 1907, Alois Alzheimer reported the first case of the illness that bears his name, a 51-year-old woman with progressive cognitive decline and behavioral changes associated with distinctive neuropathological features [1]. Much time has passed since then, however. Until recently, there has been widespread nihilism in the clinical approach to the demented patient. Most physicians believed that there was little reason to spend money or effort diagnosing diseases with no known risk factors, preventive measures or treatments [2]. However, a multidisciplinary effort has transformed this field, leading to a better understanding of the clinical presentation, risk factors, pathogenesis and treatment of the other degenerative dementias [3].

Epidemiology

AD generally begins in late life, most frequently after the age of 60 years, although in rare cases the disorders appear before the age of 50 years. The prevalence of AD is approximately 6% to 8% of all persons older than 65 years, and it doubles every 5 years after the age of 60 years, so that nearly 30% of the population older than 85 years has AD [4]. The incidence of AD increases with age, and it is estimated at 0.5% per year from age 65 to 69, 1% per year from age 70 to 74, 2% per year from age 75 to 79, 3% per year from age 80 to 84, and 8% per year from age 85 onward [5]. Disease progression is gradual and continuous, although plateaus may occur, with an average duration from the onset of symptoms until death of 8–10 years.

Several types of risk factors have been studied in different cross-sectional studies, but few have proved to be clearly related to the development of AD (see Table 1). The most well-accepted risk factor for AD is age. The elevated prevalence of AD in the more advanced ages, with the detection of senile plaques (SPs), neurofibrillary tangles (NFTs), and a reduction in choline acetyltransferase (ChAT) activity in cerebral cortex in some elderly persons without an history of dementia, has led some authors to think that AD is merely an exaggeration of the aging process [6]. Nevertheless, epidemiological data indicate that the increase in AD prevalence

Table 1. Risk factors for Alzheimer's disease

Established
– Old age
– Family history of dementia
– APOE genotype

Likely
– Female sex
– Low educational level
– Family history of Down's syndrome
– Head injury
– History of hypothyroidism
– Vascular risk factors
– No use of estrogen replacement therapy or NSAID drugs

with age reaches a plateau near the age of 95, suggesting that AD is an "age-related disease," and not an "aging-related disease" [7]. A variable proportion of subjects (from 4% to 30%, according to sample selection and diagnostic criteria) shows cognitive changes as they grow older, which are mainly characterized by a slowing of the information processes or increasing reaction time, without interference with functional status [8, 9]. Some of these are likely to be AD patients in the early stage of the disease (the annual incidence of dementia in these group is 1.5% for people 65–69 years old, and 3.6% for people 75–81 years old), but almost 90% are normal elderly with a stable and benign cognitive decline [10].

A family history of AD in a parent or sibling increases the risk of developing the illness by three to four times [11]. Most cases of AD are late onset (over 65 years) and are sporadic, but a few cases are early (less than 65 years) and are clustered within families. In 1987 a locus on chromosome 21 was found to be associated with a very aggressive form of AD with a strong familial transmission (autosomal dominant inheritance) [12]. In 1991 a mutation within the gene encoding the beta-amyloid protein precursor (β-APP) was identified [13]. This gene encodes the precursor of the beta-amyloid protein deposited in the plaques found in AD brains. However, these mutations proved to be a very rare cause of the disease: only 2%–3% of all published cases of familial AD (FAD) and 5%–7% of reported cases of early-onset FAD have been identified carrying these mutations [14]. In these families, the onset of AD is invariably in the late forties or early fifties and penetrance is nearly 100% [15]. In 1992, it was demonstrated that a second gene capable of causing AD must be located on chromosome 14 [16]. In 1995 the gene was identified and it has been named presenilin 1, or PS-1 [17]. It codes for a novel protein that is not only expressed in brain cells, but in a wide variety of other tissues; the function of this protein is not yet known; however, it has 467 amino acids and is thought to span the cell membrane 8 times [18]. To date, about 43 different mutations have been identified in about 50 families of different ethnic origin [18, 19]. Although mutations in the PS-1 gene are a more common cause of early onset AD than mutations in the β-APP gene, they probably only account for about 50% of familial early-onset AD, with a range of 28–56 years of age of onset of AD and a penetrance of 100% [19]. A gene related to the PS-1 gene has subsequently been identified on chromosome 1 and named presenilin 2, or PS-2 [20]. Two different PS-2 mutations have been reported, and the age at onset ranges between 40 and 90 years [21]. The PS-2 gene is a rare cause of AD [21]. While the physiological function of the presenilins is unknown, presenilins might cause AD by increasing amyloid deposition in the brain, although indirect evidence points to an effect on tangle formation [22, 23]. PS-2 might be involved in the apoptosis process [18].

The known genetic causes account for less than 2% of AD cases [24]. What is not rare (at least 25% of all cases of AD) is a history of dementia in one first-degree relative [21]. This observation led to the conclusion that in addition to FAD with an autosomal dominant inheritance, other forms of "familial AD" exist, in which genetic factors play a role in combination with environmental factors [21]. In 1993, the association between the apolipoprotein E ε4 allele (APOE ε4) and the risk of late-onset AD was described [25]. Apolipoprotein E (APOE) is a plasma protein involved in cholesterol transport and encoded by a gene on chromosome 19 [26]. APOE is the only apolipoprotein produced in the brain, where it has a role in damage repair [27]. There are three common alleles (ε2, ε3, ε4), with different frequencies (in normal Caucasian populations around 5%–10%, 75%–85%, and 10%–15%, respectively) [28]. Corder et al. reported a close association between the APOE ε4 and late-onset AD: members of ε4 carrier families developed AD more often (91% for ε4+/ε4+ vs 47% for ε4+/ε4- and 20% for ε4-/ε4-) and earlier (at age 68 vs 76, and 84 years, respectively) [29]. Cross-sectional studies later confirmed the higher frequency of the ε4 allele in the late onset-and early-onset sporadic AD form [21]. A meta-analysis of population-based studies shows that the risk of AD was increased for persons with genotypes ε2/ε4 (OR = 2.6), ε3/ε4 (OR = 3.2), and ε4/ε4 (OR = 14.9), whereas the ORs were decreased for the genotypes ε2/ε2 (OR = 0.6), and ε2/ε3 (OR = 0.6) [30]. A recent population-based study of incident AD confirms that APOE ε4 is an important genetic risk factor for AD, but suggests that it accounts for a fairly small fraction of the disease occurrence; in fact, persons with APOE ε4/ε4 or ε3/ε4 had 2.27 times the risk of incident AD compared with those with ε3/ε3 genotype [31]. Not everyone possessing the ε4 allele will develop AD, and many who lack the allele will also develop the illness [21]. These findings indicate that, in ε4 carriers, AD phenotype can be expressed due to the converging effects of the ε4 allele with other genetic or environmental factors [18]. The APOE genotype also appears to modulate the age of onset of AD in APP mutation families, but not in PS-1 mutation families [18]. Several studies have shown that ε4 allele is a risk factor not only for AD, but also for incident vascular dementia and poor recovery after head trauma or stroke [28, 32, 33]. Recently, after a genomic screening in late-onset familial AD, a locus on chromosome 12 was identified as a new AD susceptibility locus [34]. The gene is a susceptibility gene, and it raises the risk of developing the disease, even though it still remains to be determined how great the role of the chromosome 12 gene actually is in developing late-onset AD [35].

Other possible risk factors for AD include a previous head injury, female sex, and lower education level [36, 37]. Several studies have shown that a first-degree relative of a person with Down's syndrome or Parkinson's disease is at increased risk for developing AD; however, this association has not turned up in other studies [38]. Some researchers support a past history of hypothyroidism, psychiatric disorders, and cardiovascular disease as a risk factors of AD, but others disagree [11, 39]. Recent observations suggest that cerebrovascular disease may play an important role in determining the occurrence and severity of the clinical symptoms of AD [40]. To date, non-conclusive evidence links metals, such as zinc or aluminum, to the development of AD [38]. Possible protective factors include the use of estrogen replacement therapy and non-steroidal anti-inflammatory drugs [2]. It is becoming clear that AD can be caused by a number of different factors, both environmental and genetic, that interact differently in different people.

Neuropathology

The finding of distinctive neuropathological alterations in the brain is essential to pose the diagnosis of definitive AD in a patient presenting the characteristics symptoms of the illness [41]. The key neuropathological features of AD are a marked reduction in the population of neurons (particularly in the hippocampus, substantia innominata, locus ceruleus, and temporoparietal and frontal cortex), and the appearance of NFTs and SPs [42]. NFTs develop within the pyramidal neuronal soma as filamentous inclusions and may extend into dendrites. NFTs consist of highly insoluble, paired helical filaments (PHFs) in addition to 15-nm-wide straight filaments [43]. PHFs consist of 4–8 100-μm protofilaments containing proteins that are immunologically related to normal cytoskeletal proteins [42]. In particular, a major component of NFTs is the microtubule-associated protein tau, which has an important role in maintaining microtubule assembly and integrity [44]. In AD, the tau protein is in an abnormally phosphorylated state, which impairs its ability to bind to microtubles, which results in collapse of the neuronal cytoskeleton [45]. The NFT densities of AD brains are highest in the medial temporal lobe (uncus, amygdala, hippocampus and parahippocampal gyrus) with a moderate distribution in the association cortex of the parietal and frontal lobes [42]. The lesion counts were found to be minimal in the primary somatic and visual sensory areas [46]. Neuronal loss is also severe in regions in which NFT formation is prominent [47]. NFTs are commonly found within larger neurons, but smaller neurons may also be affected [42].

SPs are foci of enlarged axons, synaptic terminals and dendrites, associated with extracellular β-A4 amyloid [48]. They appear as roughly spherical areas 10–150 nm in diameter. The initial step in the formation of plaques is thought to be amyloid deposition. The β-A4 peptide is a post-translationally modified product of a larger amyloid precursor protein (APP). APP proteins are integral membrane glycoproteins containing a large extracytoplasmic domain, a transmembrane region, and a small cytoplasmic sequence and are expressed in almost all tissues and cell lines. Under physiological conditions, 42 aminoacid amyloid peptide isoforms can aggregate to form insoluble filaments about 7–9 nm wide. These consist of anti-parallel beta pleated sheets. The peptides found in neuritic plaque cores are a heterogeneous mixture of several isoforms, with a variable length of 39–43 amino acids [49]. What converts the normal β-A4 soluble protein into the insoluble fibrils of β-amyloid is unclear. Amyloid deposition seems be the "central event" in the etiology and pathogenesis of AD even though the relationship between β-amyloid and NFTs is not totally clear [50]. SPs are recognized as accumulations of amyloid with abnormal neurites in the aged brain and also NFTs accompany normal aging; however, the average concentration of plaques in the cortex is significantly greater in AD than in normal aging or in other disorders [42]. The disease specificity of NFTs appears to be lower than SPs [50].

There is a lot of evidence that inflammatory and immune mechanisms are involved in tissue destruction in AD. Acute phase proteins are elevated in the serum and are deposited in SPs, activated microglial cells that stain for inflammatory cytokines accumulate around SPs, and complement components, including the membrane-attack complex, are present around dystrophic neurites and NFTs [51].

Standardized neuropathological protocols that involve a semi-quantitative approach to assess the frequency of SPs, with correction for the patient's age, NFTs, and other changes, have been proposed and tested [52].

In the 1970s it was demonstrated that cholinergic blockade (scopolamine) causes memory loss in normal human subjects similar to that seen in the course of AD [53]. This early observation focused the AD research on the cerebral cholinergic system. In the 1960s it had been found that concentration of acetylcholinesterase (AChE) was reduced in the brains of patients affected by AD, and a few years later substantial depletion of cortical cholinergic innervation in AD was demonstrated [54]. Successive studies using AChE histochemistry or ChAT immunohistochemistry found a widespread loss of cortical cholinergic innervation in AD. Although the loss of AChE- and ChAT-positive cholinergic fibers is found in all cortical laminae and has been demonstrated to be substantially widespread, studies using AChE histochemistry or ChAT immunohistochemistry have shown greater loss within the superficial laminae (II-III) and regional variability of cortical cholinergic innervation in AD, with greater loss (> 75% reduction) in the temporal lobe areas (including areas 20, 21, 22, and 28 of Brodmann); intermediate loss (40%–75%) in the frontal and parietal association areas as well as the insula and temporal pole; and less than 30% loss in the anterior cingulate gyrus, primary motor, primary somatosensory, and primary visual cortex. Cholinergic fiber density also appears to be reduced within the dentate gyrus, hippocampal formation, amygdala and subcortical structures [55, 56]. A loss of neurons in the nucleus basalis of subjects affected by AD had been reported in the 1960s [57]. A 75% decrease in the number of hypercromic nucleus basalis (Ch4) neurons in AD patients was then confirmed [54]. An association between cholinergic system alterations and Alzheimer's neuropathological lesions has also been found. Early reports hypothesized that NFTs result from the degeneration of cholinergic fibers and that neuronal loss in the nucleus basalis is correlated with the presence of SPs in the anatomically related cortical areas [50]. Recently, it has been suggested that APP processing is regulated by neuronal activity via muscarinic M1 receptors and that cholinergic differentiation in AD brain might be associated with increased amyloid formation [58]. These reports provided the basis for the so-called "cholinergic theory of AD". However, further neurotransmitter systems are affected in AD and involve neurons containing excitatory amino acids, serotonin, noradrenalin, and somatostatin [59]. The deficit of the glutamatergic system in the early stage of AD has been previously shown and seems to be strongly associated with the severity of dementia [60]. The impairment of serotonergic and noradrenergic neurotransmission is known to be involved in determination of non-cognitive symptoms (depression, psychotic and behavioral disturbances) of AD, but their deficits are likely to account for part of the dysfunction of cholinergic memory processes [59].

Clinical Course

AD begins insidiously. The first cognitive symptom is usually a deficit in memory, and this is often the presenting feature [61]. In the first stage, episodic memory is predominately affected, with early loss of memory for everyday events. In contrast, short-term memory tends to be preserved early in the disease. Language deficits sometimes appear early, especially in early-onset AD, but more often occur in the middle stages. Difficulty in naming objects or in choosing the right words to express an idea (anomia) are frequently the initial language symptoms [62]. Visuospatial deficits are often manifested by impairment of topographical memory, when patients easily get lost. Deficits in other cognitive abilities, such as praxia, judgment, abstract reasoning, attention, and calculation, appear during the progression of the disease [63]. Occasionally, patients have a more focal onset, with aphasia, visual disorientation, or apraxia.

Non-cognitive symptoms are almost always present, with important consequences for the quality of life of patients and caregivers [64]. Personality changes are the most frequent features in AD patients (about 70% have apathy); 60% of patients exhibit agitation (this is a composite symptom, including aggressive behavior and persistent vocalization); about 40% manifest depressive features, anxiety, irritability, dysphoria, and aberrant motor behavior; delusion and hallucinations are present in 30%–60% of AD patients; about 5% have euphoria [65, 66]. Neurovegetative abnormalities affecting sleep, appetite, and the libido are also common in AD and, when present, constitute considerable management problems [67]. Table 2 shows the frequency of behavioral disturbances in the different stages of AD in a sample of patients evaluated at the Alzheimer Unit, IRCCS "San Giovanni di Dio Fatebenefratelli" in Brescia (Italy) [66]. The occurrence of abnormal motor behavior, agitation, apathy, and psychotic symptoms increases with the progression of the illness, whereas euphoria frequency decreases.

Table 2. Frequency of non-cognitive symptoms observed in a sample of 102 AD patients consecutively assessed at the Alzheimer Unit, IRCCS "San Giovanni di Dio Fatebenefratelli" in Brescia, Italy, in relation to stages of dementia (measured through the Clinical Dementia Rating Scale, by Hughes et al. [142])

	CDR 1 Mild AD (n = 16)	CDR 2 Moderate AD (n = 54)	CDR 3 Severe AD (n = 32)
Hallucinations*	3 (18.7%)	15 (27.8%)	17 (53.1%)
Delusions	6 (37.5%)	23 (42.6%)	14 (43.8%)
Depression	9 (56.3%)	38 (70.4%)	15 (46.9%)
Anxiety	4 (25.0%)	25 (46.3%)	17 (53.1%)
Euphoria**	7 (43.7%)	7 (13.0%)	3 (9.4%)
Disinhibition	4 (25.0%)	11 (20.4%)	9 (28.1%)
Apathy	14 (87.5%)	46 (85.2%)	29 (90.6%)
Irritability	7 (43.7%)	31 (57.4%)	25 (78.1%)
Agitation	9 (56.3%)	32 (59.3%)	23 (71,9%)
Aberrant motor behavior*	9 (56.2%)	29 (53.7%)	26 (81.8%)

Non-cognitive symptoms are assessed through the Italian version of the Neuropsychiatric Inventory [143]
* $P < 0.05$, chi-square test mild vs severe
** $P < 0.01$, chi-square test mild vs severe

The cognitive deficits and the non-cognitive symptoms, combined with comorbid conditions, cause progressive functional impairment. The demonstration of a decline in occupational or social functioning is essential to make a diagnosis of dementia syndrome (DSM-IV, [68]). Instrumental activities of daily living, such as managing finances, transportation, and communication, are impaired before basic activities, such as using the toilet, dressing, bathing, and mobility [69].

The above features and the characteristic progression have been incorporated into clinical criteria (the NINCDS/ADRDA criteria; see Table 3), which provide different levels of probability of diagnosis [41]. These criteria have proven valid for clinical and research purposes: for patients diagnosed as having probable AD according to the NINCDS/ADRDA criteria, the diagnosis is confirmed at autopsy in more than 90% [70]. Misdiagnosis is most likely to occur early in the course of illness and in patients with prominent behavioral disturbances or other atypical features [71].

Early in the history of AD clinicians and researchers recognized that AD is not a single disease, but a complex syndrome, with many subtypes and variety of patterns in its manifestations. The heterogeneity of the disease is demonstrated in many of its aspects: age at onset, duration, clinical course, types and patterns of cognitive and non-cognitive symptoms, response to treatments and neuropathological findings [72]. Different variables are associated with a distinct pattern of symptoms: age at onset, demographic factors (education, familial and social network), premorbid personality, early development of neuropsychiatric or extrapyramidal signs, co-occurrence of somatic diseases [73, 74].

Table 3. The NINCDS/ADRDA clinical criteria for Alzheimer's disease. (Adapted from [41])

- **Criteria for the clinical diagnosis of *probable Alzheimer's disease***

- Presence of dementia established by clinical examination and documented by neuropsychological tests
- Deficits in at least two areas of cognition
- Progressive deterioration of memory and other cognitive functions
- No clouding of consciousness
- Onset between ages 40 and 90
- Absence of systemic disorders or other brain diseases that could account for the dementia

The diagnosis supported by:
- Progressive deterioration of specific cognitive function
- Impaired activities of daily living and altered pattern of behavior
- Family history of dementia
- Normal lumbar puncture, EEG, and evidence of cerebral cortical atrophy on CT scan with progression documented by serial observation

Features consistent with the diagnosis:
- Plateaus in the course of the disease
- Associated psychiatric symptoms
- Neurological signs, including motor signs, such as increased muscle tone, myoclonus, or gait disorders, especially in advanced disease
- Seizures in advanced disease
- Normal CT scan

Diagnosis of Alzheimer's disease unlikely if:
- Sudden onset
- Focal neurological signs, such as hemiparesis, sensory loss, visual field deficit, and incoordination early in the course of the disease
- Seizures or gait disturbances early in the disease

- **Criteria for the clinical diagnosis of *possible Alzheimer's disease***

- In the presence of atypical onset, presentation, or clinical course
- In the presence of a systemic disease sufficient to produce dementia, but not considered to be the cause of the disease
- In the presence of a single progressive cognitive deficit

- **Criteria for diagnosis of *definite Alzheimer's disease***

- Clinical criteria for probable Alzheimer's disease
- Histopathological evidence of the disorder

Investigations

The first step in the diagnostic workup is determining whether cognitive impairment exists and if that cognitive impairment meets the criteria for dementia [75]. If dementia is identified, the second step consists of the evaluations essential to determine the etiology of the dementia and to stage its severity. The diagnosis of AD must be primarily of inclusion, not exclusion, as is often supposed [2]. The patient and an informant should be interviewed, and the history should focus on the general medical condition, with attention to drug use, psychiatric illness, chronic diseases (such as hypertension, cardiovascular diseases, cerebrovascular diseases, metabolic disorders, neurological diseases), recent trauma or surgical interventions, the onset and the evolution over time of cognitive deficits, the presence and severity of non-cognitive symptoms. The family history of dementia, stroke, depression or related conditions should be evaluated. A skillfully taken history may reveal deficits in many areas of cognitive function and disclose the functional status of the patient. Variations in common social, familial activities, or hobbies may suggest incipient intellectual deterioration.

A comprehensive physical examination, including a neurological examination and mental status testing, is essential in all patients. In early AD neurological examinations are usually normal. Some patients may develop extrapyramidal features early (and require a differential diagnosis with Lewy body dementia); in the advanced stages of the disease myoclonic jerks or seizures may occur; gait or balance deficits are frequent in the more advanced stages [62, 76]. Mental-status screening should include an assessment of cognitive and affective states and should use standardized instruments such as the Mini-Mental State Examination, or the Blessed-Information Concentration Test [77, 78]. Many variables influence the performance at psychometric evaluation: age, sex, language, socio-economic status, consciousness and attention, affective status, sensorial deficits of the responders, the attitude of the examiner and his ability to instruct the patient and to understand answers [79, 80]. For the more well-known tests cut-off points are available; when age- and education-corrected scores are used, these cut-off points are useful for the screening of patients with suspect dementia, but test scores do not establish a diagnosis of dementia by themselves, nor do they determine the etiology of the dementia syndrome [81, 82].

The evaluation of functional status plays a key role in the assessment of AD. Although a history of decline in functional ability is generally derived from the patient or caregiver, scales describing the activities of daily living may obtain this information. The systematic use of a standardized assessment instrument increases accuracy and ensures comprehensiveness, allows evaluating the modification of performance over time, and in response to treatment or rehabilitative interventions [83]. Moreover, formal assessment of functional status can determine what kind of assistance may be needed in home care or nursing home programs. Standardized scales are interviewer-administered and usually assess the basic activities of daily living (such as dressing, bathing, using the toilet, feeding, continence), and instrumental activity of daily living (such as using a telephone, getting to places beyond walking distance, shopping, preparing meals, doing housework, doing handyman work, doing laundry, taking medicine, managing money) [84]. The source of information is generally a caregiver, and only a weak correlation exists between proxy functional status rating and performance directly observed [85]. Therefore, particularly in mild and moderate dementia, performance-based assessing scales are widely used to obtain more reliable measures of functional status [86].

Table 4. Standard assessment protocol for the evaluation of demented patients in use at the Alzheimer Unit, IRCCS "San Giovanni di Dio Fatebenefratelli" in Brescia, Italy

---

Personal and family history

Somatic and neurological examination

Mental status
– Mini-Mental State Examination (as a screening tool) [77]
– Extensive neuropsychological testing (related to MMSE score or clinical features) [144]

Non-cognitive symptoms
– Neuropsychiatric Inventory [143, 145]
– Geriatric Depression Scale

Insight
– Anosognosia questionnaire [146]

Functional status
– Barthel Index for Basic Activities of Daily Living [147]
– Instrumental Activities of Daily Living [148]
– Assessment of Functional Status [149]
– Physical Performance Test [150]
– Bedford Alzheimer Nursing Severity Scale [151, 152]
– Tinetti scale for gait and balance [153]

Physical health
– Somatic symptoms [154]
– Chronic diseases (Greenfield Index) [155]

Staging measure
– Clinical Dementia Rating Scale [142, 156]

Laboratory investigations
– Standard laboratory examinations [87]
– EEG, ECG
– Chest X-ray radiography
– Morphological neuroimaging (TC or MRI) if never performed
– Functional neuroimaging (SPECT or PET) in selected cases
– CSF examination in selected cases

Caregiver's assessment for educational, psychological, legal, or social needs

---

Non-cognitive symptoms, depression, and problems in gait and balance may be assessed using standardized instruments (Table 4 shows the standard assessment protocol in use at the Alzheimer Unit, IRCCS "San Giovanni di Dio Fatebenefratelli" in Brescia, Italy). Recommendations for the instrumental examinations of demented persons are available and include laboratory examinations (complete blood count, tests for thyroid function, determination of vitamin B_{12} and folic-acid levels, sedimentation rate, BUN, creatinine, electrolytes, calcium, liver function tests, syphilis serology and urinalysis), chest X-ray and electrocardiogram [75, 87]. The utility and cost-effectiveness of standard laboratory tests in the evaluation of dementia patients have been questioned [61]. The frequency of potentially reversible etiologies has been examined by several authors and ranges between 5% and 30% [88, 89]. Recent data have demonstrated that routine laboratory examinations change the diagnosis in 9% of patients classified as AD and change patient management in 13% of cases [90]. Several studies have been conducted to identify biological markers to confirm AD diagnosis during life: lower cerebrospinal fluid (CSF) levels of APP, β-amyloid or other APP fragments, altered plasma levels of Aβ, elevated CSF levels of tau protein, and biological abnormalities in peripheral cells, mainly fibroblasts (such as alterations in transduction systems, alterations in amyloid precursor protein metabolism, in calcium homeostasis, or in oxidative metabolism) have all been found in AD patients [91, 92]. Tests for the detection of tau protein and Aβ in CSF are now commercially available, but their value compared to other diagnostic approaches and in mixed clinical populations is not well established [93]. Actually, the routine use of these tests is not recommended [2, 94].

The explosion in knowledge of the genetics of AD has led to important clinical and ethical questions regarding the possibility of introducing "genetic testing" in the assessment of AD patients [95]. Predictive genetic testing in well-characterized early-onset AD families with known mutations is appropriate, if requested, but must be confidential and accompanied by pretest and post-test counseling to help subjects understand what the test can predict and help them cope with the disclosure of their risk [96]. The real impact of the APOE genotype in determining the risk of developing AD in healthy subjects is not completely clear, and opinions on the value of APOE as a diagnostic adjunct in the differential diagnosis of patients with dementia need further investigation [97]. APOE genotyping is not recommended for individuals without dementia, and the use of this test in patients already presenting with dementia may prove useful, but this is still under investigation [21, 96, 98].

Computed tomography (CT) or magnetic resonance imaging (MRI) of the brain are used to exclude structural lesions (such as neoplasms, vascular abnormalities, subdural hematoma, hydrocephalus) that may determine the dementia syndrome [99]. Brain imaging may reveal aspecific findings, such as periventricular leukoaraiosis, or be almost normal. Widespread cortical atrophy is a characteristic finding in AD, but may also be present in normal aged subjects and cannot be used as a reliable diagnostic marker [75]. Many researchers have examined the usefulness of simple ratings or quantitation of regional atrophy (in particular in temporal lobes structures) in the diagnosis of AD, particularly in the early stages of the illness, but the results are not conclusive [100].

Functional imaging studies, such as single-photon emission CT (SPECT) or positron emission tomography (PET), may show the characteristic parietal or temporal deficits in AD, even if atypical features, such as frontal deficits, are not infrequent [101]. The clinical use of functional imaging is limited to atypical presentations or very mild patients.

Electroencephalography (EEG) is not routine, but may be helpful in distinguishing depression or delirium from AD, in assessing suspected Creutzfeldt-Jakob disease, or metabolic dementia, and in revealing seizures. The EEG may be normal in AD or show a diffuse slowing of all cerebral activity [75].

Treatment

The treatment of AD is multimodal and is guided by the stage of illness. It is focused on the specific symptoms manifested by the patient (Table 5). While no current therapy can reverse the progressive cognitive and functional decline in AD, several pharmacological substances and psychosocial techniques have been proven to be effective in controlling behavioral symptoms, slowing cognitive and functional decline and improving the quality of life of patients and caregivers [2, 102].

The widespread denial of physicians, as well as of caregivers, of the possibility of effectively treating AD arises frequently from the difficulty of defining the goals of the care plan. The traditional approach to the treatment of AD is focused on the cognitive symptoms of the disease; however, AD produces changes not only in cognition, but also in behavior, neurological function, neurovegetative state, ability to perform skills and activities, and quality of life. Moreover, AD affects the physical and psychological well-being of the caregiver and poses social, financial, and political problems [56]. A clear definition of the target of treatment and the identification of appropriate measures to detect the efficacy of interventions are essential for both clinical and political purposes [103]. In fact, the efforts to reach a better quality of life and the need to contain costs in health care determine a dramatic growth in outcome research, also in geriatric care [104]. A definition of the outcome of a health service is an important part of the improvement in the quality of care process [105]. Different variables may be used as outcome indicators in dementia care, referring to the patients [such as modification of the natural history of the disease, improvement in cognitive or functional status, reduction of non-cognitive symptoms, improvement in physical health, and in psychological well-being (depression, anxiety), satisfaction and quality of life], to the caregiver [such as physical or psychological health (depression, anxiety, guilt), social interactions, and quality of life], or to society as a whole (reduction in economic and social costs, reduction of the rate of institutionalization) [103, 106]. Therefore, outlining realistic outcomes for the patient and caregiver is the first step in the treatment of AD.

Patients with dementia require an individualized and multimodal treatment plan, which evolves with the progression of the illness to address newly emerging issues [102]. Successful patient management must be based on a solid alliance

Table 5. The multimodal treatment of AD

- Establish and maintain an alliance with the patient and family
- Provide specific psychosocial and rehabilitative interventions
- Use medications active on cognitive and functional impairment
- Treat non-cognitive symptoms
- Assess and treat comorbid conditions
- Prevent and treat complications
- Develop a treatment plan and follow-up schedule

with the patient and the family and requires a well-defined case manager that might be, in relation to cultural characteristics and peculiarities of the health-system structure, the primary physician or a specialist, such as geriatrician, neurologist, or psychiatrist [2, 107].

Since the advent of new drugs for the treatment of AD, a variety of different forms of environmental and rehabilitative interventions has proven effective for different target symptoms of the disease [108]. Changes in the environment may enhance cognitive deficits (throughout the use of cues or memory aids, the proper use of light and colors) and reduce behavioral symptoms [109]. Stimulation-oriented programs such as art therapy, recreational therapy, or occupational therapy may improve mood and quality of life of patients [110]. Validation therapy, reminiscence therapy, and sensory integration are used to reduce behavioral symptoms and improve mood [111, 112]. Cognition-oriented approaches (such as reality-orientation therapy, reactivating occupation therapy, and memory-stimulation programs) have proven to be useful in improving cognition and functional status [113, 114].

Nowadays strategies for primary therapy of AD are substantially limited to drugs to improve central cholinergic neurotransmission [115]. The cholinergic hypothesis of Alzheimer's disease has strongly influenced research on learning and memory, and on the possible pharmacological treatments of AD over the last decade. The acetylcholinesterase inhibitors (AChEIs) are the most studied and most often class of drugs used for the treatment of dementia patients. They prevent the hydrolysis of released acetylcholine, increasing the efficiency of cholinergic transmission [116]. In 1986 a paper was published discussing the possible role of tetrahydroaminoacridine (tacrine, a molecule that was synthesized in 1945, and observed to be an AChEI in 1953) in long-term treatment of AD patients [117]. The authors observed a symptomatic improvement and suggested the usefulness of tacrine in the palliative therapy of patients with AD. Even if a number of studies reported tacrine to give clinically significant improvement of AD patients, the clinical efficacy of this drug is limited only because of the frequent side effects (30%–50% of patients stop the treatment), so it is relevant in only a limited percentage of cases (20%–40%) [118, 119]. After tacrine, a number of AChEIs have been studied that have fewer side effects because of their greater selectivity for the cerebral enzyme. Nowadays donepezil is the second cholinergic drug approved by the FDA. In many European countries donepezil and rivastigmine are authorized for treatment of mild-to-moderate AD patients. The effectiveness of donepezil has been demonstrated in two randomized double-blind placebo-controlled trials [120]. Donepezil has a longer duration of inhibitory action and greater specificity for brain tissue than tacrine; moreover, donepezil is better tolerated than tacrine and does not cause hepatotoxicity [116]. The drug has a recommended starting dose of 5 mg/day, which may be increased to 10 mg after 1 month [2]. Recently, rivastigmine, an AChEI, has been approved in Europe for the treatment of AD, at the dosage of 4.5–6 mg b.i.d. [121]. Tacrine, donepezil, and rivastigmine (likewise the other cholinesterase-acting drugs) should be prescribed only for patients with a diagnosis of probable AD in the mild-to-moderate stage; use for other conditions (i.e., Lewy body dementia, more severe patients) should only be attempted within specialist services [122]. It has been suggested that therapy with AChEI improves overall behavioral disturbances and functional abilities in the activities of daily living and that it might reduce the need for institutionalization of AD patients, indicating a strongly economically positive outcome [65, 115]. However, it is not clear if cholinergic replacement treatment can prolong the duration of illness, increasing costs for services. Furthermore, at this moment, there are no scientific data on the effects of stopping cholinergic replacement therapy with AChEI, making the decision to stop therapy difficult [121].

The cholinergic agonists would be expected to activate the postsynaptic muscarinic receptor sites directly. In man, five acetylcholine receptors (mAChR, M1-M5) have been cloned. M1-type receptors are predominant in the cerebral cortex and hippocampus. They may have an important role in cognitive processes and they seem to be preserved in AD [123]. Actually, selective M1 agonist are under study that have no peripheral or central side effects mediated by M2 or M3 receptors and M2 antagonists inhibiting the effects of presynaptic AChR, such as xanomeline, SB202026, and milameline [115].

Clinical trials of other agents to improve cognitive function in AD are ongoing. These include estrogens, nonsteroidal anti-inflammatory agents, and botanic agents, such as gingko bilboa and huperzine A [2, 124]. Currently, these agents cannot be recommended for treatment of AD, because the evidence of benefits is only indirect and the data are insufficient. The results of large-scale clinical trials using antioxidants (vitamin E and selegiline) have shown decreased rates of functional decline compared with placebo [125]. On the basis of these data, and in consideration of the low toxicity of the drug, vitamin E may be used in moderate AD patients to delay progression of the disease [102]. Others agents with a less clearly defined mechanism of action, such as hydergine, acetyl-l-carnitine, nimodipine, nootropil, are widely used, but have not been associated with a any-clinically important improvement in AD [126].

Several strategies for the development of disease-modifying treatment are under investigation: neurotrophic factors, drugs modulating protein processing, antioxidants, anti-inflammatory drugs, glutamate antagonists, and heavy metal shelters [127].

Behavioral problems cause significant morbidity, interfere with the performance of activities of daily living, and often have a greater impact on the quality of life of the family and caregivers than cognitive impairment. Treatment is often challenging and requires education of the family members/caregivers, and communication between all persons involved in patient care, in addition to the standard pharmacological and non-pharmacological interventions [128].

Symptoms considered responsive to pharmacological intervention include physical and verbal agitation, depression, and delusional and psychotic symptoms.

Psychosis only needs to be treated if it substantially interferes with the patient, caregiver or family. The preferred treatment is environmental manipulation because neuroleptics are the only effective medication, and they may have rele-

Table 6. Environmental strategies to minimize behavioral disturbances in AD patients

Psychosis
- Ignoring false accusations
- Keeping lights or music on if reduced sensory input triggers the psychosis
- Having a regular activity program to increase motor and social activity
- Distracting the patients from their focus of concern by shifting subjects or changing location
- Providing familiar objects for the patient in common locations
- Providing them something familiar and reassuring to hold such as their wallet or purse
- Comforting and reassuring the patient through touch and tone of voice

Agitation
- Avoiding events that precipitate the behavior
- Removing the precipitating and perpetuating stimuli
- Distracting the patient
- Providing emotional support

vant side effects or worsen the dementia [129]. Identifying the source of the psychosis and correcting or eliminating it is best (Table 6 shows some examples of strategies to minimize behavioral disturbances in AD patients). When drugs are indicated, risperidone (0.2–2 mg b.i.d.), clozapine (12.5–50 mg b.i.d.) and thioridazine (25-75 mg b.i.d.) are the first choice [130, 131].

Agitation may occur in persons with brain damage to structures affecting perception and inhibitory control, especially frontal and temporal lobes, or new onset or relapse of a medical illness [128]. Treatment should first be directed at identifying precipitants or undiagnosed medical illnesses. Environmental strategies are shown in Table 6. In the pharmacological treatment of agitation, antipsychotics are not necessarily the best choice; sedating serotoninergic agents, such as trazodone (50–200 mg t.i.d.), non-sedating selective serotoninergic reuptake inhibitors, low-dose beta blockers, such as propranolol (10–30 mg t.i.d.), carbamazepine at low doses (as low as 50 mg twice a day to 200 mg t.i.d.), buspirone (5–20 mg b.i.d. to t.i.d.) are preferred [102]. Low doses of selective D2-receptor antagonists will minimize extrapyramidal side effects; short-acting benzodiazepines, such as oxazepam or temazepam (15–30 mg), should be considered as a second line of treatment [132].

AD patients with depressive syndrome should be considered for treatment, even if they do not meet the criteria for a depressive syndrome [2]. Behavioral approaches (psychotherapy, activity groups, exercise, pets) may be useful [133]. When medical treatment is indicated, the choice of the drugs should be based on the patient's general medical conditions and the behavioral symptoms coexisting. If insomnia coexists, then an antidepressant that helps sleep (such as trazodone, or nortryptiline) should be given 1–2 h before bedtime. If no sleep problems exist, then one can use one of the selective serotonin reuptake inhibitor antidepressants in the morning (fluoxetine, paroxetine, or sertraline) [134]. Venflaxine and bupropion are used in selected cases. Tricyclic antidepressant are effective in the treatment of depression, but have significant anticholinergic activity (especially amitriptyline, imipramine, and clorimipramine) [102]. Other types of behavioral symptoms (such as wandering, sexual behaviors, persistent vocalization, and abnormal eating behavior) may be difficult to manage, and little is known about effective treatment [128, 135].

Caregivers play a key role in the management of dementia and represent the main social buffer to the early institutionalization of patients. Since they are psychologically affected by the heavy burden of care-giving, the patient's behavioral disturbances being the most distressing, the use of supportive and educational programs has been regarded as the most reliable way to reduce distress and decrease the risk of institutional placement of demented patients [136]. Caregiver interventions have used different strategies, the elements of which can be categorized into three broad focuses: psychological, educational, and social (see Table 7) [137]. Patients and family must be educated regarding the illness, possible hazards (driving, wandering, living alone), environmental factors that control symptoms, available treatments, source of care and supports, legal, ethical and financial issues.

Despite the progress in the development of new treatment strategies for AD, many patients and caregivers receive inadequate care [2]. One reason is that the health service organizations for the care of patients with AD are not well defined in many countries. In recent years different health services have been designed to diagnose dementia more accurately and to improve the care of patients [138]. The

Table 7. The elements of interventions on AD caregivers

Psychological
- Support
- Ventilation, group process, sharing, acknowledging, learning, mutual support, recognition of universality
- Cognitive
- Counseling, insight therapy, cognitive therapy, relaxation training, stress management
- Emotional impact-stress, anger, grief, guilt
- Self-care
- Interpersonal relationship and communication

Educational
- Information
- Improving home care skills
- Developing therapeutic skills, problem-solving
- Behavioral techniques
- Planning emergencies, legal, financial

Developing a support system
- Personal, family
- Community
- Professional

aims of these services (diagnostic evaluation, treatment/referral, case management, research, training and continuing education, rehabilitation programs, special programs of care), their target subjects (outpatients, inpatients, nursing home residents), organization, and costs are specific and substantially different, but still not thoroughly studied [139]. To identify the causes of memory impairment in the elderly, memory clinics have been developed in addition to geriatric and psychogeriatric services [140]. In North America memory clinics are linked to Alzheimer's Disease Centers for the assessment of patients with cognitive impairment. These centers provide training and education of patients and relatives, support services, diagnostic treatment and rehabilitation programs. In 1994, in Italy, the regional government of Lombardia (northern Italy, 8.5 million inhabitants) decided to set up a comprehensive care network of services called "Piano Alzheimer," the aim of which was to take care of patients in different phases of AD [141]. On the basis of the results of a pilot study, 20-bed Special Care Units (SCUs) in 40 nursing homes (corresponding to 3.5% of the total nursing home beds in the region) were opened and reserved for demented patients with a high level of behavioral disturbances [113]. Nine Regional Alzheimer Centers, and day care centers were also realized. The Regional Alzheimer Centers were designed to provide comprehensive assessment of medical, psychological and social problems of the demented elderly, to provide them with therapy, rehabilitation, counseling, and social, legal and ethical support, to provide the formation of physicians and nurses, and to establish research protocols. The cost-effectiveness analysis demonstrated that, with 11% increase in costs, considerable improvement in cognitive, functional, and behavioral symptoms can be attained, with a reduction in physical and pharmacological restraints and with improvement in the well-being of patients and caregivers [113].

References

1. Maurer K, Volk S, Gerbaldo H, Auguste D (1997) Alzheimer's disease. Lancet 349: 1546-1549
2. Small GW, Rabins PV, Barry PP, Buckholtz NS, DeKosky ST, Ferris SH, Finkel SI, et al (1997) Diagnosis and treatment of Alzheimer disease and related disorders. Consensus Statement of the American Association for Geriatric Psychiatry, the Alzheimer's Association, and the American Geriatrics Society. JAMA 278: 1363-1371
3. Miller B (1997) Clinical advances in degenerative dementias. Br J Psychiatry 171: 1-3
4. Hofman A, Rocca WA, Brayne C, Breteler MM, Clarke M, Cooper B, Copeland JR, Dartigues JF, da Silva Droux A, Hagnell O (1991) The prevalence of dementia in Europe: a collaborative study of 1980-1990 findings. Eurodem Prevalence Research Group. Int J Epidemiol 20: 736-48
5. Hebert LE, Scherr PA, Beckett LA, Albert MS, Pilgrim DM, Chown MJ, Funkenstein HH, Evans DA (1995) Age specific incidence of Alzheimer's disease in a community population. JAMA 273: 1354-1359
6. Drachman DA (1994) If we live long enough, will we all be demented? Neurology 44: 1563-1565

7. Ritchie K, Kildea D (1995) Is senile dementia "age-related" or "ageing-related"? – evidence from meta-analysis of dementia prevalence in the oldest old. Lancet 346: 931-934

8. Larrabee GJ, McEntee WJ (1995) Age-associated memory impairment: sorting out the controversies. Neurology 45: 611-614

9. Christensen H, Henderson AS, Jorm AF, Mackinnon AJ, Scott R, Korten AE (1995) ICD-10 mild cognitive disorder: epidemiological evidence on its validity. Psychol Med 25: 105-120

10. Hanninen T, Hallikainen M, Koivisto K, Helkala EL, Reinikainen KJ, Soininen H, Mykkanen L, Laakso M, Pyorala K, Riekkinen PJ Sr (1995) A follow-up study of age-associated memory impairment: neuropsychological predictors of dementia. J Am Geriatr Soc 43: 1007-1015

11. Fratiglioni L (1993) Epidemiology of Alzheimer's disease. Issues of etiology and validity. Acta Neurol Scand 145: 1-70

12. St. George-Hyslop P, Tanzi RE, Polinsky RJ (1987) The genetic defect causing familial Alzheimer's disease maps on chromosome 21. Science 235: 885-889

13. Goate A, Chartier-Harlin MC, Mullan M (1991) Segregation of a missense mutation in the amyloid precursor protein gene with familial Alzheimer's disease. Nature 349: 704-706

14. Sorbi S (1993) Mulecular genetics of Alzheimer's disease. Aging Clin Exp Res 5: 417-425

15. Mullan M, Crawford F, Buchanan J (1994) Technical feasibility of a genetic testing for Alzheimer's disease. Alzheimer Dis Assoc Disord 8: 102-115

16. Schellenberg G, Bird T, Wijsman E, et al (1992) Genetic linkage evidence for a familial Alzheimer's disease locus on chromosome 14. Science 258: 668-671

17. Sherrington R, Rogaev EI, Liang Y, et al (1995) Cloning a gene bearing a missense mutations in early-onset familial Alzheimer's disease. Nature 375: 754-760

18. Lendon CL, Ashall F, Goate AM (1997) Exploring the etiology of Alzheimer disease using molecular genetics. JAMA 277: 825-831

19. Tanzi RE, Kovacs DM, Kim TW, Moir RD, Guenette SY, Wasco W (1996) The presenilin genes and their role in early-onset familial Alzheimer's disease. Alzheimer Dis Rev 1: 91-98

20. Levy-Lahad E, Wijsman EM, Nemens E, et al (1995) A familial Alzheimer's disease locus on chromosome 1. Science 269: 970-973

21. Frisoni GB, Trabucchi M (1997) Clinical rationale of genetic testing in dementia. J Neurol Neurosurg Psychiatry 62: 217-221

22. Hardy J (1996) New insights into the genetics of Alzheimer's disease. Ann Med 28: 255-258

23. Dickson DW, Liu WK, Yen SH (1996) An antibody to an amino-terminal domain of presenilin-1 immunostains neurofibrillary tangles in Alzheimer's disease. Neurobiol Aging 17: S18

24. Farrer LA (1997) Genetics and the dementia patient. Neurologist 3: 13-30

25. Strittmatter WJ, Saunders AM, Schmechel D, et al (1993) Apolipoprotein E: high avidity binding to β-amyloid and increased frequency of type 4 allele in late-onset familial Alzheimer disease. Proc Natl Acad Sci USA 90: 1977-1981

26. Saunders AM, Strittmatter WJ, Schmechel D, et al (1993) Association of apolipoprotein E allele epsilon 4 with late onset familial and sporadic Alzheimer's disease. Neurology 43: 1467-1472

27. Nathan BP, Bellosta S, Sanan DA, Weisgraber KH, Pitas RE (1994) Differential effects of apolipoprotein ε3 and ε4 on neuronal grow in vitro. Science 264: 850-852

28. Frisoni GB, Calabresi L, Geroldi C, Bianchetti A, D'Acquarica AL, Trabucchi M, Govoni

S, Franceschini G (1994) Apolipoprotein E ε4 allele in Alzheimer's disease and vascular dementia. Dementia 5: 240-242

29. Corder EH, Saunders AM, Strittmatter WJ, et al (1993) Gene dose of apolipoprotein E type 4 allele and the risk of Alzheimer's disease in late onset families. Science 261: 921-923

30. Farrer LA, Cupples LA, Haines JL, Hyman B, Kukull WA, Mayeux R, et al (1997) Effects of age, sex, and ethnicity on the association between apolipoprotein E genotype and Alzheimer's disease. A meta-analysis. JAMA 278: 1349-1356

31. Evans DA, Beckett LA, Field TS, Feng L, Albert MS, et al (1997) Apolipoprotein E ε4 and incidence of Alzheimer disease in a community population of older persons. JAMA 277: 822-824

32. Slooter AJC, Tang MX, Dujin CM, Stern Y, Ott A, et al (1997) Apolipoprotein E ε4 and the risk of dementia after stroke. JAMA 277: 818-821

33. Nicol JAR, Roberts GW, Graham DJ (1995) Apolipoprotein E ε4 allele is associated with deposition of amyloid β-protein following head injury. Nat Med 1: 135-137

34. Pericak-Vance MA, Bass MP, Yamaoka LH, Gaskell PC, Scott WK, et al (1997) Complete genomic screen in late-onset familial Alzheimer disease. Evidence for a new locus on chromosome 12. JAMA 278: 1237-1241

35. Stephenson J (1997) Reserchers find evidence of a new gene for late-onset Alzheimer disease. JAMA 277: 775

36. Mortimer JA, van Duijn CM, Chandra V, Fratiglioni L, Graves AB, Heyman A, Jorm AF, Kokmen E, Kondo K, Rocca WA, et al (1991) Head trauma as a risk factor for Alzheimer's disease: a collaborative re-analysis of case-control studies. EURODEM risk factors research group. Int J Epidemiol 20[Suppl 2]: S28-S35

37. Stern Y, Gurland B, Tatemichi TK, Tang MX, Wilder D, Mayeux R (1994) Influence of education and occupation on the incidence of Alzheimer's disease. JAMA 271: 1004-1010

38. Rocca WA (1994) Frequency, distribution, and risk factors for Alzheimer's disease. Nurs Clin North Am 29: 101-11

39. Breteler MM, van Dijon CM, Chandler V, Fratiglioni L, Graves AB, He-man A, Corm AF, Kokune E, Kondo K, Mortimer JA, et al (1991) Medical history and the risk of Alzheimer's disease: a collaborative re-analysis of case control study. EURODEM risk factors research group. Int J Epidemiol 20[Suppl 2]: S36-S42

40. Snowdon DA, Greiner LH, Mortimer JA, Riley KP, Greiner PA, Markesbery WR (1997) Brain infarction and the clinical expression of Alzheimer disease. The Nun Study. JAMA 277: 813-817

41. McKhann G, Drachman D, Folstein M, Katzman R, Price D, Stadlan EM (1984) Clinical diagnosis of Alzheimer's disease: report of the NINCDS-ADRDA work group under the auspices of Department of Health and Human Services Task Force on Alzheimer's disease. Neurology 34: 939-944

42. Terry RD, Masliah E, Hansen LA (1994) Structural basis of the cognitive alterations in Alzheimer disease. In: Terry RD, Katzman R, Bick KL (eds) Alzheimer disease. Raven, New York, pp 179-196

43. Goedert M (1993) Tau protein and the neurofibrillary pathology of Alzheimer's disease. Trends Neurosci 16: 460-465

44. Goedert M, Jakes R, Spillantini MG, Crowther RA (1994) Tau protein and Alzheimer's disease. In: Hyams JS, Lloyd CW (eds) Microtubules, modern cell biology, vol 13. Wiley-Liss, New York, pp 183-200

45. Lee VMY, Balin BJ, Otvos L Jr, Trojanowski JQ (1991) A 68: a major subunit of paired helical filaments and derivatized form of normal tau. Science 251: 675-678

46. Esiri MM, Pearson RCA, Steele JE, Bowen DM, Powell TPS (1990) A quantitative study

of the neurofibrillary tangles and the choline acetyltransferase activity in the cerebral cortex and the amygdala in Alzheimer's disease. J Neurol Neurosurg Psychiatry 53: 161-165

47. Mann DMA (1989) The pathogenesis and progression of the pathological changes of Alzheimer's disease. In: Davies DC (ed) Current problems in neurology. II. Alzheimer's disease: towards an understanding of the aetiology and pathogenesis. John Libbey, London, pp 43-53

48. Zemlan FP, Vogelsong GD, McLaughlin L, Dean GE (1994) Alzheimer's paired helical filaments: amyloid precursor protein epitope mapping. Brain Res Bull 33: 387-392

49. Robakis NK (1994) Beta amyloid and amyloid precursor protein chemistry, molecular biology and neuropathology in Alzheimer disease. In: Terry RD, Katzman R, Bick KL (1994) Alzheimer disease. Raven, New York, pp 317-325

50. Blennow K, Cowburn RF (1996) The neurochemistry of Alzheimer's disease. Acta Neurol Scand Suppl 168: 77-86

51. Aisen PS, Davis KL (1994) Inflammatory mechanism in Alzheimer's disease: implications for therapy. Am J Psychiatry 151: 1105-1113

52. Mirra SS, Gearing M, Nash F (1997) Neuropathologic assessment of Alzheimer's disease. Neurology 49[3 Suppl]: S14-S16

53. Drachman DA, Leavitt J (1974) Human memory and the cholinergic system: a relationship to aging? Arch Neurol 30: 113-121

54. Whitehouse PJ, Price DL, Struble RG, Clark AW, Coyle JT, DeLong MR (1982) Alzheimer's disease and senile dementia: loss of neurons in the basal forebrain. Science 215: 1297-1239

55. Geula C, Mesulam MM (1989) Cortical cholinergic fibers in aging and Alzheimer's disease: a morphometric study. Neuroscience 33: 469-481

56. Rossor M, Garrett NJ, Johnson AL, Mountjoy CQ, Roth M, Iversen LL (1982) A postmortem study of the cholinergic and GABA systems in senile dementia. Brain 105: 313-330

57. Pilleri G (1966) The Kluver Bucy syndrome in man – a clinicoanatomical contribution to the function of the medial temporal lobe structures. Psychiatr Neurol 152: 65

58. Nitsch RM (1996) From acetylcholine to amyloid: neurotransmitters and the pathology of Alzheimer's disease. Neurodegeneration 5: 477-482

59. Palmer AM (1996) Neurochemical studies of Alzheimer's disease. Neurodegeneration 5: 381-391

60. Palmer AM, Gershon S (1990) Is neuronal basis of Alzheimer's disease cholinergic or glutamatergic. FASEB J 4: 2745-2752

61. Geldmacher DS, Whitehouse PJ (1996) Evaluation of dementia. N Engl J Med 335: 330-336

62. Rossor M (1993) Alzheimer's disease. BMJ 307: 779-782

63. Bianchetti A, Rozzini R, Trabucchi M (1994) Demenze. In: De Leo D, Stella A (eds) Manuale di psichiatria dell'anziano. Piccin, Padova, pp 343-371

64. Binetti G, Bianchetti A, Padovani A, Lenzi G, De Leo D, Trabucchi M (1993) Delusions in Alzheimer's disease and multi-infarct dementia. Acta Neurol Scand 88: 5-9

65. Cummings JL (1997) Changes in neuropsychiatric symptoms as outcome measures in clinical trials with cholinergic therapies for Alzheimer disease. Alzheimer Dis Assoc Disord 11[Suppl 4]: S1-S9

66. Trabucchi M, Bianchetti A (1996) Delusions. Int Psychogeriatr 8[Suppl 3]: 383-386

67. Bianchetti A, Scuratti A, Zanetti O, Binetti G, Frisoni GB, Magni E, Trabucchi M (1995) Predictors of mortality and institutionalization in Alzheimer's Disease patients one year after discharge from an Alzheimer's dementia Unit. Dementia 6: 108-112

68. American Psychiatric Association (1994) Diagnostic and statistical manual of mental disorders, 4th edn. American Psychiatric Association, Washington DC

69. Zanetti O, Bianchetti A, Frisoni GB, Rozzini R, Trabucchi M (1993) Determinants of disability in Alzheimer's disease. Int J Geriatr Psychiatry 8: 581-586

70. Gearing M, Mirra SS, Hedreen JC, Sumi SH, Hansen LA, Heyman A (1995) CERAD part X. Neuropathology confirmation of the clinical diagnosis of Alzheimer's disease. Neurology 45: 461-466

71. Rasmusson DX, Brandt J, Steele C, Hedreen JC, Troncoso JC, Folstein MF (1996) Accuracy of clinical diagnosis of Alzheimer disease and clinical features of patients with non-Alzheimer disease neuropathology. Alzheimer Dis Assoc Dis 10: 180-188

72. Teri L, McCurry SM, Edland SD, Kukull WA, Larson EB (1995) Cognitive decline in Alzheimer's disease: a longitudinal investigation of risk factor for accelerated decline. J Gerontol 50A: M49-M55

73. Strauss ME, Lee MM, Di Filippo JM (1997) Premorbid personality and behavior symptoms in Alzheimer disease. Arch Neurol 54: 257-259

74. Kitwood T (1993) Person and process in dementia. Int J Geriatr Psychiatry 8: 541-545

75. Corey-Bloom J, Thal LJ, Galasko D, Folstein M, Drachman D, Raskind M, Lanska DJ (1995) Diagnosis and evaluation of dementia. Neurology 45: 211-218

76. Geroldi C, Frisoni GB, Bianchetti A, Trabucchi M (1997) Drug treatment in Lewy body dementia. Dement Geriatr Cogn Disord 8: 188-197

77. Folstein MF, Folstein SE, McHugh PR (1975) "Mini-Mental State": a practical method for grading the cognitive state of patients for the clinician. J Psychiatr Res 12: 189-198

78. Blessed G, Tomlinson BE, Roth M (1968) The association between quantitative measures of dementia and of senile change in the cerebral gray matter of elderly subjects. Br J Psychiatry 114: 797-811

79. Cipolotti L, Warrington EK (1995) Neuropsychological assessment. J Neurol Neurosurg Psychiatry 58: 655-664

80. Frisoni GB, Rozzini R, Bianchetti A, Trabucchi M (1993) Principal lifetime occupation and MMSE score in the community dwelling elderly. J Gerontol Soc Sci 48: 310-314

81. Cronin-Golomb A, Corkin S, Rosen TJ (1993) Neuropsychological assessment of dementia. In: Whitehouse PJ (ed) Dementia, vol 40 (Contemporary Neurology Series). Davis, Philadelphia, pp 130-164

82. Magni E, Binetti G, Bianchetti A, Rozzini R, Trabucchi M (1996) Mini-Mental State Examination: a normative study in italian ederly population. Eur J Neurol 3: 1-5

83. Reuben DB (1993) Use and abuse of assesement instruments. In: Osterwail D, Reuben DB, Rozzini R, Rubenstein LZ, Trabucchi M (eds) New frontiers in geriatric medicine. Kendall, Padova, pp 17-26

84. Reuben DB, Solomon DH (1989) Assessement in geriatrics: of caveats and names. J Am Geriatr Soc 37: 570-572

85. Zanetti O, Bianchetti A, Trabucchi M (1995) The puzzle of functional status in mild and moderate Alzheimer's disease: self-report, family report, and performance-based assessement. Gerontologist 35: 148

86. Guralnik JM, Branch LG, Cummings SR, Crib JD (1989) Physical performance measures in aging research. J Gerontol 44: 141-146

87. NIH Consensus Conference (1987) Differential diagnosis of dementing diseases. JAMA 23: 3411-3416

88. Clarfield AM (1988) The reversible dementia: do they reverse? Ann Int Med 109: 476-486

89. Weytingh MD, Bossuyt PMM, van Crevel H (1995) Reversible dementia: more than 10% or less than 1%? A quantitative review. J Neurol 242: 446-471

90. Chui H, Zhang Q (1997) Evaluation of dementia: a systematic study of the American Academy of Neurology Practice Parameters. Neurology 49: 925-935

91. Growdon JH (1995) Advances in the diagnosis of Alzheimer's disease. In: Iqbal K, Mortimer JA, Winblad B, Wisniewski HM (eds) Recent advances in Alzheimer's disease and related disorders. Wiley, Chichester, pp 139-154

92. Gasparini L, Racchi M, Binetti G, Trabucchi M, Solerte B, Alkon D, Etcheberrigaray R, Gibson G, Blass J, Paoletti R, Govoni S (1998) Peripheral markers in testing pathophysiological hypotheses and diagnosing Alzheimer's disease. FASEB J 12: 17-34

93. Geldmacher DS, Whitehouse PJ (1997) Differential diagnosis of Alzheimer's disease. Neurology 48[Suppl 6]: S2-S9

94. Green RC, Clarke VC, Thompson NJ, Woodard JL, Letz R (1997) Early detection of Alzheimer disease: methods, markers, and misgivings. Alzheimer Dis Assoc Disord 11[Suppl 5]: S1-S5

95. Post SG (1994) Genetics, ethics, and Alzheimer disease. J Am Geriatr Soc 42: 782-786

96. Post SG, Whitehouse PJ, Bisntock RH, Bird TD, Eckert SK, et al (1997) The clinical introduction of genetic testing for Alzheimer disease: an ethical perspective. JAMA 277: 832-836

97. Frisoni GB, Bianchetti A, Trabucchi M, Govoni S (1996) Diagnostic usefulness of apolipoprotein E ε4 in the diagnosis of dementia. J Neurol Neurosurg Psychiatry 60: 699-700

98. Relkin NR (1996) National Institute on Aging/Alzheimer's Association Working Group. Apolipoprotein E genotyping in Alzheimer's disease. Lancet 347: 1091-1095

99. Jagus WJ, Eberling JL (1991) MRI, CT, SPECT, PET: their use in diagnosing dementia. Geriatrics 46: 28-35

100. Frisoni GB, Bianchetti A, Geroldi C, Trabucchi M, Beltramello A, Weiss C (1994) Measures of temporal lobe atrophy in probable Alzheimer's disease. J Neurol Neurosurg Psychiatry 57: 1438-1439

101. Frisoni GB, Pizzolato G, Govoni S, Bianchetti A, Ferlin G, Trabucchi M (1994) Single photon emission computed tomography with [99mTc]-HM-PAO and [123I]-IBZM in Alzheimer's Disease and dementia of frontal type. Acta Neurol Scand 89: 199-203

102. Rabins P, Blacker D, Bland W, Bright-Long L, Cohen E, Katz I, et al (1997) Practice guideline for treatment of patients with Alzheimer's disease and other dementias of late life. Am J Psychiatry 154: 1-39

103. Lussignoli G, Frisoni GB, Bianchetti A, Trabucchi M (1998) Measurement of outcomes in Alzheimer's disease. In: Vellas B, Fitten J, Frisoni GB (1998) Research and practice in Alzheimer's disease. Serdi, Paris, pp 437-446

104. Kane RL (1995) Improving the quality of long-term care. JAMA 273: 1376-1380

105. Brook RH, McGlynn EA, Cleary PD (1996) Maeasuring quality of care. N Engl J Med 335: 966-970

106. Ramsay M, Winget C, Higginson I (1995) Review: measures to determine the outcome of community services for people with dementia. Age Aging 24: 73-83

107. Zanetti O, Bianchetti A, Trabucchi M (1995) Il geriatra e la gestione del paziente demente. Giorn Geront 43: 343-349

108. Bianchetti A, Zanetti O, Trabucchi M (1997) Non pharmacological treatment in Alzheimer's disease. Funct Neurol 12: 215-217

109. Coons DH (1991) The therapeutic milieu: concepts and criteria. In: Coons DH (ed) Specialized dementia care units. John Hopkins University Press, Baltimore

110. Rovner BW, Steel CD, Shumely Y, Folstein MF (1996) A randomized trial of dementia care in nursing home. J Am Geriatr Soc 44: 7-13

111. Burnside L, Haight B (1994) Reminiscence and life review: therapeutic interventions for older people. Nurs Pract 19: 55-61
112. Feil N (1992) Validation: the Feil method. Feil, Cleveland
113. Bianchetti A, Benvenuti P, Ghisla KM, Frisoni GB, Trabucchi M (1997) An Italian model of Dementia Special Care Unit: results of a pilot study. Alzheimer Dis Assoc Disord 11: 53-56
114. Zanetti O, Binetti G, Magni E, Rozzini L, Bianchetti A, Trabucchi M (1997) Procedural memory stimulation in Alzheimer's disease: impact of a training programme. Acta Neurol Scand 95: 152-157
115. Knopman DS, Morris JC (1997) An update on primary drug therapies for Alzheimer disease. Arch Neurol 54: 1406-1409
116. Wilcock GK (1996) Current approach to the treatment of Alzheimer disease. Neurodegeneration 5: 505-509
117. Summers WK, Majorski LV, Marsh GM, Tachiki K, Kling A (1986) Oral tetrahydroaminoacridine in long term treatment of senile dementia of Alzheimer type. N Engl J Med 315: 1241-1245
118. Davis KL, Thal LJ, Gamzu ER, et al (1992) A double-blind , placebo-controlled multicenter study of tacrine for Alzheimer's disease. N Engl J Med 327: 1253-1259
119. Farlow M, Gracon SI, Hershey LA, Lewis KW, Sadowsky CH, Dolan-Ureno J (1992) A controlled trial of tacrine in Alzheimer's disease. JAMA 268: 2523-2529
120. Nightingale SL (1997) Donepezil approved for treatment of Alzheimer disease. JAMA 277: 10
121. Geroldi C, Bianchetti A, Trabucchi (1997) Manipulation of the cholinergic system. Funct Neurol 12: 187-191
122. Lovestone S, Graham N, Howard R (1997) Guidelines on drug treatments for Alzheimer's disease. Lancet 350: 232-233
123. Levely A (1996) Muscarinic acetylcholine receptor expression in memory circuits. Implications for treatment of Alzheimer disease. PNAS 93: 13541-13546
124. Skolnick AA (1997) Old Chinese herbal medicine used for fever yields possible new Alzheimer disease therapy. JAMA 277: 776-778
125. Sano M, Ernesto C, Thomas RG, et al (1997) A controlled trial of selegiline, alpha-tocopherol, or both as treatment for Alzheimer's disease. N Engl J Med 336: 1216-1222
126. Riedel WJ, Jolles J (1996) Cognition enhancers in age-related cognitve decline. Drugs Aging 8: 245-274
127. Aisen PS, Davis KL (1997) The search for disease-modifying treatment for Alzheimer's disease. Neurology 48[Suppl 6]: S35-S41
128. Teri L, Rabins P, Whitehouse PJ, Berg L, Reisberg B, Sunderland T, et al (1992) Management of behavior disturbane in Alzheimer disease: current knowledge and future directions. Alzheimer Dis Assoc Disord 6: 77-88
129. Marchello V, Boczko F, Shelkey M (1995) Progressive dementia: strategies to manage new problems behaviors. Geriatrics 50: 40-43
130. Salzman C, Vaccaro B, Lieff J, Weiner AS (1995) Clozapine in older patients with psychosis and behavioral disturbances. Am J Geriatr Psychiatry 3: 26-33
131. Goldberg RJ, Goldberg J (1997) Risperidone for dementia-related disturbed behavior in nursing home residents: a clinical experience. Int Psychogeriatr 9: 65-68
132. Tariot PN, Schenider LS, Katz IR (1995) Anticonvulsivant and other non-neuroleptic treatment of agitation in dementia. J Geriatr Psychiatry Neurol 8[Suppl 1]: S28-S39
133. Teri L (1994) Behavioral treatment of depression in patients with dementia. Alzheimer Dis Assoc Disord 8[Suppl 3]: 66-74

134. Knesper DJ (1995) The depression in Alzheimer's disease: sorting pharmacotherapy, and clinical advice. J Geriatr Psychiatry Neurol 8[Suppl 1]: S40-S51

135. Dodds P (1994) Wandering: a short report on coping strategies adopted by informal careers. Int J Geriatr Psychiatry 9: 751-756

136. Zanetti O, Magni E, Sandri C, Frisoni GB, Bianchetti A, Trabucchi M (1996) Determinants of burden in an Italian sample of Alzheimer's patient caregivers. J Cross-Cultural Gerontol 11: 17-27

137. Burgeois MS, Shulz R, Burgio L (1996) Interventions for caregivers of patients with Alzheimer's disease: a review and analysis of content, process, and outcomes. Int J Aging Hum Dev 43: 35-92

138. Berg L, Buckwalter KC, Chafets PK, et al (1991) Special care units for persons with dementia. J Am Geriatr Soc 39: 1229-1236

139. Sloane PD, Lindeman DA, Phillips C, Moritz DJ, Koch G (1995) Evaluating Alzheimer's special care units: reviewing the evidence and identifying potential sources of study bias. Gerontologist 35:1 03-111

140. Philpot MP, Levy R (1987) A memory clinic for the early diagnosis of dementia. Int J Ger Psychiatry 2: 195-200

141. Trabucchi M, Bianchetti A, Zanetti O (1995) An italian network for the care of demented patients. Gerontologist 35[Suppl 1]: 90

142. Hughes CP, Berg L, Danziger WL (1982) A new clinical scale for the staging of dementia. Br J Psychiatry 140: 566-572

143. Binetti G, Magni E, Rozzini L, Bianchetti A, Trabucchi M, Cummings JL (1995) "Neuropsychiatric Inventory": validazione italiana di una scala per la valutazione psicopatologica della demenza. Giorn Geront 43: 864-865

144. Binetti G, Magni E, Cappa S, Padovani A, Bianchetti A, Trabucchi M (1993) Neuropsychological heterogeneity in mild Alzheimer's disease. Dementia 4: 321-326

145. Cummings JL, Mega M, Gray K, Rosemberg-Thompson S, Carusi DA, Gornbei J (1994) The Neuropsychiatric Inventory: comprehensive assessment of psychopathology in dementia. Neurology 44: 2308-2314

146. Migliorelli R, Teson A, Sabe L, et al (1995) Anosognosia in Alzheimer's disease: a study of associated factors. J Neuropsychol Clin Neurosci 7: 338-344

147. Mahoney FI, Barthel DW (1965) Functional evaluation: the Barthel Index. Md Med J 14: 61-65

148. Lawton MP, Brody EM (1969) Assessement of older people; self-maintaining and instrumental activities of daily living. Gerontologist 9: 179-186

149. Lowenstein DA, Amigo E, Duara R, Guterman A, Hurwitz D, Berkowitz N, Wilkie F, Weinberg G, Black B, Gittelman B, Eisdorfer C (1989) A new scale for the assessement of functional status in Alzheimer's disease and related disorders. J Gerontol 44: 114-121

150. Reuben DB, Siu AL (1990) An objective measure of physical function of elderly outpatients. J Am Geriatr Soc 38: 1105-1109

151. Volicer L, Hurley AC, Lathi DC, Kowall NW (1994) Measurement of severity in advanced Alzheimer's disease. J Gerontol 49: M223-M226

152. Bellelli G, Frisoni GB, Bianchetti A, Trabucchi M (1997) The Bedford Alzheimer Nursing Severity Scale for the Demented: validation study. Alzheimer Dis Assoc Dis 11: 71-77

153. Tinetti M (1986) Performance-oriented assessment of mobility problems in elderly patients. J Am Geriatr Soc 34: 119-126

154. De Leo D, Frisoni GB, Rozzini R, Bernardini M, Dello Buono M, Trabucchi M (1991) The Profile of Elderly Quality of Life (PEQOL): a quick package to assess general health conditions in old age. 5th Congress of the International Psychogeriatric Association, Rome

155. Greenfield S, Blanco DM, Elashoff RM, et al (1987) Development and testing of a new index of comorbidity. Clin Res 335-346
156. Heyman A, Wilkinson WE, Hurwitz BJ, Helms MJ, Haynes BA, Utley CM, Gwyther LP (1987) Early-onset Alzheimer's disease: clinical predictors of institutionalization and death. Neurology 37: 980-984

Lacunes in – the Knowledge of – Vascular Dementia

P. Martinez-Lage and V. Hachinski

Introduction

A recent article on clinical-neuropathological correlations in multi-infarct dementia (MID) questions many assertions regarding the frequency and importance of MID [1]. Aiming to define neuropathological criteria for the diagnosis of MID, the authors searched for cases among the files of ten neuropathology departments in major medical centers involved in the CERAD program. After reviewing autopsies from patients with dementia, neuropathologists from four centers could not find even a single case in which cerebral infarction was "the only explanation of the clinical dementia." The other six neuropathologists detected six valid cases ... among 1929 patients with dementia! Even considering overly strict inclusion criteria and the biased interest of the ten centers in Alzheimer's disease (AD), the quick conclusion is that "multi-infarct dementia unaccompanied by neuropathological evidence of Alzheimer's disease is rare." However, a second careful analysis of these unexpected numbers raises a number of questions regarding the concept of MID or vascular dementia (VD). At the pathological level the absence of commonly agreed upon diagnostic criteria is of concern. Lack of agreement may lead to a biased underestimation of the contribution of vascular lesions to cognitive impairment and dementia. The presence of senile plaques with or without neurofibrillary tangles in the brain of a patient with dementia and vascular lesions automatically excludes a diagnosis of VD in favor of AD. Paradoxically, when similar changes are present in a cognitively intact subject, establishing a diagnosis of AD may be too bold. This margin of uncertainty should be kept also for patients with vascular lesions, so that their relevance in the pathophysiology of cognitive decline and dementia can be investigated and clarified. Even when a case is labelled as mixed dementia, the responsibility for the clinical picture is heavily weighted towards the plausible degenerative rather than the vascular component. Observation, quantification and collection of neuropathological data may be laborious, but it is the interpretation and transformation of anatomical findings into clinical correlates that entails the highest difficulty and demands scientific rigor, inspiration, and open-mindedness. One of the most important sources of confusion and misunderstanding probably resides in the way clinical information is collected and classified.

In the study quoted, neuropathologists were requested to look for cases with a clinical diagnosis of dementia. Again, and always considering the methodological limitations and biases, the quick conclusion is that cerebrovascular disease alone does not cause dementia. Another explanation might be that dementia in these patients would have been sufficiently explained in life by the accumulation of strokes or neuroradiological findings so that autopsy would not be sought to the same extent as in patients with degenerative, radiologically silent diseases. An alternative explanation, however, is that cerebrovascular patients with cognitive impairment might have not received a diagnosis of dementia. This could be the case if cognitive changes are overlooked and not evaluated or if cognitive syndromes in these patients do not fit the concept and criteria for dementia. This brings up one of the most important caveats in the understanding and investigation of VD. Is the current concept of dementia suitable for patients with cognitive decline from vascular causes? The definition of dementia, as it is generally understood, has been extracted from the clinical picture caused by AD and includes cognitive deficits, such as memory impairment, which may not necessarily be present in a cognitively impaired patient on a cerebrovascular basis. On the other hand, a diagnosis of dementia requires a degree of severity important enough to restrict social or occupational functioning of the individual. This probably requires a significant amount of brain-damage that in most instances may prove irreversible. Can VD be treated as is usually claimed? The answer is "unlikely," if dementia has developed after the damage has been done. But if the question is rephrased into: Can cognitive decline from vascular causes be prevented and its progression stopped? The answer is probably "yes," insofar as it is detected early [2].

The simplest definitions in the field of VD hypotheses from its clinical manifestations to its neuropathological substrate are permanently controversial. Many concepts have been based on unproven hypotheses. Consequently, concepts and definitions regarding its epidemiology, pathophysiology, etiopathogenic classification, neuroradiological manifestations, and preventive or therapeutic strategies lack a solid base [3]. There are still too many "lacunes" in the knowledge and understanding of how cerebrovascular disease causes cognitive impairment (and ultimately dementia), how cognitive impairment can be detected and correctly diagnosed, and the appropriate measures to prevent or treat it.

Throughout this chapter, the authors will use the term vascular cognitive impairment (VCI), as recently proposed by Hachinski and Bowler [4] to substitute for that of VD. Far from defining a new entity, this concept should be understood as a new tool, an alternative way to approach the controversies in this field. VCI will encompass all degrees of cognitive decline, from the earliest stages in which risk factors or mild cerebrovascular disease is present with no cognitive changes to the stage of established dementia. This idea of a continuum, a progression that can potentially be prevented, is inherent to the concept of VCI in which the word "vascular" refers to all possible mechanisms of ischemic cerebrovascular disease.

Epidemiology: Is VCI Frequent?

Research in the field of dementia has probably transcended the "epidemiological stage." The literature is full of papers on the prevalence and incidence of dementia, which are widely known and fairly stable throughout many populations all over the world. Understandably, epidemiological research is now oriented towards genetic epidemiology, investigation of risk factors and early detection. While this is applicable to AD, the literature on the epidemiology of VD is scarce [5]. Prevalence figures for VD are highly variable across studies, and such variability cannot be explained by ethnical or geographical differences only, but rather by methodological disparity. Strictly speaking, it could be argued that while the prevalence of VD can be roughly estimated, the exact figures remain unsettled. This is particularly true if the contribution of vascular causes to cognitive decline and dementia in the general population, rather than VD itself, is to be estimated.

Variations in the prevalence figures of VD even for geographically close populations is easily explainable on methodological grounds. The concept of prevalence is based on the premise that a population can be divided into cases and controls or non-cases. This is particularly complicated for a disease such as VD, as it is chronic, in need of reliable diagnostic markers or diagnostic criteria, and it shares many features with other conditions that affect the same age groups, such as stroke or AD [6]. The reliability of the diagnosis of VD in the setting of extensive population-based studies is inevitably poor, as no neuroradiological information is available. In the absence of CT or MRI examinations, the diagnosis of VD is necessarily based on the evolution or chronological pattern of cognitive deficits (acute onset, step-wise deterioration), as well as in the presence of clinically recognizable cerebrovascular disease [history of stroke or transient ischemic attack (TIA)]. The former assertion that VD starts abruptly and progresses in steps has never been consistently proven, and there is enough evidence in the literature that patients with autopsy-proven VD may lack a history of vascular events [7]. Consequently, the frequency or prevalence of VD may be significantly underestimated. Hébert and Brayne [5] have recently reviewed ten prevalence studies on VD. All were performed on representative samples of the reference populations; they applied the same diagnostic criteria (DSM-III) and reported specific data on VD. Prevalence rates varied from 1.2% to 5.6%. The two studies using CT evaluation reported higher prevalence figures. Similarly, in the study by Skoog et al. [8], in which they obtained a prevalence rate of 13.9% in a population of octogenarians, ten patients received a diagnosis of VD only after the CT was evaluated.

Another important source of bias leading to confusion in the estimation of the contribution of cerebrovascular conditions to the development of cognitive decline and dementia results from how a diagnosis of mixed dementia is interpreted. When cerebrovascular disease accompanies AD, the crucial question is whether the patient would be demented had the vascular component been absent. While the answer to this question is difficult to reach in the epidemiological setting, the diagnostic classification of these patients is important. For instance, Skoog et al. [8] included mixed dementia cases as VD, obtaining a prevalence rate of 13.9% for VD. In contrast, in the Rotterdam study, where mixed dementia cases were included as AD, the prevalence of VD was 1% [9]. These numbers are obviously misleading if the contribution of vascular causes to the prevalence of dementia is to be estimated. If VD patients and those with mixed dementia in which the vascular component plays a significant role are grouped together, the frequency and relevance of cerebrovascular disease as a cause of dementia may prove to be as important as that of AD.

The prevalence of VCI remains to be established, as cases with early cognitive changes are not detected in population-based studies. The role that vascular factors play as a cause of mild or moderate cognitive decline has been highlighted by numerous recent studies. Hypertension and diabetes are significant negative predictors of cognitive function in non-demented subjects in population-based studies, and this effect appears to be independent of the presence of stroke [10]. The deleterious effect of hypertension on cognitive function has been known for years and has been related to the presence of white matter changes. Additional epidemiological studies have shown that the presence of electrocardiographic evidence of myocardial infarction, peripheral artery disease, or presence of carotid atheromatous changes doubles the proportion of subjects scoring 23 or less in the Mini-Mental State Examination (MMSE) [11]. History of stroke increases this proportion by three. In stroke-free subjects, diabetes and hypercholesterolemia were independent correlates of abstract reasoning/visuospatial deficits and memory dysfunction, respectively [12]. In patients with stroke the appearance of cognitive impairment, but frank dementia, has received significant attention in the last few years. Some degree of cognitive impairment may be detected in up to 60% of stroke patients. However, the proportion depends on which diagnostic tool is applied and the level at which cut-off points are placed. Pohjasvaara and colleagues [13] found that 61.7% of 486 stroke patients had deficits in at least one cognitive domain, 34.8% in one or two domains and 26.8% in three cognitive areas or more, as measured by the modified MMSE (3MS) and other brief tests. Using a comprehensive neuropsychological battery, Grace and colleagues [14] found that 46% of their stroke patients were cognitively impaired, that is, scoring two standard deviations below the norm in at least two cognitive domains. Interestingly, more than half of these patients scored higher than 24 on the Mini-Mental, and 31% obtained more than 79 on the 3MS, hence questioning the sensitivity of these cognitive scales. Tatemichi and colleagues [15] used similar criteria and required failure (scoring below the 5th percentile of the control sample) in at least 4 of the 17 items of their neuropsychological battery, which evaluated memory, orientation, language, visuoespatial ability, abstract reasoning, and attention. Cognitive impairment was detected in 35% of their 227 stroke patients. Remarkably, cognitive impairment had a significant effect on functional impairment and was a significant predictor of dependent living even after adjusting for physical disability.

The importance of cerebrovascular disease as a cause of dementia is supported by the cumulative evidence that stroke is a highly significant risk factor for dementia. Almost 30% of stroke patients develop dementia in the first 3 months after stroke. The risk is nine times higher than that in the general population [16].

Surprisingly, the type of dementia that stroke patients develop is both VD and AD (mixed dementia?), and in many cases no new cerebrovascular events precede dementia. In population-based longitudinal studies, the incidence of dementia in patients with stroke at study entry is higher by nine times, and the incidence of AD is twice that of the general population [17]. Some authors have suggested the term "post-stroke dementia" to include all possible mechanisms of dementia after stroke [18].

Epidemiological studies on VCI await the definition of valid diagnostic criteria to include all cases of mild, moderate or severe cognitive decline. In the meantime, recently published results from the Canadian Study of Health and Aging (CSHA), in which 10,000 elderly subjects were evaluated and screened for the presence of cognitive impairment, suggest that the role of vascular factors may be more important than currently thought [19]. Using the Modified MMSE, 18.1% of the population showed some degree of cognitive impairment. Of these, a total of 861 subjects presented cognitive decline without dementia and of these, 149 (17.3%) had a vascular cause for their cognitive impairment. VCI was the second cause of cognitive decline without dementia in this study, after age-associated memory impairment. If patients with VD and mixed dementia (AD with a vascular component) are considered, it can be said that 4.8% of the population evaluated in the CSHA had some degree of VCI.

What Are the Risk Factors for VCI?

Despite the intensive and extensive literature published on this matter, risk factors for VD remain unknown [20]. This is not surprising as the two main requirements for the investigation of risk factors cannot easily be met in the case of VD: a clear separation of cases and non-cases and a reliable definition of the presence or absence of a specific factor. Since no reliable diagnostic criteria are available for VD, samples of cases and controls cannot be defined and compared. In this respect there is no agreement as to whether VD patients should be compared with neurologically normal control or rather with non-demented cerebrovascular disease patients or both. However, the important question is whether risk factors for VD are the same as those for stroke [21]. In this respect some studies have compared stroke patients with and without dementia regarding the presence or absence of traditional risk factors. In the series of Ladurner and colleagues [22] only hypertension was found to be more frequent among demented stroke patients than non-demented patients. Investigating patients with multiple infarcts, Loeb and colleagues [23] reported a higher prevalence of hypertension, cardiopathy and diabetes in those with dementia. In the Stroke Data Bank Cohort, only myocardial infarction and prior stroke were significantly associated with dementia [24]. In a prospective study Tatemichi and colleagues [25] found a significant association of dementia and age, education, race, history of prior stroke, and diabetes. Gorelick and colleagues [26] investigated patients with two or more infarcts and found that age, hypertension, proteinuria, myocardial infarction, smoking and systolic blood pressure were significant predictors of dementia. Both hypertension and proteinuria lost significance when analyses were adjusted for educational level. Surprisingly, the presence of high systolic blood pressure showed a "protective" effect. Studies on population-based surveys are scarce. Using incident cases with a high rate of autopsy confirmation, Yoshitake and colleagues [27] found age, systolic blood pressure, history of stroke and alcohol abuse to be significant predictors on a multivariate analysis. Diabetes, diastolic blood pressure, and increased hematocrit were significant only on univariate analysis. History of hypertension, alcohol abuse, heart conditions, and exposure to pesticides/fertilizers or liquid plastics and rubbers, as well as low educational level showed a significant effect in analysis performed in the CSHA database [28]. In this study, diagnosis of VD followed ICD-10 criteria and no neuroimaging study was required. Patients with cognitive decline and no dementia were excluded from the control group, but also from the case group. Reports from the population-based study of Rochester have described the significant effect of age, male sex, recurrent stroke, and mitral valve prolapse on incident dementia [17]. Results from all these studies are too varied to draw any conclusions, as might be expected from the different methodologies applied.

Further progress in the understanding of risk factors for VCI should take into account some crucial methodological points. First, the definition of the presence or absence of a particular factor should consider if the condition, hypertension for instance, has been treated and successfully controlled or not. Second, cases with mild cognitive impairment should be adequately detected, as otherwise they might be included in a control group and artificially reduce the role of a particular factor. Third, the current criteria for the definition of VD does not differentiate its etiopathogenic categories. This poses a crucial problem since patients with different diseases that might hypothetically respond to different risk factors, such as cardioembolic infarcts or lacunar disease, are grouped together under the same diagnosis (VD). Last, but not least, most studies have paid attention to traditional vascular risk factors: hypertension, diabetes, cardiac disease, smoking, alcohol intake, etc; however, the role of other conditions such as hypoxic-ischemic states [29], hypotension [30], immunological abnormalities [31], increased fibrinogen [32], increased platelet activation [33], or genetic factors [34] are gaining acceptance as potential significant predictors of dementia in cerebrovascular disease patients. Polymorphism of the apolipoprotein E has been claimed to be a risk factor for VD, but anatomopathological studies do not seem to confirm this effect, except for mixed dementia [35]. The urgent need for further research in this area is highlighted by the fact that despite intensive control of vascular risk factors for several decades, the incidence of VD has not been significantly reduced.

The Diagnosis of VD According to Current Criteria: Is it Accurate? Is it Valid?

At this point, the reader will no doubt easily anticipate the answer to both questions. The validity and reliability of current diagnostic criteria merit a number of considerations. It is not only that neuropathological validation has not been performed, or that their reproducibility has been questioned in several reports, it is that some à priori limitations can be anticipated.

There are two types of diagnostic criteria sets available for clinical practice or research purposes. All sets are based on the same diagnostic algorithm: first, a diagnosis of dementia must be established; second, a degree of cerebrovascular disease must be present; third, a cause-effect relationship between both processes must be established. The criteria from the Diagnostic and Statistical Manual-version IV (DSM-IV) [36] and those from the International Classification of Diseases, version 10 (ICD-10) [37] are included as general diagnostic tools and outline general principles for the diagnosis of dementia and its different types, including AD, VD, and others. In both sets of criteria, dementia is defined by the presence of memory impairment plus deficits in other cognitive areas, and these deficits must be sufficiently severe to interfere with social or occupational functioning. With regard to VD, history of TIAs or stroke, presence of focal signs, and laboratory data will provide evidence of underlying cerebrovascular disease. The etiological relationship to the cognitive deficits is left to the responsibility and subjectivity of the clinician. The criteria from the National Institute of Neurological Disorders and Stroke–Association Internationale pour la Recherche et L'Enseignement en Neurosciences (NINDS-AIREN) [38] and the criteria from the State of California AD Diagnostic and Treatment Centres (ADDTC) [39] differ in that they were specifically designed and operationalized for the diagnosis of VD.

NINDS-AIREN criteria maintain a similar definition of dementia. Memory impairment and interference with activities of daily living are both required. Confirmation of underlying cerebrovascular disease requires the presence of focal signs consistent with stroke and (not or) radiological evidence, such as large arterial infarcts, multiple lacunes, or extensive white matter disease. The relationship of both processes is then established on the basis of temporal coincidence (dementia appears in the 3 months following a clinical stroke) or if dementia follows a typical chronological pattern (abrupt onset, fluctuating or stepwise course). In cases of dementia with focal signs in which the radiological or pathochronological criteria are missing, the NINDS-AIREN criteria establish a category of "possible" VD.

The ADDTC criteria have been less widely accepted despite the fact that they offer two original and interesting modifications. First, the pattern of cognitive impairment is not subject to strict limits and rules and any combination of neuropsychological deficits is allowed in the definition of dementia as far as it is not restricted to "a single narrow category." This seems important for a process such as cerebrovascular disease that may involve any part of the brain. For confirmation of the presence of vascular brain disease, two situations are considered. In patients with a clinical history of at least two strokes, radiological evidence of at least one infarct outside the cerebellum is required. On the other hand, if there is radiological evidence of two or more ischemic strokes, a clinical history of stroke or the presence of focal signs is not considered necessary. The temporal coincidence of dementia and stroke is considered relevant for those patients with only one cerebrovascular event, and no specific chronological pattern in the evolution of dementia is required. The ADDTC criteria also contemplate a category of "possible" VD for demented patients with a history of a single stroke not chronologically related, as well as for those patients with a Binswanger's syndrome who present with vascular risk factors and extensive white matter changes on neuroimaging.

A critical review of these criteria leads to a number of considerations at the three levels of definition of dementia, confirmation of cerebrovascular disease and establishment of their etiopathogenic relationship. First, none of the criteria considers early cognitive changes. Only advanced cognitive impairment that is severe enough to affect activities of daily living is contemplated. Subsequently, their applicability for prevention and treatment is seriously limited. Another relevant aspect regarding the definition of dementia is the requirement of memory impairment. Vascular brain disease can affect any cognitive domain, depending on the type, extension and location of lesions, and still spare memory functions, contrary to AD in which amnesia is a prominent and early feature. Establishing limits to the pattern of neuropsychological manifestations may leave a number of patients undiagnosed and restrict their access to preventive or therapeutic measures. In this respect the definition of dementia proposed by the ADDTC criteria seems more adequate, as they leave an open field for research projects in which different patterns of vascular disease may be correlated with different combinations of neuropsychological deficits. Second, the requirements proposed by the NINDS-AIREN criteria to establish the presence of cerebrovascular disease may be excessively restrictive. Both the presence of focal signs and radiological evidence are required. The concept of focal signs may be difficult for the clinician that wonders whether aphasia and apraxia are considered focal signs. On the other hand, if only pyramidal, sensory, or visual signs are counted, it is conceivable that vascular disease may well affect cognition without involving the corticospinal tract, the optic radiations or the sensory pathways. The ADDTC criteria are more permissive, as long as radiological or (not and) clinical findings provide evidence of two or more ischemic infarcts. Third, concerning the etiological connection between dementia and vascular brain disease, as previously suggested, the pathochronological requirements imposed by the NINDS-AIREN criteria have not been specifically tested. Additionally, the literature provides a number of examples of autopsy-proven cases of VD in which cognitive deficits develop in a slowly progressive fashion [1]. The ADDTC criteria leave this question unresolved and apply the chronological criteria for cases with only a single stroke. Finally, all sets of criteria suffer from a serious omission, as none of them provides a proposal for an etiopathogenic classification of VD, while it is generally accepted that dementia from vascular causes responds to different etiologies [40].

None of the criteria sets has been validated on neuropathologically confirmed cases, and the interater reliability of some has been questioned [41]. Remarkably, the different sets are not interchangeable and the frequency of VD in the same population may vary from 6% to 32%, depending on the applied criteria [13, 42].

In a population of 167 elderly patients with dementia, Wetterling et al. [43] found 45 cases of VD according to DSM-IV criteria and 21 according to the ICD-10 research criteria. Only 12 patients met NINDS-AIREN criteria, while 23 subjects would have been diagnosed according to ADDTC criteria. The authors emphasized the poor concordance among the different criteria sets, as only 5 patients fulfilled all criteria sets.

Operationalized criteria are necessary to provide means of communication and understanding among the scientific community, as well as to develop a useful tool for research. However, current criteria are based on an unproven hypothesis regarding clinical and radiological manifestations and prevent further progress in the understanding and early detection of VD. Additionally, communication is limited in the absence of a rational etiopathogenic classification. For patients diagnosed with VD according to any criteria, it is impossible to know what specific disease is causing dementia. New criteria to diagnose VCI should be the result of prospective studies on patients with early cerebrovascular disease, from those with only risk factors to those with established stroke, in which functional, neuropsychological, and neuroimaging data are gathered systematically and analyzed without preformed ideas. A selected comprehensive neuropsychological battery administered to these patients would define cognitive impairment according to agreed cut-off values in each test without emphasizing memory or any other cognitive domain. Short screening tests would then be developed and validated from these neuropsychological tests. The MMSE is widely used, but its sensitivity for detecting cognitive decline in cerebrovascular patients may be too low [14]. Patients could be classified according to possible or probable etiopathogenic criteria. Patterns of cognitive deficits could then be defined for each group of cerebrovascular diseases. Cases with cognitive decline, from early changes to dementia, could then be compared to those without changes, particularly with regard to clinical and neuroimaging features. Results from such an analysis could lead to the formulation of rational and reliable criteria for the diagnosis, early detection and differentiation of VCI from other forms of cognitive decline such as AD. Using the current criteria, a differential diagnosis between AD and VD is theoretically impossible on a clinical basis, as dementia is identically defined in both cases. However, if cases are detected early and cognitive deficits are thoroughly characterized, both diseases should be clearly distinguishable. AD patients may show impairment of orientation, memory and language while VD patients tend to be more impaired in executive functions such as attention, self-regulation, planning, and fine motor coordination [44].

Neuropathology: New Concepts in the Vascular Mechanisms of Cognitive Impairment

The anatomopathological characteristics of focal vascular damage to the brain in patients with VD have been extensively reviewed [45–48]. Both parenchymal and blood vessel wall lesions are varied and heterogeneous (Table 1), and different pathological entities usually coexist. Classical lesions include macroscopic arterial infarcts, lacunar infarcts and white-matter rarefaction.

Blood Vessel Pathology

Vessel wall lesions in patients with VD include atherosclerosis of the large or medium-sized cerebral arteries and small-vessel disease [48]. *Atherosclerotic changes* with accumulation of lipids, extracellular matrix components and fibroblast proliferation are mainly distributed in the extracranial trunks, the circle of Willis, and the leptomeningeal arteries. Also aortic atheromatous changes are gaining significance as a possible source of artery-to-artery emboli. Two different pathological entities are encompassed under the term "small-vessel disease": arteriolosclerosis and amyloid angiopathy. *Arteriolosclerosis* is characterized by the loss of muscular and elastic components of small arteries and arterioles, which are substituted by hyaline material and collagenous strands (hyalinosis, fibrohyalinosis). Lipid-laden macrophages (lipohyalinosis) and fibrinoid necrosis may also be present in some vessels. This type of angiopathy has been classically described in chronic hypertension, but may be present in normotensive subjects. Almost identical changes are described in patients with hereditary forms of VD. In *amyloid angiopathy* a congophilic proteinaceous material is deposited in the wall of meningocortical arterioles, inducing media degeneration. It has been described in both sporadic and familial cases and is a common finding in patients with AD. Other forms of *inflammatory or noninflammatory angiopathies* may involve small vessels in rare and selected cases of VD (Table 1).

Parenchymal Lesions

Arterial macroscopic infarcts may involve cortical regions and extend into the subcortical underlying areas in the territory of a single medium- to large-sized artery or the watershed areas of two arterial territories. Coagulation necrosis and cavitation characterize these lesions, which in most cases respond to an atherothrombotic or cardioembolic mechanism. *Lacunes* are defined as small cavitated infarcts of less than 1.5 cm distributed in the territory of small penetrating arteries and arterioles of the vascular centrencephalon and white matter. A local occlusive mechanism is usually postulated, but lacunes may also be caused by microembolism or even represent small, healed hemorrhages. When multiple lacunes coexist in the same brain, the term "lacunar state" is usually applied. Lacunar infarcts are to be distinguished from other small cavitaries, that is, membrane-covered lesions that surround a small artery, correspond to enlarged Virchow-Robin spaces and probably represent brain atrophy. Such a distinction is important as the term *état criblé* (cribiform state), which describes the macroscopic appearance of a brain with multiple small subcortical or white-matter cavitary lesions, is frequently confused and taken as a synonym of a lacunar state. A microscopic description is necessary to separate the two entities. Another type of small lesions not to be confused with lacunes is represented by discrete areas of

Table 1. Neuropathological findings in the brain and blood vessels in patients with VD. Potential etiopathogenic mechanisms

Brain vascular lesions

- Cortical infarcts

  Large/medium-sized arterial infarcts
  　Atherothrombosis
  　Cardioembolism
  　Coagulopathies

  Cortical microinfarcts (granular cortical atrophy)
  　Small-vessel disease
  　Amyloid angiopathy
  　Multiple microemboli/microthrombosis
  　Angiitis obliterans

  Watershed infarcts
  　Hypoperfusion states
  　Severe carotid stenosis/occlusion (?)

  Other ischemic lesions
  Selective ischemic necrosis:

- Subcortical infarcts

  Lacunes
  　Arteriolosclerosis
  　Microthrombosis

  Deep infarcts
  　Atherothrombosis
  　Embolic
  　Amyloid angiopathy (familial?)
  　Hereditary arteriopathies (CADASIL)

  Watershed infarcts
  　Hypoperfusion/hypoxic states
  　Severe carotid stenosis/occlusion (?)

- White-matter vascular damage

Selective ischemic necrosis:		Lacunes/infarcts	Astrocytic reaction
Laminar necrosis	Global anoxia	Spongiosis	Small-vessel disease
Hippocampal sclerosis	Mild focal ischemia?	"Incomplete infarcts"	Ischemia/hypoxia
Incomplete infarction?	Hypoperfusion states	Perivascular demyelination	Increased permeability
			Increased fibrinogen

Vessel wall disease

- Large artery disease

  Atherosclerosis (aortic, carotid, cerebral arteries)
  Giant cell or other inflammatory arteritis

- Small-vessel disease

Arteriolosclerosis	Hypertensive	
	Non-hypertensive	Hereditary (CADASIL)
		Diabetic (?)
		Other
Amyloid angiopathy	Sporadic, familial	
	AD-associated	

  Inflammatory/non-inflammatory arteritis

demyelination that surround a small white-matter vessel. Unlike Virchow-Robin spaces, these lesions are not covered by a membrane and are usually non-cavitated. The mechanism by which these lesions developed is under discussion and both ischemia and increased vascular permeability with perivascular protein deposits have been postulated. The relevance of *white-matter lesions* as pathological correlates of "arteriosclerotic dementia" was initially highlighted by Binswanger and Alzheimer. Numerous studies have described areas of gliosis/astrocytosis, extensive demyelination, as well as areas of diffuse vacuolization and edema, loss of oligodendrocytes, and cavitated infarcts (lacunes) in the white matter of patients with VD. The pathophysiology of these changes and whether they respond to ischemia or other vascular mechanisms are matters for continuing discussion.

Other Neuropathological Findings

Apart from these classical ischemic lesions, other neuropathological findings in patients with VD include granular cortical atrophy, laminar necrosis and hippocampal sclerosis. *Granular cortical atrophy* is the term used to describe the macroscopic appearance of the brain when multiple, widespread small cortical infarcts coexist in the same region. Small cortical vessel disease in the form of amyloid angiopathy, multiple microembolism, or angiitis obliterans have been described in association with these lesions. *Laminar necrosis* appears as a destructive lesion selectively involving neurons in the vulnerable layers of the cerebral cortex and seems to be common in the boundary zones of two different vascular territories. The term *hippocampal sclerosis* refers to a selective loss of neurons in sector CA1 and subiculum of the hippocampal formation accompanied by a severe gliotic reaction, which has been reported as a frequent finding in aged subjects with dementia [49]. These changes occur in patients with global cerebral anoxia-ischemia and, remarkably, they may occur as a delayed phenomenon [50]. A vascular mechanism has been claimed for this type of neuropathological finding, but whether this is true for all cases of hippocampal sclerosis has yet to be determined, as other neurodegenerative findings are associated in some cases [51]. Dickson et al. [49] found 13 cases among 81 autopsies performed on subjects who were 80 years of age or older and had been recruited in longitudinal studies of aging and dementia. While risk factors for cerebral hypoperfusion were common, clinical cardiovascular measures were not significantly different from patients without hippocampal sclerosis. Four patients had other neuropathological findings such as ballooned neurons or argyrophilic grains suggestive of a neurodegenerative disorder. In the series of Knopman et al. [52] 10 out of 14 patients with so-called dementia without distinctive histopathological features had hippocampal sclerosis, together with neurodegenerative findings in frontoparietal cortices. After a thorough review of clinical and pathological data from eight patients with hippocampal sclerosis, seven of whom had a cardiac disorder, Corey-Bloom et al. [51] concluded that "the relationship to cardiovascular disease remains unclear." They also found a decrease in neocortical synaptic density in these patients. Whether there is a connection between this finding and previous episodes of ischemia-hypoxia is also unclear.

These two last entities, laminar necrosis and hippocampal sclerosis, may represent examples of a broad concept labelled "incomplete infarction" but for which the term "selective ischemic necrosis" or "selective vulnerability" seems more appropriate. The term "incomplete infarction" was originally used by Lassen [53] to describe focal ischemic lesions with necrosis but not "emollision." A few years later it was applied to designate vascular changes in the white matter (loss of myelin, axons and oligodendrocytes, glial reaction) of presumed ischemic origin and comparable to those occurring in the "transitional zone surrounding many complete infarctions." More recently, the term has been reviewed [54] and recovered to describe areas of brain damage, including the cortex, caused by reversible ischemia of moderate severity. Histologically, these lesions are characterized by selective neuronal death with little or no glial reaction. Such neuropathological changes are well known to occur in relation to global cerebral ischemia, but they also appear in the periphery of classical cavitary infarcts. Experimental and clinical evidence suggests that they can occur in cases of focal ischemia in the absence of necrotic lesions.

Together with both classical and less conspicuously described lesions, further research in the neuropathological correlates of VCI should take into account the data provided by recent advances in the understanding of the biological phenomena that accompany brain ischemia.

Experimental studies have shown that cerebral ischemia leads to DNA fragmentation and protein synthesis inhibition and has been interpreted as a sign of "programmed" cell death or apoptosis [55]. The question of whether such a process might lead to delayed, or even progressive, cell loss remains open. Of especial interest are the profound changes that ischemia induces in the modulation and regulation of the expression of certain gene families such as the "immediate early genes" (c-*fos*, c-*jun*) and those related to the stress response [56]. Such changes at the gene-expression level occur in both global and transient or prolonged focal ischemia models and have been related to mechanisms of excitotoxicity, neuronal death, as well as to mechanisms of neuroprotection. Remarkably, in models of focal ischemia, changes in gene expression may explain delayed neuronal changes in remote areas outside the cerebral ischemic focus.

Of exceptional relevance is the increasingly recognized role that non-neuronal cell populations play in the production of tissue damage induced by ischemia [57]. Oligodendrocytes are particularly sensitive to ischemia, which might explain the development of white-matter changes with ischemic insults that are otherwise insufficient to cause cavitary infarctions [58]. On the other hand, astrocytes and microglia are rapidly activated by ischemic insults and breakdown of the blood-brain barrier. Astrocytes are actively involved in several neuroprotective mechanisms such as the maintenance of extracellular fluid homeostasis, scavenging of oxygen-free radicals, reuptake of excytotoxic neurotransmitters, or release of trophic factors. Distortion of these responses or overactivation and proliferation of astrocytes surviving an ischemic insult may result in an inflammatory response with overexpression of nitric oxide synthetase, sustained production of nitric oxide, release of glutamate and cytokines, overproduction of free radicals, and phagocytic behavior, thereby contributing to neuronal death postischemia. A similar reaction involves the microglia with overexpression and production of mediators of the inflammatory response and acute phase proteins [59]. Interesting is the fact that one of the proteins overexpressed is the amyloid precursor protein (APP) that places cerebral ischemia as a potential triggering factor of β-amyloid deposition and development of neuropathological changes of the

Alzheimer type. In experimental models of focal cerebral ischemia, the microglial response occurs not only at the primary ischemic lesion, but also in other ipsilateral and contralateral brain regions days or even weeks after the original insult.

How Does Cerebrovascular Disease Cause Dementia?

The pathological anatomy and etiopathogenesis of ischemic brain damage is multiple and varied. How these processes alone or in combination produce loss of cognitive functions is a question of open debate despite decades of research and discussion [12]. Unfortunately, most studies, either with anatomopathological or neuroimaging approaches, have failed to separate different etiopathogenic entities and have grouped patients under the excessively broad and abstract diagnosis of dementia. In general, two hypotheses have traditionally been held to explain how a focal process affects a global capacity such as cognition or intelligence: the volumetric hypothesis, based on a model of functional reserve of the brain and the topographical hypothesis that stresses the complex field of the anatomy of intelligence. However, data from different studies have provided contradictory results regarding the effect of volume, localization or even the type of lesions in the production of dementia. Topographical or anatomical correlations may be applied for cases in which those cognitive functions considered as localized, that is, with a concrete cortical representation, are predominantly affected. However, it is the impairment of disperse cognitive functions such as attention and concentration, executive functions and memory processes, that most specifically defines cognitive decline and dementia. The traditional "localization of function" models are limited in explaining how the brain sustains these cognitive domains. Instead, complex large-scale neurocognitive network models integrating cortical areas, subcortical nuclei and white matter connecting pathways have been postulated [60]. In this setting (1) components of a single function are represented in different interconnected locations throughout the brain; (2) individual cortical or subcortical areas may belong to different overlapping networks and participate in several complex functions; (3) lesions confined to a single region may result in multiple deficits; (4) impairment of an individual complex function usually requires simultaneous involvement of several regions to be severe and persisting; (5) the same cognitive function may be affected by a lesion in different brain areas. According to this simplified, but useful model, cerebrovascular disease can affect cognition by destroying neuronal components of the cognitive networks both in the cortex or subcortical relay nuclei and by lesioning their connections (white-matter disease). Both size and location of infarcts are important and no assumptions should be made as to what cognitive functions are to be impaired. The role of destructive processes such as selective necrosis, apoptosis, or glial inflammatory reactions, which may certainly be radiologically or even anatomopathologically silent, should be further investigated. Whether brain atrophy, which has repeatedly been associated with the presence or risk of dementia in cerebrovascular patients, is the macroscopic expression of these processes is a hypothesis to be tested.

Structural Neuroimaging in VCI

That VCI cannot be diagnosed in the absence of conclusive evidence of cerebrovascular disease requires little discussion. Such evidence can be obtained from two sources: clinical history and examination or neuroimaging investigations. Collecting clinical information regarding cognitive complaints, as opposed to motor or sensory symptoms, may be extremely difficult. Establishing the type of onset or pattern of evolution of complaints such as poor memory, disorientation, and personality changes is often impossible. Even with reliable informants or caregivers the absence of an acute onset or a stepwise progression does not rule out a vascular origin. In stroke patients the onset of cognitive symptoms may easily be perceived after the acute event, but this may also happen in relation to other circumstances such as general anesthesia or a relative's death. Even a fluctuating course of the dementia may be observed in conditions other than VD such as diffuse Lewy body disease or rarely in AD. With regard to clinical examination, the importance of focal motor, sensory, and visual signs has received significant attention, but the fact that cognitive deficits, if well characterized, constitute focal signs themselves has been overlooked. Examination of cognition requires time and experience, and the pattern of intellectual deficits is not always easy to establish. In such a situation it is difficult to ascribe symptoms and signs to a particular neuroradiological finding. The role of neuroimaging techniques in detecting vascular lesions in patients with cognitive decline is evident, but information provided by clinical and radiological sources must be carefully contrasted.

Cortical and Subcortical Infarcts

Neuroimaging studies in patients with VCI or VD may show any kind of infarcts that appear as hypodense areas on CT and hyperintense/hypointense areas on T_2-weighted/T_1-weighted MRI images [61]. The size and distribution of lesions, together with clinical manifestations, may help in the differential etiological diagnosis. Medium-sized or large infarcts involving the cortical areas in an end-artery distribution usually respond to atherothrombotic or cardioembolic mechanisms. The presence of multiple wedge-shaped cortical infarcts or evidence of hemorrhagic transformation (which can be demonstrated in the chronic phase by the low-signal intensity of ferritin and hemosiderin on T_2-weighted MRI images) may reinforce the latter. Evidence of hemorrhagic infarction may also suggest an underlying amyloid angiopathy. Cortical infarcts involving the posterior frontal lobe and posterior parieto-occipital regions are compatible with watershed or border zone infarction and might suggest a hemodynamic origin. Subcortical lesions include lacunar infarcts, but also atheroembolic and border zone infarcts that may be distinguished according to size, distribution and other clinical and laboratory data. Lacunar infarcts appear as small lesions, less than 1.5 cm in size, and are distributed in the territory of deep penetrating arteries (white matter, thalami, basal ganglia, brainstem). Dilated Virchow-Robin spaces that show behavior similar to that of the CSF on the different MRI sequences should not be confused with lacunes that appear hyperintense with respect to the CSF on proton density images. All these data will surely help to classify patients with VCI according to possible and probable etiologies. However, neuroradiological findings should be adequately compared to appropriately collected clinical and neuropsychological information. Otherwise, neuroimaging techniques would only be useful in detecting vascular lesions, rather than assisting in the diagnosis of cognitive impairment from vascular lesions. For the former purpose, the ischemic scale has proven to be acceptably sensitive and extremely inexpensive.

A number of studies have analyzed CT and MRI findings in patients with VD, as compared to patients with cerebrovascular disease but no dementia [20]. Some authors have emphasized the presence of bilateral lesions, while others highlight the importance of dominant hemispheric infarction, and still others report no effect of infarct location. The importance of lacunar over cortical infarcts is reported in some studies, but not all. Interestingly, most studies coincide when reporting cerebral and ventricular atrophy as significantly associated with dementia. In one longitudinal survey cerebral atrophy proved to be a significant predictor of subsequent dementia among patients with stroke [24].

Unfortunately, the majority of these studies have been performed in patients with a diagnosis of established dementia and are hardly applicable to the diagnosis of early VCI or to estimate the risk of progressive cognitive decline in patients with mild or moderate cerebrovascular disease. Prospective studies should investigate what CT or MRI patterns correlate with what cognitive deficits in cerebrovascular patients, and longitudinal follow-up will help to determine if certain neuroimaging features such as silent infarcts, atrophy or white-matter disease may increase the risk of developing dementia.

White-Matter Vascular Damage

The "matter of white-matter lesions" has received significant attention since the introduction of CT scanning in the diagnostic evaluation of neurological patients. White-matter rarefaction, which appears as a hypodense area surrounding the ventricles and spreading into the subcortical white matter, was soon found to be related to the presence of cognitive decline, dementia and a higher incidence of vascular risk factors [62]. Many authors hurried to diagnose patients with this finding as Binswanger's disease (BD) patients, and an "epidemic" of this previously rare disease invaded the literature [63]. However, the pathological correlates were varied and heterogeneous and included not only lacunar infarcts, areas of demyelination, and associated arteriolosclerosis, but also dilated perivascular spaces, hydrocephalus and ependimitis granularis. This heterogeneity and the initial confusion regarding the etiology and clinical significance of white-matter changes were discussed by Hachinski and colleagues who suggested the term "leukoaraiosis" (LA) to describe this neuroradiological finding. The introduction of MR imaging brought more confusion and misunderstanding. Due to its high sensitivity, MRI detected some changes in the white matter in the majority of healthy subjects older than 60. Not surprisingly, LA detected on MRI did not correlate with cognitive dysfunction or vascular risk factors [64]. The low specificity of MRI was subsequently improved and periventricular lesions were distinguished from deep white-matter changes separated from the ventricles, and semiquantitative or quantitative grading approaches were applied [65]. Pathological studies confirmed that small areas of LA surrounding the frontal horns corresponded to a normal anatomical structure, the subcallosal fascicle. Similarly, thin areas adjacent to the body of the lateral ventricle ("rims" or "halos") and the frontal/occipital horns ("caps") were the expression of ependimitis granularis, caused by ependymal disruption and periventricular gliosis. When periventricular changes extended into the subcortical white matter, vascular changes such as arteriolosclerosis and ischemic demyelination were present. Deep white-matter changes were classified as punctate, patchy, patchy-confluent and diffuse [66] and were related to perivascular demyelination, spongiosis, gliosis, decreased oligodendrocyte densities, and lacunar or other infarcts [67]. Small vessel disease was present in most cases [68]. In some instances, punctate lesions corresponded to dilated perivascular Virchow-Robin spaces [69]. With regard to VD, several studies showed that moderate-to-severe degrees of LA were present in 50% to 80% of cases compared to 20% in healthy controls or 25%-30% of AD cases [62]. The presence of LA on CT has been related with an extra risk of future stroke in patients with TIAs or minor stroke [70], as well as with cerebrovascular progressive disease in patients with symptomatic lacunes [71].

As has been outlined before, the etiopathogenesis of LA remains a question of open debate. Ischemia and infarction are undoubtedly present in many cases. However, some authors have suggested that small vessel disease could lead to increased permeability of the vessel wall and leakage of serum proteins into the surrounding white-matter, leading to an astrocytic reaction, gliosis, spongiosis and further ischemia [46, 72]. The role of hypotension and short-lived episodes of ischemia has been emphasized by some groups [30]. In this respect, it is well known that oligodendrocytes are highly vulnerable to mild ischemia [58]. A mechanism of distal filed ischemia has been suggested in cases with white-matter changes associated with amyloid angiopathy that typically affects meningocortical and not white-matter vessels. In summary, there may be different mechanisms leading to white-matter changes, but all of them share a vascular origin. Subjects with moderate-to-severe LA, especially if involving the deep white matter, show cognitive deficits in executive functions and speed of mental processing, as well as a higher incidence of vascular risk factors such as hypertension and less frequently diabetes or cardiac disease. Increased fibrinogen or factor VIIc activity [9], hypotension [73], and genetic factors [34] have also been implicated and require further investigation.

Neuroimaging Criteria for VD

Neuroimaging criteria have been recently proposed to assist in the diagnosis of VD. Pullicino and colleagues [74] designed a rating scale ranging from 0 to 3 to estimate the severity of vascular disease as seen on CT. This scale was based on the number of infarcts, total volume of infarcted tissue (less or more than 100 ml), severity of white-matter lesions (graded from 0 to 4, depending on the extension throughout the brain and whether they are restricted to the periventricular area or extend beyond it to the subcortical area), as well as a measure of brain atrophy (ventricular index). They classified CT scans from patients with VD, AD, and stroke with focal neuropsychological deficits according to this scale and found that the proportion of patients with VD increased significantly with the increasing degree of severity on CT. However, this analysis was not significant when only cerebrovascular patients were included. Using the most severe grade of 3 the sensitivity and specificity of the scale were 41% and 92%, respectively. Similarly, using grade 0, which includes patients with no infarcts and mild white matter changes, yielded a specificity of 56% and a sensitivity of 94%. Patients with VD in this study were diagnosed according to NINDS-AIREN criteria so the analysis was subjected to the limitations imposed by the use of these diagnostic rules. It is conceivable that separation of patients with different types of infarcts (lacunes versus macroscopic infarcts, for instance) might have improved the validity of the scale.

Categorization of VCI. A Proposed Encounter of Neuropathology, Etiopathogenesis and Clinical/Neuroradiological Manifestations

It has been previously pointed out that none of the currently accepted diagnostic criteria proposes a systematic classification of vascular "dementias" according to etiopathogenic mechanisms. This is one of the big paradoxes in the literature on vascular neurology, as almost any text on the subject includes a more or less appropriate classification of VD. That VD or VCI are the result of ischemic infarction requires little discussion; however, this can only be understood as the common final pathway or common pathological substrate where multiple pathophysiological and etiopathogenic mechanisms concur. An analogy with dementia from infectious diseases is easy to establish. A study of clinical manifestations, risk factors, and a clinical trial in which demented AIDS patients would be grouped with Whipple's disease or Creuztfeldt-Jakob patients would not be acceptable despite the fact that they all represent infectious dementias. A similar situation is creating confusion in the field of VD.

Proposals for classification of the different types of VD are abundant in the literature [12, 30, 38, 61, 75-80]. However, most of them mix anatomical and pathophysiological criteria, and those that propose clinically recognizable categories are limited and incomplete. Some classifications are based on potential etiologies, but they consist of exhaustive lists of possible causes for stroke and are of limited value. Some concepts are still confusing and arbitrarily used. The term MID is used with ambiguous meanings. On the one hand, some authors apply it to differentiate dementia caused by multiple lesions from cases of single-infarct dementia. Other authors use the term to separate dementia with multiple well-established ischemic infarcts from "hemodynamic" dementia, forgetting that this type of dementia is caused by multiple infarcts too. This term was originally coined to emphasize the fact that dementia from vascular causes is also caused by multiple infarcts and not chronic ischemia [81], and for a long time it was used to designate all types of ischemic VD. Nowadays its use is restricted to cases of dementia from multiple macroscopic infarcts. Another example of the confusion that terminology creates is represented by the terms Binswanger's disease and lacunar state, which are proposed as different in all available classifications, while they probably represent the same entity, as will be later discussed.

Any classification of VCI should be based on etiology and pathophysiology. The different subtypes resulting from such categorization should be supported by pathological data and, more importantly, should be clinically recognizable. The question arises as to whether a *pathologic diagnosis* can be reached from clinical and paraclinical data, and how this can be related to a particular *etiologic diagnosis*. The traditional clinical method of neurology may prove to be the best method to classify VCI.

The first step consists in the recognition of cognitive symptoms and signs and their combination into clinical syndromes. A thorough characterization of the different patterns of cognitive deficits would facilitate an approximation to a topographical and, subsequently an etiopathogenic diagnosis. In this respect two main types of clinical situations may be distinguished: those in which cognitive deficits are restricted to one cognitive area and those involving two or more intellectual functions.

Syndromes Restricted to One Area of Cognition

Localized Cognitive Functions

That cerebrovascular disease can cause intellectual deficits confined to a single cognitive area has been known since the work of Broca. Disturbances of language (aphasia), reading/writing (alexia/agraphia), praxis (apraxia), recognition (agnosia, neglect) will usually respond to cortical lesions, which may be easily detected on CT or MRI. Subcortical lesions in the thalamus or basal ganglia may also manifest with isolated language or hemineglect disturbances [82]. Atheroembolic or cardioembolic mechanisms will be responsible in the majority of cases. In some cases of isolated transcortical aphasia or visuospatial disturbances, neuroimaging techniques may show watershed infarction and suggest a hemodynamic mechanism [83]. Rare cases will present in which no vascular lesion will be detected, and instead focal atrophy may be the only radiological correlate. Whether these cases represent examples of incomplete infarction, laminar necrosis or focal accumulation of cortical microinfarcts requires further research. Hypothetically, a differential diagnosis with focal degenerative cortical atrophies may be established. The value of clinical information regarding type of onset and progression, presence of vascular risk factors, as well as information provided by functional neuroimaging techniques will need to be tested. Some authors may question the value of this diagnostic category. In many instances focal neuropsychological deficits will remain stable, but in some cases cognitive decline may continue to progress and lead to significant functional limitation. One of the challenges for modern neurology is to recognize these patients and determine which mechanisms underlie the process.

Disperse Cognitive Functions

Isolated impairment of memory functions is rare in cerebrovascular disease [7]. Thalamic lesions and basal forebrain infarcts either lacunar or secondary to large artery embolism may cause severe memory impairment, but are usually accompanied by other deficits. Infarcts in the territory of the posterior cerebral artery may involve the mesial temporal lobe and cause amnesia, but other visuospatial or language deficits will probably be present [84]. Hippocampal sclerosis may be suspected when episodes of global anoxia or ischemia damage both hippocampi and produce amnesia [50]. The value of MR imaging to detect hippocampal sclerosis has not been determined. This might be of relevance for those cases in which the entity is suspected, but no episode of global anoxia or ischemia can be elicited by history.

Describing an "isolated" compromise of executive functions may be imprecise, as they include tasks as varied as attention, planning, programming, anticipation, set shifting, or memory-search strategies. Dysexecutive syndromes may result from different lesions with different etiopathogenic mechanisms. In each case, clinical examination, as well as neuroradiological findings will help to distinguish cortical infarcts in the territory of the anterior cerebral artery or the anterior branches of the middle cerebral artery; lacunar infarcts in the subcortical frontal white matter, thalamus, capsular genu; or deep infarcts in the caudate or basal forebrain [84]. It is important to emphasize when dealing with these types of lesions that disturbances of motivation and drive, as well as profound personality changes or affective disorders, may be more prominent than cognitive manifestations and characterize the clinical picture. Again, a differential diagnosis with frontal lobe dementia should be easily established on a neuroradiological basis.

Focal Lesions Causing Syndromes Involving More Than One Cognitive Domain

It is well known that certain whimsically placed infarcts may cause significant impairment in several cognitive areas, which may be severe enough to be labeled as dementia. Paramedian unilateral or bilateral thalamic infarcts may give way to a number of combinations of severe amnesia, aphasia, hemineglect, apathy, loss of drive and dysexecutive frontal-like syndromes [85]. In most instances the underlying mechanism is an arterial or cardiac embolism, but the importance of thalamic lacunes as correlates of dementia has also been emphasized [45]. Apathy, personality changes and memory impairment may also be caused by lacunar infarcts in the caudal capsular genu [86]. Aphasia, alexia, visuoconstructional impairment and Gerstmann syndrome are the classical manifestations of the angular gyrus syndrome caused by arterial macroscopic infarcts in the posterior territory of the medial cerebral artery. This syndrome has been described to behave clinically as AD except for the absence of memory impairment [61]. As previously mentioned, embolic or atherothrombotic occlusion of the posterior cerebral artery may provoke different combinations of visual agnosia, alexia, agraphia, and memory loss. Basal forebrain infarcts either in relation with rupture of anterior communicating artery aneurysms or due to occlusion of the deep penetrating arteries are associated with severe memory loss and personality changes [82]. Caudate infarcts in the context of small deep penetrating artery disease are associated with frontal-like syndromes, aphasia and neglect syndromes [87]. As can be seen, the examples are varied and the etiopathogenic mechanisms heterogeneous. In most instances the abrupt onset of cognitive deficits will lead to the diagnosis. In some cases, however, the clinical picture may be insidious and resemble degenerative processes. In these cases, a thorough characterization of the neuropsychological pattern will allow appropriate clinicoradiological correlations to be established. Dementia caused by a single strategically placed infarct is included as a separate category in the majority of classifications of VD available in the literature. While such diagnoses may be academically laudable, they provide little information regarding etiology and pathophysiology.

Syndromes Involving Multiple Cognitive Areas

From the diagnostic perspective, patients with cognitive changes in multiple domains that cannot be related to a single infarct will represent the most challenging category of VCI. In all probability they will also represent the most frequent type of VCI – at least in a memory clinic setting and at the primary care level. Detailed prospective studies are needed in which carefully collected clinical and neuropsychological data are correlated with information provided by neuroimaging and other laboratory investigations so that adequate etiopathogenic categories be defined. While results from such studies are available, it can be hypothesized that most patients will fall into one of the following three clinical forms according to the pattern of neuropsychological deficits.

Cortical Forms of VCI

These patients will present deficits in cognitive areas with a clear cortical representation such as language, praxis, visuospatial and visuoconstructional functions, or calculation. Although generally considered a disperse function, memory has an important representation in the temporal lobes and hippocampal formation so that certain amnesic deficits, especially those related to impaired registration, storage or learning, may be considered typically cortical. In a substantial proportion of patients with multiple cortical cognitive deficits, neuroimaging studies will reveal the presence of cortical macroscopic infarcts. These probably respond to atherothrombotic or cardioembolic mechanisms. Clinicoradiological correlation will then be easily established. In other cases infarcts may be present, but in areas that would not explain the neuropsychological deficits, and a number of patients will show no vascular lesions. As has been previously outlined, some vascular lesions such as laminar necrosis or cortical microinfarcts may be radiologically silent. Consequently, a diagnosis of VCI should not be automatically rejected when CTs or MRIs are normal or show only mild cerebrovascular disease. It is in these two situations that a diagnosis of either VCI or AD must be made with extreme caution. In this respect, it is likely that detection of early cases will allow a better definition of differential traits between the two diseases. If diagnosis is delayed until dementia ensues, and especially if dementia is defined in the same way, neuropsychological characteristics will prove to be similar. However, in the early stages, detailed and thorough neuropsychological evaluation might potentially detect distinct patterns. Prominent memory deficits may favor a diagnosis of AD, but aphasia or visuoconstructional disturbances with little or no memory loss would make it unlikely. A differential diagnosis with degenerative focal cortical atrophies (progressive aphasia, posterior cortical atrophy, progressive apraxia) should always be taken into account. The presence of focal atrophy, as suggested, may also reveal selective ischemic necrosis or laminar necrosis in the watershed territories. In these cases, presence of vascular risk factors, history of orthostatic hypotension, syncope or other hypoperfusional states, as well as other causes of focal transient ischemia might provide clues for the diagnosis. Investigation with SPECT using neuronal markers such as iomazenil may prove to be useful to detect "incomplete" infarcts, as has been recently reported [88]. Similarly, perfusional deficits in the hippocampal, cingulate and anterior thalamic regions may favor the diagnosis of AD [89]. Detection of old hemorrhages on MRI or a family history of lobar hemorrhages and dementia may suggest a diagnosis of amyloid angiopathy. Except for infrequent familial cases, amyloid angiopathy is accompanied by AD in most instances. Even in this situation detection of modifiable risk factors may be crucial (see below). Prospective studies are obviously needed in this field before accurate criteria can be formulated. These studies should focus on early rather than demented cases.

Subcortical Forms of VCI

Symptoms and signs of dysexecutive deficits, disturbances of motivation and drive, attentional deficits, personality changes and affective disorders are the predominant features of subcortical cognitive syndromes [61]. Memory loss may also be prominent but, as opposed to cortical forms of amnesia, patients may show good performance in cued recall or assisted learning tasks [90]. Memory performance may improve if attentional deficits are corrected during the examination or even if affective disorders are adequately treated. Impairment of motor control, including gait disturbances, slowness, rigidity, tremulousness, or pseudobulbar syndrome, are usually present [32]. Most published classifications of VD have related these forms of subcortical dementia to small-vessel disease and two entities are usually described: "Binswanger's disease" (BD) and lacunar states. This traditional and generally accepted notion needs reconsideration. First, atheroembolic infarcts, and not only lacunes or white matter changes, may cause subcortical deficits. Second, BD and lacunar states might well represent the same clinical and radiological entity. BD is characterized by the presence of white-matter changes and small subcortical lacunar infarcts, while lacunar states that cause cognitive decline are frequently accompanied by white-matter rarefaction [63]. Both entities coincide in clinical manifestations, risk factors and neuroradiological findings, and it is not surprising that some authors have described them as "closely related" [91] or even "probably identical" [65].

A diagnosis such as BD may be misleading in terms of etiology and pathogenesis. There are several processes that may cause subcortical cognitive deficits with white-matter changes and small subcortical infarcts or lacunes. Hypertension is probably the most frequent, but other entities such as amyloid angiopathy, hereditary forms of small-vessel disease, or coagulation disorders with increased fibrinogen should be taken into account, especially in normotensive patients [32]. This is probably the most important consideration. The syndrome that combines clinical strokes, subcortical cognitive deficits, gait disturbance, pyramidal and extrapyramidal signs, pseudobulbar syndrome with radiological findings of white-matter rarefaction, multiple lacunar infarcts and ventricular dilation that progresses either gradually or in a stepwise fashion with periods of stabilization, is in fact a syndrome with different etiologies and pathogeneses. If this is kept in mind, terms like Binswanger's syndrome or subcortical VCI might be more appropriate as long as they are followed by a statement about etiology. Hypotension and other processes leading to global brain ischemia or anoxia have also been related to subcortical cognitive deficits [30]. The role of hypotension as an etiopathogenic factor for vascular dementia has been highlighted by Sulkava and Erkinjuntti [92]. Therefore, cardiac arrhythmia, hypotension, anesthesia, sleep apnea and other disorders inducing cerebral hypoperfusion-hypoxia, associated or not with other occlusive vasculopathies, should be added to the list of causes or contributing factors of the syndrome called "Binswanger's disease." Cerebral hypoxic and ischemic injury resulting from systemic illnesses is in fact a significant risk factor for the development of dementia after stroke [29]. Another entity to be added to the list of causes of subcortical forms of VCI is the hereditary form called cerebral autosomal dominant arteriopathy with subcortical strokes and ischemic leukoencephalopathy (CADASIL) [93]. The disease is characterized by the combination of migraine with aura, mood disturbances, TIAs and strokes at an unusually young age, pseudobulbar palsy, gait disturbances with pyramidal signs and subcortical dementia with predominance of frontal-like symptoms in the members of the same family. Except for the absence of hypertension or other vascular risk factors, the clinical features are strikingly similar to those described in patients given a diagnosis of BD. The resemblance persists in MRI findings, as patients with CADASIL show small infarcts and extensive areas of abnormal signal in the periventricular and deep white matter, as well as in the basal ganglia and brainstem [94]. Histologically, apart from white-matter changes and lacunes, small arteries in the leptomeninges, white-matter and deep gray structures show concentric thickening with splitting of the lamina elastica interna, hyalinosis of the media, and deposits of a granular PAS-positive material replacing the smooth muscle cells. Similar changes have been described in muscle and skin vessels in some patients. The prevalence of this disease is unknown, but it may be more frequent than initially presumed. There were 25 families reported in Europe by 1993 [95] and this number doubled in 2 years [93]. Genetic linkage studies had located the responsible gene in chromosome 19, and the critical region has already been mapped with mutations described in the Notch-3 gene [96]. Genetic tests will probably be available soon and this will allow us to investigate whether cases of so-called BD are in fact "CADASIL" sporadic patients.

Subcortical cognitive syndromes should be identified syndromically and classified etiologically. Terms like BD or subcortical VCI would be acceptable when followed by a statement about possible or probable etiopathogenic mechanisms. A distinction of hypertensive versus nonhypertensive forms could be a first useful step towards a better categorization and understanding of this frequent form of VCI and its potential treatment and prevention.

White-Matter Forms of VCI (The Loss of Brain Connectivity)

Numerous studies have provided enough evidence to support the idea that vascular white-matter damage as seen on CT or MRI causes cognitive impairment. In many of these studies, patients with discrete vascular lesions such as lacunes or small infarcts were excluded and the effect of LA on cognition could still demonstrated. Subjects with white-matter hyperintensities and "no other focal MRI changes" perform poorly on tasks measuring speed of mental processing such as the trail-making test or an assembly procedure or the Purdue pegboard test while memory, attention, and conceptualization performance are similar to the performance of subjects without leukoaraiosis [65]. Other studies have reported low scores on attention, speed of information processing, and tests measuring frontal lobe functions among patients with severe white-matter lesions [97]. Ylikoski and colleagues [98] found significant negative correlation between the severity of leukoaraiosis and attention and mental-processing speed scores, but not with memory, verbal or constructional functions. It may be interpreted that isolated white-matter changes may induce similar but more restricted or less severe deficits than subcortical nuclei lesions. Pathological studies have also shown that white-matter disease may occur in the absence of lacunes and subsequently respond to different etiopathogenic mechanisms in some patients [72]. If this is so, the clinical and radiological distinction of both entities is important as risk factors, and therapeutic measures might differ.

Classification of the different types of VCI on a clinical and etiopathogenic basis is obviously complex and laborious. The categorization proposed herein (Table 2) is hypothetical, necessarily incomplete and modifiable as research advances in this field. Two categories have not been included, as they can only be retrospectively defined. One category should include high-risk patients not cog-

Table 2. Categorization of VCI. Cognitive syndromes and plausible neuropathological/neuroimaging findings (see Table 1 for potential etiopathogenic mechanisms)

Clinical picture			Neuropathology/neuroradiology findings
• Circumscribed cognitive deficits			
	Localized functions		Cortical infarcts
			Cortical microinfarcts
			Watershed infarction
			Laminar necrosis?
	Disperse functions	Memory	Selective vulnerability (hippocampal sclerosis)
			PCA infarcts
			Subcortical infarcts (thalamus, basal forebrain)
			Lacunes (thalamus, basal forebrain?)
		Executive function	Cortical infarcts/microinfarcts (frontal)
			Lacunes (frontal white matter, basal ganglia, thalamus, capsular genu)
			Subcortical infarcts
			White-matter disease
• Multiple cognitive deficits			
	With focal lesions	Angular gyrus	Cortical infarct
		PCA infarcts	Cortical infarct
		Capsular genu	Lacunes/subcortical infarcts
		Basal forebrain	Lacunes/subcortical infarcts
		Thalamus	Lacunes/subcortical infarcts
		Cortical VCI	Multiple cortical infarcts/microinfarcts
			Watershed infarction
			Selective vulnerability (laminar necrosis?)
	Multiple lesions (?)	Subcortical VCI	Lacunes (lacunar state)
			Lacunes (lacunar state) + white-matter disease
			Watershed infarction
			Subcortical infarcts
		White-matter VCI	Vascular white-matter disease

nitively impaired and a second category, with multiple subtypes, would include cases in which different etiopathogenic mechanisms are combined. An international databank of patients with different kinds and grades of cerebrovascular disease in which clinical, radiological and neuropsychological information is systematically gathered through consensuated standardized forms would probably constitute the best scenario to achieve this complex task.

Vascular Changes in AD: New Concepts on Mixed Dementia

The concept of mixed dementia was first used by Tomlinson and colleagues [99]. When studying brains of old, demented people, they found a few cases in which both vascular lesions and senile plaques/neurofibrillary tangles coexisted. In these patients the extent and severity of both processes clearly exceeded that observed in their non-demented control brains, and any of the two findings, vascular or degenerative, was considered sufficient to have caused dementia. Interestingly, they also found five patients with dementia in which vascular and degenerative changes coexisted, but the degree of severity for both was in the upper limit of that observed in controls. The density of neurofibrillary pathology was significantly low. They postulated that the synergistic effect of both pathologies was necessary to produce dementia in these subjects and applied the term "probable mixed dementia." In other words, they hypothesized that these patients would probably have remained free of dementia had one of the two components been absent.

In the last three decades the concept of mixed dementia has received little attention despite the fact that it represents the third-most-frequent cause of dementia in most clinical and epidemiological series. Patients with clinical pictures resembling AD in which either clinical history or neuroimaging techniques reveal some degree of cerebrovascular disease may have received this diagnosis regardless of whether the vascular component was relevant or not in the resultant dementia. This question has been often avoided and hidden under the disguise of terminology. "Possible AD," "AD with cerebrovascular disease" or "dementia after stroke" are terms frequently used but whose accuracy has seldom been tested experimentally.

A significant amount of evidence from studies reported in the last years on the association of vascular pathology and AD is highly suggestive of significant pathophysiological interactions in the production of dementia in a substantial proportion of patients with AD. Even a common etiopathogenic origin of both processes has been insinuated. Epidemiological research on population-based studies has revealed that high blood pressure [100], diabetes [101] or atrial fibrillation [102] are significant risk factors for dementia, not only VD, but also AD. In samples of stroke patients followed longitudinally, the incidence of AD is significantly higher than that in the general population [16, 17]. Data from anatomopathological studies are also remarkable. Numerous autopsy series have demonstrated the presence of ischemic infarcts in up to 30% of patients with a pathological (not clinical) diagnosis of AD (not mixed dementia) [103]. In a group of patients with similar AD changes at autopsy, those who also presented vascular lesions had presented dementia in life much more frequently than those without infarcts (93 versus 53%). The risk of having presented dementia increased by a factor of 20 in relation to ischemic infarcts. Of note, the vascular lesions in most cases included only small infarcts in the thalamus, basal ganglia or frontal white matter [104]. That "pure" AD and AD with vascular lesions may behave differently from the clinical and anatomopathological perspective has also been suggested. When infarcts are present, the density of neurofibrillary tangles is significantly lower, and in contrast to what happens in "pure" AD, cognitive scores do not correlate with this pathological marker [105]. Another difference involves the age of onset that may be surprisingly delayed at least six years in patients with vascular lesions as compared to "pure" AD cases (Bowler et al., unpublished). This probably reflects the fact that early AD tends to be more severe and rapidly progressive, whereas in the 1970s and 1980s Alzheimer changes are milder, and whether they manifest clinically may depend in part on the occurrence of cerebral infarcts.

The idea of a common etiology for both vascular lesions and Alzheimer changes is hypothethical, and only anecdotal evidence has been provided in patients with global cerebral ischemia who at autopsy show accumulation of β-amyloid in the boundary zones of two arterial territories. In addition, amyloid angiopathy and white-matter changes are frequently associated with AD. It is difficult to establish whether AD with vascular lesions represents a separate entity or not and whether it can be clinically recognizable on neuropsychological or neuroimaging investigations. It is important, however, to recognize cases in which modifiable risk factors may be detected and adequately treated. For instance, in patients with AD and moderate-to-severe amyloid angiopathy, the risk of having an ischemic infarct at autopsy increases by a factor of three; if the patient is hypertensive, this risk increases 14 times [103].

Conclusions and Perspectives: the Definition of VCI

Cerebrovascular disease with or without clinical stroke is a frequent cause of cognitive impairment and dementia. VD has been traditionally considered treatable and preventable, but many questions are still unclear regarding its risk factors. More importantly, early detection is not possible as no diagnostic criteria exist. Current diagnostic criteria for VD are only fulfilled by patients with advanced cognitive decline that matches a specific neuropsychological pattern. For these reasons, the concept of VCI was formulated as a necessary new approach to this important medical challenge. Cognitive decline from vascular causes should be identified in its initial stages, diagnosed on the basis of reliable clinical and radiological criteria and classified according to its multiple etiologies.

As has been suggested [2], an international data bank of prospectively collected and longitudinally followed patients will certainly be the best setting to achieve this task. Clinical data on age, symptoms and mode of onset, history of cerebrovascular events, pattern of progression, as well as demographic data regarding level of education, employment, physical and intellectual activities should be systematically collected. Results from clinical neurological and cardiovascular examinations as well as measures of functional and physical disability should be gathered on standardized forms. Information on risk factors should be simple and complete, including traditional and putative risk factors, as well as information on whether they are treated and effectively controlled. Exhaustive questionnaires on family history would be essential. Quantitative and comparable data on cognitive functions would be mandatory. While valid short screening tests are selected or developed, a comprehensive battery of neuropsychological tests are desirable. Neuroimaging data should provide information on number, type and location of infarcts. White-matter changes should be classified according to type, location, and degree of severity. Simple consensual measures of brain atrophy should be obtained. Genetic testing for apolipoprotein E genotypes has been recommended for any research project on cognitive decline and dementia. DNA should be stored, as new susceptibility genes will soon be discovered. Laboratory tests including SPECT, echocardiogram, carotid ultrasound, Holter ECG, blood-pressure monitoring, autonomic function tests, tests for sleep apnea, vasculitis screening and others may prove useful in selected cases. The combination of clinical and laboratory data would allow classification of patients according to possible or probable etiopathogenic mechanisms. By doing so it will be possible to discern what cognitive deficits are caused by what cerebrovascular process and reliable diagnostic criteria will be formulated. Longitudinal follow-up will help to detect and understand the factors that determine further progression of cognitive deficits and ultimately dementia. The rationale for future therapy and prevention of VCI will be established.

Acknowledgements. Dr. Pablo Martinez-Lage was supported by grant (BAE 95/551) from the Fondo de Investigaciones Sanitarias, Ministry of Health, Government of Spain. Dr. Vladimir Hachinski is the first recipient of the Trillium Clinical Scientist Award, given by the Ontario Ministry of Health.

References

1. Hulette C, Nochlin D, McKeel D, Morris JC, Mirra SS, et al (1997) Clinical-neuropathologic findings in multi-infarct dementia: a report of six autopsied cases. Neurology 48: 688-672
2. Hachinski VC (1992) Preventable senility: a call for actions against vascular dementias. Lancet 340: 645-648
3. Hachinski VC (1994) Vascular dementia: a radical redefinition. Dementia 5: 130-132
4. Hachinski VC, Bowler JV (1993) Vascular dementia. Neurology 43: 2159-2160
5. Hébert R, Brayne C (1995) Epidemiology of vascular dementia. Neuroepidemiology 14: 240-257
6. Jorm AF (1990) The epidemiology of Alzheimer's disease and related disorders. Chapman and Hall, London

7. Bowler J, Hachinski VC (1996) History of the concept of vascular dementia: two opposing views on current definitions and criteria for vascular dementia. In: Prohovnik I, Wade J, Knezevic S, Tatemichi T, Erkinjuntti T (eds) Vascular dementia: current concepts. Wiley, Chichester, pp 1-24

8. Skoog I, Nilsson R, Palmertz B, et al (1993) A population-based study of dementia in 85-year-olds. N Engl J Med 328: 153-158

9. Breteler MMB, van Swieten JC, Bolts ML, et al (1994) Cerebral white matter lesions, vascular risk factors, and cognitive function in a population-based study: the Rotterdam Study. Neurology 44: 1246-1252

10. Martinez-Lage P, Manubens JM, Martinez-Lage JM, et al (1996) Vascular risk factors and cognitive performance in a non-demented elderly population. Neurology [Suppl] 1: A289

11. Breteler MMB, Claus JJ, Grobbee DE, et al (1994) Cardiovacular disease and distribution of cognitive function in elderly people: the Rotterdam Study. BMJ 308: 1604-1608

12. Tatemichi TK, Sacktor N, Mayeux R (1994) Dementia associated with cerebrovascular disease, other degenerative diseases, and metabolic disorders. In: Terry RD, Katzman R, Bick KL (eds) Alzheimer's disease. Raven, New York, pp 123-166

13. Pohjasvaara T, Erkinjuntti T, Vataja R, Kaste M (1997) Dementia three months after stroke. Baseline frequency and effect of different definitions of dementia in the Helsinki Stroke Aging. Memory Study (SAM) cohort. Stroke 28: 785-792

14. Grace J, Nadler JD, White DA, et al (1995) Folstein vs Modified Mini-Mental State Examination in geriatric stroke. Stability, validity, and screening utility. Arch Neurol 52: 477-484

15. Tatemichi TK, Desmond DW, Stern Y, et al (1994) Cognitive impairment after stroke: frequency, patterns, and relationship to functional disabilities. J Neurol Neurosurg Psychiatry 57: 202-207

16. Tatemichi TK, Desmond DW (1996) Epidemiology of vascular dementia. In: Prohovnik I, Wade J, Knezevic S, Tatemichi T, Erkinjuntti T (eds) Vascular dementia: current concepts. Wiley, Chichester, p 40

17. Kokmen E, Whisnant JP, O'Fallon WM, et al (1996) Dementia after ischemic stroke: a population-based study in Rochester, Minnesota. Neurology 46: 154-159

18. Pasquier F, Leys D (1997) Why are stroke patients prone to develop dementia? J Neurol 244: 135-142

19. Rockwood K, Ebly E, Hachinski H, et al (1997) Presence and treatment of vascular risk factors in patients with vascular cognitive impairment. Arch Neurol 54: 33-39

20. Martinez-Lage P, Hachinski VC (1998) Multi-infarct dementia. The vascular causes of cognitive impairment and dementia. In: Barnett HJM, Mohr JP, Stein BM, Yatsu FM (eds) Stroke: pathophysiology, diagnosis, and management, 3rd edn. Churchill Livingstone, New York

21. Moncayo J, Bogousslavsky J (1996) Vascular dementia: persisting controversies and questions. Eur J Neurol 3: 299-308

22. Ladurner G, Iliff LD, Lechner H (1982) Clinical factors associated with dementia in ischaemic stroke. J Neurol Neurosurg Psychiatry 45: 97-101

23. Loeb C, Gandolfo C, Bino G (1988) Intellectual impairment and cerebral lesions in multiple cerebral infarcts. A clinical-computed tomography study. Stroke 19: 560-565

24. Tatemichi TK, Foulkes MA, Mohr JP, et al (1990) Dementia in stroke survivors in the Stroke Data Bank Cohort. Prevalence, incidence, risk factors, and computed tomographic findings. Stroke 21: 858-866

25. Tatemichi TK, Desmond DW, Paik M, et al (1993) Clinical determinants of dementia related to stroke. Ann Neurol 33: 568-575
26. Gorelick PB, Brody J, Cohen D, et al (1993) Risk factors for dementia associated with multiple cerebral infarcts. A case-control analysis in predominantly African-American hospital-based patients. Arch Neurol 50: 714-720
27. Yoshitake T, Kiyohara W, Kato I, et al (1995) Incidence and risk factors of vascular dementia and Alzheimer's disease in a defined elderly Japanese population: the Hisayama Study. Neurology 45: 1161-1168
28. Lindsay J, Hebert R, Rockwood K (1997) The Canadian Study of Health and Aging. Risk factors for vascular dementia. Stroke 28: 526-530
29. Moroney JT, Bagiella E, Desmond DW, et al (1995) Global hypoxic-ischemic events increase the risk of dementia after stroke. Ann Neurol 38: 290
30. Brun A (1994) Pathology and pathophysiology of cerebrovascular dementia: pure subgroups of obstructive and hypoperfusive etiology. Dementia 5: 145-147
31. Lopez OL, Rabin BS (1995) Alteraciones inmunológicas de la demencia vascular. In: Lopez Pousa S, Manubens JM, Rocca WA (eds) Epidemiología de la demencia vascular. Controversias en su tratamiento. JR Prous, Barcelona, pp 91-98
32. Caplan LR (1995) Binswanger's disease-revisited. Neurology 45: 626-633
33. Iwamoto T, Kubo H, Takasaki M (1995) Platelet activation in the cerebral circulation in different subtypes of ischemic stroke and Binswanger's disease. Stroke 26: 52-56
34. Bowler JV, Hachisnki VC (1994) Progress in the genetics of cerebrovascular disease: inherited subcortical arteriopathies. Stroke 25: 1696-1698
35. Betard C, Robitaille Y, Gee M, et al (1994) ApoE allele frequencies in Alzheimer's disease, Lewy body dementia, Alzheimer's disease with cerebrovascular disease and vascular dementia. Neuroreport 5: 1893-1896
36. American Psychiatric Association (1994) Diagnostic and statistical manual of mental disorders, 4th edn. American Psychiatric Association, Washington DC
37. World Health Organization (1993) The ICD-10 classification of mental and behavioural disorders. Diagnostic criteria for research. World Health Organization, Geneva
38. Román GC, Tatemichi TK, Erkinjuntti T, et al (1993) Vascular dementia: diagnostic criteria for research studies. Report of the NINDS-AIREN International Workshop. Neurology 43: 250-260
39. Chui HC, Victoroff JI, Margolin D, et al (1992) Criteria for the diagnosis of ischemic vascular dementia proposed by the State of California Alzheimer's disease diagnostic and treatment centers. Neurology 42: 473-480
40. Wallin A, Blennow K (1993) Heterogeneity of vascualr dementia: mechanisms and subgroups. J Geriatr Psychiatry Neurol 6: 177-185
41. Lopez OL, Larumbe MR, Becker JT, et al (1994) Reliability of NINDS-AIREN criteria for the diagnosis of vascular dementia. Neurology 44: 1240-1245
42. Verhey FRJ, Lodder J, Rozendaal N, et al (1966) Comparison of seven sets of criteria used for the diagnosis of vascular dementia. Neuroepidemiology 15: 166-172
43. Wetterling T, Kanitz R-D, Borgis K-J (1996) Comparison of different diagnostic criteria for vascular dementia (ADTCC, DSM-IV, ICD-10, NINDS-AIREN). Stroke 27: 30-36
44. Villardita C (1992) Alzheimer's disease compared with cerebrovascular dementia. Neuropsychological similarities and differences. Acta Neurol Scand 87: 299-308
45. del Ser T, Bermejo F, Portera A, et al (1990) Vascular dementia. A clinicopathological study. Neurol Sci 96: 1-17
46. Munoz DG (1991) The pathological basis of multi-infarct dementia. Alzheimer Dis Assoc Disord 5: 77-90

47. Garcia JH, Brown GG (1992) Vascular dementia: neuropathologic alterations and metabolic brain changes. J Neurol Sci 109: 121-131
48. Olsson Y, Brun A, Englund E (1996) Fundamental pathological lesions in vascular dementia. Acta Neurol Scand Suppl 168: 31-38
49. Dickson DW, Davies P, Bevona C, et al (1994) Hippocampal sclerosis: a common pathological feature of dementia in very old (≥80 years of age) humans. Acta Neuropathol 88: 212-221
50. Petito CK, Feldemann E, Pulsinelli WA, et al (1987) Delayed hipoocampal damage in humans following cardiorespiratory arrest. Neurology 37: 1281-286
51. Corey-Bloom J, Sabbagh MN, Bondi MW, et al (1997) Hippocampal sclerosis contributes to dementia in the elderly. Neurology 48: 154-160
52. Knopman DS, Mastri AR, Frey WHII, et al (1990) Dementia lacking distinctive histologic features: a common non-Alzheimer degenerative dementia. Neurology 40: 251-256
53. Lassen NA (1982) Incomplete cerebral infarction-focal incomplete ischemic tissue necrosis not leading to emollision. Stroke 13: 522-523
54. Garcia JH, Lassen NA, Weiller C, et al (1996) Ischemic stroke and incomplete infarction. Stroke 27: 761-765
55. Bredesen DE (1995) Neural apoptosis. Ann Neurol 38: 839-851
56. Matsushima K, Schmidt-Kastner R, Hakim AM (1996) Genes and cerebral ischemia: therapeutic perspectives. Cerbrovasc Disord 6: 119-127
57. Gehrmann J, Banati RB, Wiessnert C, et al (1995) Reactive microglia in cerebral ischaemia: an early mediator of tissue damage? Neuropathol Appl Neurobiol 21: 277-289
58. Yamanouchi H (1991) Loss of white matter oligodendrocytes and astrocytes in progressive subcortical vascular encephalopathy of Binswanger type. Acta Neurol Scand 83: 301-305
59. Higashi T, Nishi S, Nakai ZA, et al (1995) Regulatory mechanisms of stress response in mammalian nervous system during cerebral ischaemia or after heat shock. Neuropathol Appl Neurobiol 21: 471-483
60. Mesulam MM (1990) Large-scale neurocognitive networks and distributed processing for attention, language, and memory. Ann Neurol 28: 597-613
61. Cummings JL, Benson DF (1992) Vascular dementias. In: Cummings JL, Benson DF (eds) Dementia: a clinical approach. Butterworth-Heinemann, Boston, pp 153-176
62. Meyer JS, Kawamura J, Terayama Y (1992) White matter lesions in the elderly. J Neurol Sci 110: 1-7
63. Román GC (1987) Senile dementia of the Binswanger type. A vascular form of dementia in the elderly. JAMA 258: 1782-1788
64. Chimowitz MI, Awad IA, Furlan AJ (1989) Periventricular lesions on MRI. Facts and theories. Stroke 24: 7-12
65. Schmidt R, Fazekas F, Offenbacher H, et al (1993) Neuropsychologic correlates of MRI white matter hyperintsities: a study of 150 normal volunteers. Neurology 43: 2490-2494
66. Fazekas F, Schmidt R, Offenbacher H, et al (1991) Prevalence of white matter and periventricular magnetic resonance hyperintesities in asymptomatic volunteers. J Neuroimaging 1: 27-30
67. Grafton ST, Sumi SM, Stimac GK, et al (1991) Comparison of postmortem magnetic resonance imaging and neuropathologic findings in the cerebral white matter. Arch Neurol 48: 293-298
68. Leifer D, Buonanno FS, Richardson EP (1990) Clinicopathologic correlations of cranial magnetic resonance imaging of periventricular white matter. Neurology 40: 911-918
69. Hachinski VC, Munoz DG (1991) Leuko-araiosis: an update. Bull Clin Neurosci 56: 24-33

70. Van Swieten JC, Kappelle LJ, Algra A, et al (1992) Hypodensity of the cerebral white matter in patients with transient ischemic attacks or minor stroke: the influence on the rate of subsequent stroke. Ann Neurol 32: 177-183

71. van Zagten M, Boiten J, Kessels F, Lodder J (1996) Significant progression of white matter lesions and small deep (lacunar) infarcts in patients with stroke. Arch Neurol 53: 650-655

72. Erkinjuntti T, Enavente O, Eliasziw M, et al (1996) Diffuse vacuolization (spongiosis) and arteriolosclerosis in the frontal white matter occurs in vascular dementia. Arch-Neurol 53: 325-332

73. Tarvonen-Schröder S, Röyttä M, Räihä I, et al (1996) Clinical features of leuko-araiosis. J Neurol Neurosurg Psychiatry 60: 431-436

74. Pullicino P, Benedict RHB, Capruso DX, et al (1996) Neuroimaging criteria for vascular dementia. Arch Neurol 53: 723-728

75. Meyer JS, Ropgers RL, Mortel KF (1989) Multi-infarct dementia: demography, risk factors, and therapy. In: Ginsberg MD, Dietrich WD (eds) Cerebrovacular diseases. Raven, New York, pp 199-207

76. Mirsen T, Hachinski VC (1988) Epidemiology and classification of vascular and multi-infarct dementia. In: Meyer JS, Lechner H, Marshal J, Toole JF (eds) Vascular and multi-infarct dementia. Future, Mount Kisco, pp 61-76

77. Loeb C (1991) Vascular dementia: terminology and classification. In: Chopra JS, Jagannathan K, Sawhney IMS, Lechner H, Szendey GL (eds) Progress in cerebrovascular disease. Current concepts in stroke and vascular dementia. Elsevier Science, Amsterdam, pp 79-88

78. Ross GW, Cummings JL (1992) Vascular dementias. In: Thal LJ, Moos WH, Gamzu ER (eds) Cognitive disorders. Pathophysiology and treatment. Dekker, New York, pp 271-289

79. Erkinjuntti T (1987) Types of multi-infarct dementia. Acta Neurol Scand 75: 391-399

80. Loeb C, Meyer JS (1996) Vascular dementia: still a debatable entity? J Neurol Sci 143: 31-40

81. Hachinski VC, Lassen NA, Marshall J (1974) Multi-infarct dementia: a cause of mental deterioration in the elderly. Lancet 2: 207-210

82. Freeman R, Bear D, Greenberg MS (1989) Behavioral disturbances in cerebrovascular disease. In: Toole JF (ed) Handbook of clinical neurology, vol 11 (55). Vascular diseases, part III. Elsevier Science, Amsterdam, pp 137-150

83. Arboix A, Martí-Vilalta JL (1992) Manifestaciones clínicas de los infartos isquémicos cerebrales de los "territorios frontera". In: Matías-Guiu J, Martinez-Vila E, Martí-Vilalta JL (eds) Isquemia cerebral global. JR Prous, Barcelona, pp 33-39

84. Mahler ME, Cummings JL (1991) Behavioral neurology of multi-infarct dementia. Alzheimer Dis Assoc Disord 5: 122-130

85. Castaigne P, Lhermitte F, Buge A, et al (1981) Paramedian thalamic and midbrain infarcts: clinical and neuropathological study. Ann Neurol 10: 127-148

86. Tatemichi YK, Desmond DW, Prohovnik I, et al (1992) Confusion and memory loss from capsular genu infarction: a thalmocortical disconnection syndrome? Neurology 42: 1966-1979

87. Caplan LR, Schmahmann JD, Kase CS, et al (1990) Caudate infarcts. Arch Neurol 47: 133-143

88. Nakagawara J, Sperling B, Lassen NA (1997) Incomplete infarction of reperfused cortex may be quantitated with iomazenil. Stroke 28: 124-132

89. Keith AJ, Jones KJ, Becker A, et al (1997) Preclinical diagnosis of Alzheimer's disease with SPECT. Neurology 48: A207

90. Cummings JL (1993) Frontal-subcortical circuits and human behavior. Arch Neurol 50: 873-880

91. Nichols FT III, Mohr JP (1986) Binswanger's subacute arteriosclerotic encephalopathy. In: Barnett HJM, Mohr JP, Stein BM, Yatsu FM (eds) Stroke: pathophysiology, diagnosis, and management. Churchill Livingstone, New York, pp 875-885

92. Sulkava R, Erkinjuntti T (1987) Vascualr dementia due to cardiac arrhythmias and hypotension. Acta Neurol Scand 76: 123-128

93. Chabriat H, Vahedi K, Iba-zizen MT, et al (1995) Clinical spectrum of CADASIL: a study of 7 families. Lancet 346: 934-939

94. Sabbadini G, Francia A, Calandriello L, et al (1995) Cerebral autosomal dominant arteriopathy with subcortical infarcts and leucoencephalopathy (CADASIL). Clinical, neuroimaging, pathological, and genetic study of a large Italian family. Brain 118: 207-215

95. Bousser MG, Tournier-Lasserve E (1994) Summary of the proceedings of the first International Workshop on CADASIL. Paris, 19-21 May 1993. Stroke 25: 704-707

96. Joutel A, Corpechot C, Ducros A, et al (1996) Notch3 mutations in CADASIL, a hereditary adult-onset condition causing stroke and dementia. Nature 383: 707-710

97. Boone KB, Miller BL, Lesser IM, et al (1992) Neuropsychological correlates of white matter lesions in healthy elderly subjects. A threshold effect. Arch Neurol 49: 549-554

98. Ylikoski R, Ylikoski A, Erkinjuntti E, et al (1993) White matter changes in healthy elderly persons correlate with attention and speed of mental processing. Arch Neurol 50: 818-824

99. Tomlinson BE, Blessed G, Roth M (1970) Observations on the brains of old demented people. J Neurol Sci 11: 205-242

100. Skoog I, Lernfelt B, Landahl S, et al (1996) 15-year follow-up study of blood pressure and dementia. Lancet 347: 1141-1145

101. Ott A, Stolk RP, Hofman A, et al (1996) Association of diabetes mellitus and dementia: the Rotterdam Study. Diabetologia 39: 1392-1397

102. Ott A, Breteler MMB, de Bruyne MC, et al (1997) Atrial fibrillation and dementia in a population based study. The Rotterdam Study. Stroke 28: 316-321

103. Olichney JM, Hansen LA, Hofstetter R, et al (1995) Cerebral infarction in Alzheimer's disease is associated with severe amyloid angiopathy and hypertension. Arch Neurol 52: 702-708

104. Snowdon DA, Greiner LH, Mortimer JA, et al (1997) Brain infarction and the clinical expression of Alzheimer's disease. JAMA 277: 813-817

105. Nagy Z, Esiri MM, Jobst KA, et al (1997) The effects of additional pathology on the cognitive deficit in Alzheimer disease. J Neuropathol Exp Neurol 56: 165-170

Dementia with Lewy Bodies

C. Ballard, C. Morris, and M. Piggott

Background

Lewy bodies are intracytoplasmic, eosinophilic intraneuronal inclusion bo-dies [1]. They contain phosphorylated and non-phosphorylated microfilaments, β-crystalline, ubiquitin and proteinases [2]. They are therefore indicators of cell response to stress and markers for the degradation and excretion of abnormal proteins from the cell. Although Lewy bodies in the substantia nigra were identi-fied as a key neuropathological element of Parkinson's disease (PD) early in the century, the importance of cortical Lewy bodies as a feature of some cases of dementia has only recently become evident. A series of single case reports and small case series appeared in the literature between the 1960s and the mid-1980s, suggesting that Lewy bodies in the cerebral cortex could be associated with a rare form of dementia [3–6]. Hospital-based postmortem series reporting consecutive dementia cases have subsequently identified cortical Lewy bodies in at least 10% of dementia sufferers [7–10].

The current chapter aims:

1. To summarise current knowledge regarding the clinical symptom profile of dementia with Lewy bodies (DLB).
2. To describe developments in the understanding of the neurochemistry, neu-ropathology and genetics of the disorder.
3. To investigate the relationship between DLB and both Alzheimer's disease (AD) and PD. A summary of the main points is shown in Table 1.

Byrne et al. [8], Crystal et al. [11] and McKeith et al. [12] identified the clin-ical syndrome by retrospective examination of hospital notes from neu-ropathologically diagnosed cases. There was broad agreement about the com-mon symptoms, which included significant cognitive impairment with fluctua-tion in severity, disturbance of consciousness, visual hallucinations, delusions, auditory hallucinations, falls, and significant parkinsonism. Two sets of diag-nostic criteria were proposed. McKeith et al. [12] suggested fluctuating confu-sion as the key symptom and required one from the list of other features, which included visual hallucinations, falls, clouded consciousness, and parkinsonism in order to make the diagnosis. Byrne et al. [13] placed an increased emphasis

Table 1. Summary of the main differences between DLB, AD and PD

	Alzheimer's disease	Parkinson's disease	Dementia with Lewy bodies
(a) Clinical symptoms			
Visual hallucinations	20%	20%–50%	>80%
Significant parkinsonism	5%–10%	100%	60%–70%
Significant cognitive impairment	100%	10%–20%	100%
Marked fluctuation of cognitive impairment	10%–20%	Unknown	>90%
Neuroleptic sensitivity	<5%	Unknown	30%–50%
(b) Neuropathology			
Plaques (per mm^2)	15	<2	15
Tangles (per mm^2)	13	0	<1
Hippocampal CA2 neurites	Absent	Occasional	Common
(c) Genetics			
Apolipoprotein E4 allele frequency	35%–45%	15%–20%	35%–40%
CYP2D6 poor metabolisers	<2%	15%–20%	15%–20%
(d) Neurochemistry			
Dopaminergic changes			
Substantia nigra neuron density	No loss	69%↓ especially in early or unmedicated cases	48%↓
Striatal D2 receptors	No change	41%↑	Not elevated
Cholinergic changes			
Temporal cortex ChAT	↓70%	↓38%	↓80%
Muscarinic receptor binding	Not elevated	↑45%	↑24%
Nicotinic receptor binding	↓54% reduced in all cortical layers	Not significantly reduced	40%↓ significantly reduced only in inner cortical layers
Substantia nigra nicotinic receptor binding	Not reduced	↓50%–70%	↓50%–60%

upon parkinsonian symptoms and did not include visual hallucinations as a core feature.

Three clinical series have now reported patients identified using the McKeith et al. [12] criteria [14–16]. The prevalence rates varied from 9.7 to 26%, illustrating that patients with this symptom profile are common in clinical practice. There are, however, no published data from prospective studies that have attempted to apply standardised criteria prospectively and validate them against postmortem diagnosis. Provisional data from the ongoing prospective study in Newcastle are however available. Eighteen out of 19 patients with an operational diagnosis of DLB ($n = 14$) or NINCDS ADRDA probable or possible AD ($n = 5$) were correctly diagnosed.

DLB shows neuropathological features of both PD and AD, such as the presence of Lewy bodies (LB), senile plaques, and neurofibrillary tangles (NFT) [9, 17]. The hallmark feature of DLB is the presence of widespread LBs throughout the neo- and archi-cortex and the presence of LB and cell loss in subcortical nuclei, particularly the substantia nigra.

Ultrastructurally, LBs have a core of granular or vesicular material, and emanating from this core are radiating 8- to 10-nm diameter fibrils. Immunocytochemical investigations have shown a major component to be phosphorylated neurofilaments, but they do not contain the phosphorylated tau protein found in NFT [18]. In common with many other neuronal filamentous inclusions, the LB appears to contain ubiquitinated epitopes, and ubiquitin immunocytochemistry is the method of choice for demonstrating cortical LBs in DLB [19]. The presence of LBs in multiple neocortical areas appears to be one of the distinguishing features of DLB from PD where cortical LBs appear to be confined to vulnerable limbic and insular cortices [9, 19].

The major distinction between DLB and AD is in the deposition of NFT, which are absent or scarce in DLB [9, 10, 17, 20]. Mean cortical tangle counts are usually one-tenth that found in AD, and often only the temporal lobe is involved [9]. The CA1 sector of the hippocampus is normally vulnerable to NFT formation in AD, but in DLB this region is often spared, showing reduced or absent NFT [17, 21]. One distinguishing feature of DLB may be that ubiquitin-positive neuritic degeneration in the CA2-3 region is specific to this disease, since this is not normally apparent in AD [17].

Certain neuropathological similarities exist between AD and DLB, with both showing the presence of senile plaques at similar densities in the neocortex in the majority of elderly cases [9]. Senile plaques in DLB are characterised by increased neuronal sprouting, as assessed by GAP-43 staining [20]. Cortical cerebral amyloid angiopathy in DLB is however only related to concurrent ageing and is not a major feature [22].

Neocortical cholinergic activity, as measured by the activity of the synthesising enzyme choline acetyltransferase (ChAT) and histochemical activity of the degredative enzyme acetylcholinesterase, is more extensively reduced in DLB than AD in the frontal, temporal, and parietal cortex [17]. In contrast, in AD losses in ChAT are more extensive in the hippocampus and entorhinal cortex. Other parameters of cholinergic transmission, including reduced [^{3}H] nicotine binding and elevated M1 muscarinic cholinergic receptor subtype, similarly differentiate these dementias in a manner consistent with the putative pattern of cognitive deficits [9].

Clinical Symptoms

Data pertaining to clinical symptoms have been collected from a number of sources, including small case series reported between the 1960s and the 1980s, larger postmortem case series [8, 12, 23–25], clinical series [14–16] and the ongoing prospective Newcastle study. In total, this includes information from 252 cases.

Most groups have identified visual hallucinations as a common symptom in patients with DLB. Sixteen per cent of patients from smaller neuropathological case series and 13%–80% of patients from larger neuropathological series experienced these symptoms. This included 63% of patients in contact with psychiatric services and 27% of patients in neurology settings. The two clinical studies which used a standardised instrument for the identification of visual hallucinations both found them to occur in more than 80% of patients. Perhaps most importantly, all seven studies that have compared patients with DLB to patients with AD found visual hallucinations to occur significantly more frequently in DLB.

The Newcastle study has provided some phenomenological information about visual hallucinations. The characteristics appear to be similar in AD and DLB patients. Complete, detailed images of normal size are seen which tend to move. The most common themes are people, animals, children and objects. The only significant difference is that the visual hallucinations are significantly more likely to 'talk' in DLB patients. There is also a great deal of similarity between these hallucinations and the visual hallucinations experienced by patients suffering from a wide variety of conditions, including PD [26] and the Charles Bonnet syndrome [27].

There are now three studies that provide prospective information regarding the course of visual hallucinations. McShane et al. [28] reported that patients with cortical LBs at postmortem experienced visual hallucinations at significantly more of their clinical follow-up assessments. Ballard et al. [29] found that patients with an operationalised diagnosis of DLB were significantly more likely to continue to have visual hallucinations during 1 year of follow-up than patients with AD. This finding is mirrored by data from the ongoing Newcastle study. Patients with DLB also experience visual hallucinations significantly more frequently and are more likely to be distressed by them [16].

The propensity to hallucinate may be exacerbated by serotonin. Hallucinating DLB subjects have increased 5-hydroxyindole acetic acid:5-HT molar ratios, indicating increased turnover of 5-HT in the frontal cortex [9]. The ratio of serotonergic to cholinergic activity is also raised in the frontal and temporal cortices of hallucinators [9]. Furthermore, higher 5-HT$_2$ receptor binding has been identified in the temporal cortex of patients with hallucinations [30].

Delusions occurred in 68% of patients from psychiatric series and 39% from neurology series, with prevalence rates varying from 20% to 80%. Auditory hallucinations also occurred commonly, with prevalence rates varying from 10% to 57%. Although most groups found both symptoms to occur more frequently in DLB patients than those with AD, the difference was only significant in three out of seven series. In addition, many of the delusions and auditory hallucinations experienced by the DLB patients were clearly secondary to their visual hallucinations. Although it is important to be aware of these symptoms as they occur commonly in DLB patients and may cause distress, they probably are not helpful in discriminating between patients with DLB and those with other dementias.

Significant depression was identified in 14% of patients from small neuropathological series, 30%–50% of patients from large neuropathological series and 12%–33% of patients in clinical cohorts, with a median of 15%–20%. A much higher rate of 50% was reported by Klakta et al. [25]. The latter series reported the prevalence of depression during the whole course of the illness, although this does not explain the size of the discrepancy. Three out of eight series have found depression to occur significantly more commonly in DLB patients than in those with AD.

There is insufficient evidence to suggest that depression is a helpful diagnostic discriminator. It is, however, clear that it occurs commonly in patients with DLB, and more information is needed about the effects, associations and outcome of depression in these individuals. Preliminary data from the Newcastle study suggest a significant association with the severity of parkinsonism, which would make intuitive sense, given the high prevalence of depression amongst patients with PD.

Frequent falls occurred in 14% of patients identified in small neuropathological series and 25%–50% of patients in large neuropathological series. Higher prevalence rates of falls were described in clinical studies, although this might be somewhat tautological, as falls are included within the diagnostic criteria. In the current Newcastle study 50% of DLB patients had 2 or more falls in the preceding year and 10% had more than 5 falls. Only two out of six studies making direct comparisons found falls to occur significantly more often in patients with DLB compared to those with AD. They are therefore not a good discriminator between the two dementias, but are nevertheless an important cause of morbidity and mortality.

Significant parkinsonism occurred in all 15 of the patients with DLB reported in the Byrne's et al. [8] series, but only 9 had these symptoms at presentation. The other patients took a mean of 4 years to develop parkinsonism. Forty-five to 70% of patients within large neuropathological series had experienced these symptoms. Again, the prevalence rates are higher in clinical series, although this is tautological given the diagnostic criteria. Most importantly, all 7 series comparing the two dementias found parkinsonism to occur significantly more often in patients with DLB.

Several of the studies report the commonly occurring parkinsonian symptoms [8, 12, 24]. These include tremor, rigidity, bradykinesia, diminishing of facial expressions, shuffling gait and impaired fine motor performance. Tremor is common, but it is very rare for it to be pill rolling in nature. The severity of these symptoms tends to be mild [24], but it is unknown whether these symptoms become worse over time. Significant parkinsonism would appear to be a good diagnostic discriminator between the two dementias.

Functional activity of the basal ganglia is the sum of interactions between several different neuronal tracts, where perturbations in one may be responsible for behavioural dysfunction. Dopamine release in the striatum is controlled in part by the cholinergic nicotinic receptor. These receptors are found in the substantia nigra pars compacta in a coincident distribution with the pigmented dopaminergic neurones and also in the dorsolateral tegmental area and the raphe nucleus. Autoradiographic [^{3}H] nicotine binding has been found to be reduced in these areas in DLB to an extent similar to that seen in PD. Since the loss of substantia nigra neurones in DLB is much less than in PD, it may be that this reduction in nicotine binding precedes degeneration of the nigrostriatal dopaminergic neurones [9]. A reciprocal relationship between dopamine and acetylcholine exists in that activation of dopamine D1 receptors at the glutamatergic corticostriatal synapses stimulates release of acetylcholine, while D2 receptors suppress the release of acetylcholine. ChAT tends to be reduced rather than elevated in the putamen in DLB, in contrast to PD [31], and this reduction in cholinergic activity is one factor mitigating against the emergence of severe extrapyramidal signs in this disease.

Fluctuation is identified as a core clinical symptom by most groups. Byrne et al. [8] found 80% of all patients to fluctuate, whereas 90% of the patients in the series by McKeith et al. [12, 23] were thought to fluctuate significantly. Although most authorities agree that fluctuation is the result of disturbances of consciousness, only 25% of the patients who fluctuated in the McKeith et al. [12, 23] series were identified as having impairment of their conscious level. Both unexplained fluctuation and disturbances of consciousness are probably good discriminators between DLB and AD, but there is clearly a need to improve the definitions of these symptoms and to improve their identification. This could be achieved by using more tangible parameters such as fluctuation in performance in activities of daily living, serial cognitive assessment and measurement of related cognitive parameters such as impairment of attention.

Recent data from Newcastle showed extensive reductions in acetylcholinesterase and ChAT activity in the cingulate cortex in DLB cases. The significance of this is uncertain, but it may underly the occurrence of attentional dysfunction [32, 33].

McKeith et al. [34] also suggested that neuroleptic sensitivity was a key feature of DLB. Seven out of 16 (54%) patients in their series taking neuroleptics experienced a severe neuroleptic sensitivity reaction. This was characterised by marked parkinsonism, deterioration in cognitive function, drowsiness and often death. Two of the patients had characteristic features of the neuroleptic malignant syndrome. All of the patients died within 4 months of neuroleptic exposure and 4 died within 1 month. Neuroleptic sensitivity was not related to the dose of the neuroleptic. Data from the ongoing Newcastle study have identified 6 out of 18 patients exposed to neuroleptics to have an excessive reaction. The symptoms were similar to those described in the earlier series, but milder and more likely to resolve with only 1 out of 6 patients dying. The mean doses of neuroleptics were much lower in the more recent series (40 mg chlorpromazine equivalents). Two of the 6 patients were taking risperidone. The concept of neuroleptic sensitivity has been criticised because it has not been clearly operationalised. This will however require standardised data from patients with other dementias exposed to neuroleptics in order to define more clearly what is a severe and unexpected neuroleptic response.

Previous analyses suggested moderate losses of dopamine and homovanillic acid in DLB, which are less severe than those seen in PD [35]. In PD symptomatic parkinsonism occurs with substantia nigra neurone losses of more than 60%. In patients with DLB there is a 60% loss of substantia nigra neurones in the neuroleptic-treated cases (both tolerant and sensitive) compared to the neuroleptic-naive cases with a mean loss of 35% [36]. These reductions in dopaminergic nigrostriatal neurones were also examined by changes in autoradiographic [^{3}H] mazindol binding to the presynaptic dopamine uptake sites. Mazindol binding was reduced in PD in both the caudate and putamen compared to controls and this reduction correlated with nigral cell loss. Mazindol binding was not significantly altered in DLB groups [36]. Dopamine D1 receptors were also unchanged [37], although D2 receptors showed differential changes, with significant elevations in the caudate and putamen of patients with PD and a tendency to lower levels in neuroleptic naive patients with DLB [36]. The neuroleptic-tolerant DLB patients had raised D2 receptors relative to neuroleptic-naive cases. The neuroleptic-sensitive cases did not have elevated D2 receptors, a finding which would have been expected, given the reduced nigrostriatal input and the administration of neuroleptic medication [36]. These considerations suggest that failure of upregulation of D2 receptors in neuroleptic-sensitive DLB is a fundamental defect predisposing to total dopaminergic blockade with typical neuroleptic drugs. Neuroleptic treatment, in both sensitive and tolerant groups, was associated with reduced dopamine and homovanillic acid concentrations in the caudate and putamen compared to neuroleptic-naive cases.

A number of studies have compared the profile of cognitive function between patients with DLB and those with AD. McKeith et al. [12] found that DLB patients were less impaired than AD patients at the time of presentation on the Mental Test Score. This is probably a reflection of the complex non-cognitive symptoms experienced by these patients, which expedites referral to specialist services. Förstl et al. [38] reported eight cases who clinically met the criteria for NINCDS ADRDA probable AD but were found to have LBs at neuropathological examination. They did not differ in their cognitive profile from patients with AD. Hansen et al. [10] reported nine patients with AD and nine patients with cortical LBs in addition to Alzheimer's-type pathology. The patients with cortical LBs were more impaired in attention and abstract thinking tasks from the WAIS R battery. Galasko et al. [24] reported a series that included six patients with cortical LBs but no Alzheimer's type pathology. These patients had a very "sub-cortical" profile of abnormalities with slower processing speed, as well as impairment of attention and planning. Sahgal et al. [39] reported 20 clinically diagnosed patients with DLB and compared them to patients with a diagnosis of AD. The DLB patients had more impaired attention and planning. Ballard et al. [40] compared 12 patients with a clinical diagnosis of DLB to patients with AD. Patients with DLB had better memory function but more impaired verbal fluency. Patients with DLB also experienced a significantly greater decline over 1 year, deteriorating 27 points on the Cambridge Assessment for Mental Disorders in the Elderly (CAMCOG) schedule compared to a deterioration of 13 points in the patients with AD. Although these data are provisional and based upon small samples, there are a number of common themes. Most impressively, there appears to be a consistent trend for DLB patients to perform less well in tests traditionally associated with the frontal lobes and sub-cortical regions, particularly tasks of attention. It is also possible that DLB patients experience more rapid cognitive decline.

Genetics of DLB

The main genetic factor influencing late-onset AD is the apolipoprotein E gene (Apo E) on chromosome 19. The ε4 allele of Apo E is associated with a fourfold increased risk of developing AD [41, 42]. The ε2 allele has been suggested to be protective in AD with ε2 positive individuals living longer and developing AD at a later age [43]. A biological basis for this effect has been suggested on the basis of interactions between Apo E and the major proteins involved in AD, including β-amyloid, cerebrovascular amyloid and tau [44].

Very few studies have been directed towards genetic determinants in DLB. Because of the association of the apolipoprotein E ε4 allele (Apo E ε4) with AD and the presence of β-amyloid in DLB, several groups have reported genotyping studies. In DLB [45, 46], the ε4 allele frequency is similar to that found in AD. In contrast, in PD [47] no apparent association is observed with Apo E ε4.

Our own studies on DLB in Newcastle have demonstrated that the Apo E ε4 allele shows a significant increase compared to the control population, but unlike late-onset AD, no reduction in the ε2 allele frequency is observed [46]. Calculated odds ratios suggest that the ε4 allele is a significant risk factor for the development of DLB (OR = 2.50; 95% CI = 1.36–4.58), although the ε2 allele does not show a protective effect (OR = 1.18; 95% CI = 0.45–3.08).

To date, the only other published investigations into the genetics of DLB are concerned with polymorphism at the cytochrome P450 IID6 (CYP2D6) locus [48]. This enzyme is involved in the metabolism of endogenous compounds and xenobiotics and is associated with the poor-metaboliser phenotype for the drug debrisoquine. This phenotype is associated with poor metabolism of several clinically relevant drugs and because of this reduced metabolism, individuals with this phenotype consequently show adverse reaction to these drugs [49].

The poor metaboliser phenotype is caused by mutation of the P450 IID6 gene, with the result that there is aberrant splicing of the pre-mRNA, causing reduced or absent expression of the protein in the liver and other tissues [50]. Several deletions or point mutations have been described that give rise to the same phenotype. These alterations in the CYP2D6 gene are readily identified by PCR methods and are in common usage. Genotypes associated with the poor metaboliser phenotype have been shown to be twice as common in PD as in the control population [51]. In DLB as with PD, there appears to be an increase in the frequency of the poor metaboliser phenotype, which is not apparent amongst patients with AD [48]. This would suggest that in some individuals with DLB there are two genetic influences, Apo E ε4 and CYP2D6 acting in tandem to produce a clinical and pathological phenotype of DLB.

Whether there are other genes that could influence the risk of developing DLB is unknown at present, though the description of disease-associated polymorphisms in the α-1 antichymotrypsin gene [52] and the presenilin 1 gene [53] in AD suggests possible targets for future research.

Treatment Issues

When considering treatment there are several main areas of importance. Treating the main dementia syndrome, treating impairment of consciousness, treating secondary psychotic symptoms, particularly visual hallucinations, treating parkinsonism and limiting the occurrence of neuroleptic sensitivity reactions.

Although there are global deficits of neurotransmitter systems, perhaps the most striking abnormality is the profound deficit of acetylcholine in the cortical and sub-cortical regions amongst DLB patients. It has been hypothesised that cholinergic centres in the reticular activating system may be important in controlling consciousness. It is well established in patients with AD that cholinergic deficits correlate closely with the overall severity of cognitive impairment. There are therefore good reasons to suppose that interventions able to replenish cholinergic function to some degree might improve both the attention and overall cognitive function. Cholinesterase inhibitors have been found to bring about small, but significant improvements in cognitive function amongst patients suffering from AD [54]. It has been suggested that the benefits may be greater in DLB patients. Levy et al. [55] reported that two of the patients responding well to the cholinesterase inhibitor THA in their trial had cortical LBs at subsequent postmortem examination. Trials of cholinesterase inhibitors and other similar drugs are probably merited in DLB patients under carefully controlled conditions.

Visual hallucinations result in marked distress amongst DLB patients. Neuroleptic drugs, which are the mainstay of antipsychotic treatment in other patient groups, may potentially cause extreme sensitivity reactions. This poses a considerable treatment dilemma. Neuroleptic agents only improve psychotic symptoms in 40% of patients with AD [56] and are hence of only limited efficacy even if they could be prescribed at effective doses. Clozapine has been used to great effect to treat psychosis amongst patients with PD and has been well tolerated in these patients [57]. There is probably a case to merit a carefully monitored open trial of clozapine for the treatment of visual hallucinations in patients with DLB. A single case report has appeared in the literature describing the effectiveness of risperidone in alleviating psychotic symptoms in DLB patients [58]. Whilst this report is encouraging, a single case report cannot be accepted as irrefutable evidence of effectiveness. Although risperidone's neuropharmacological profile may be slightly less likely to induce extra pyramidal side effects, severe neuroleptic-sensitivity reactions have been reported by McKeith et al. [59].

In the longer term there is a need to develop strategies based upon the associated neurochemical deficits. Perry et al. [9] demonstrated a strong association between cholinergic deficits with a relative preservation of 5 hydroxy-tryptamine and the occurrence of visual hallucinations. There is in addition one clinical trial of Physostigmine in three patients with AD suffering from visual hallucinations [60]. In each of these patients physostigmine was equally effective to haloperidol in reducing visual hallucinations. Cholinesterase inhibitors may offer a treatment alternative.

Perhaps the most important point to emphasise, given the sensitivity of these patients to neuropharmacological agents, is that drugs should only be prescribed for psychosis or "behavioural problems" when absolutely essential and that any treatment should only be undertaken in conditions of stringent monitoring.

Parkinsonism is usually relatively mild in these patients, but it is unclear whether it progresses. Therefore, if anti-parkinsonian treatment is given, it should be reviewed periodically, as it may not continue to be necessary. Byrne et al. [8] suggested that 10 out of 11 patients given l-dopa showed some responsiveness without any adverse side-effects. Single case reports are less clear about the benefits of l-dopa prescription. The ongoing Newcastle study has found an association between l-dopa prescription and depression, although no association was found between the occurrence of visual hallucinations or other psychotic symptoms and taking l-dopa. As there is provisional evidence that l-dopa may improve motor symptoms in some of these patients, it is probably worth considering as a treatment option if the severity of parkinsonism is sufficient to cause additional disability to the sufferer. Again, it should be prescribed under carefully monitored and the agent should be withdrawn if side-effects ensue. It is also important to obtain detailed measurement of parkinsonian symptoms prior to treatment intervention so that l-dopa is only continued if there is clear evidence of responsiveness.

References

1. Forno LS (1969) Concentric hyalin intraneuronal inclusions of Lewy type in the brains of elderly persons (950 incidental cases). Relationship to parkinsonism. J Am Geriatr Soc 17: 557-575
2. Lowe JS, Mayer RJ, London M (1996) Pathological significance of Lewy bodies in dementia. In: McKeith EG, Perry EK (eds) Dementia with Lewy bodies. Cambridge University Press, Cambridge, pp 195-203
3. Woodard JS (1962) Concentric hyaline inclusion body formation in mental disease. Analysing 27 cases. J Neuropathol Exp Neurol 21: 442-449

4. Yoshimura M (1983) Cortical changes in the parkinsonian brain: a contribution to the delineation of diffuse Lewy body disease. J Neurol 229: 17-32

5. Kosaka K (1978) Lewy bodies in cerebral cortex: report of 3 cases. Acta Neuropathol (Berl) 42: 127-134

6. Gibb WRG, Esiri MM, Lees AJ (1987) Clinical and pathological features of diffuse Lewy body disease (Lewy body dementia). Brain 100: 1131-1153

7. Burns A, Jacoby R, Levy R (1990) Psychiatric phenomena in Alzheimer's disease. Br J Psychiatry 157: 72-94

8. Byrne EJ, Lennox G, Lowe J, Goodwin-Austen RB (1989) Diffuse Lewy body disease, clinical features in 15 cases. J Neurol Neurosurg Psychiatry 52: 709-717

9. Perry RH, Irving D, Blessed G, Fairbairn A, Perry EK (1990) Senile dementia of the Lewy body type: a clinically and neuropathologically distinct form of Lewy body dementia in the elderly. J Neurol Sci 95: 119-139

10. Hansen LA, Salmon D, Galasko D, Masliah E, Katzman R, DeTeresa R, Thal L, Pay MM, Hofstetter R, Klauber M (1990) The Lewy body variant of Alzheimer's disease: a clinical and pathological entity. Neurology 40: 1-7

11. Crystal HA, Dickson DW, Lizadi HE, et al (1990) Antemortem diagnosis of diffuse Lewy body disease. Neurology 40: 1523-1528

12. McKeith IG, Perry RH, Fairbairn AF, Jabeen S, Perry EK (1992) Operational criteria for senile dementia of Lewy body type (SDLT). Psychol Med 22: 911-922

13. Byrne EJ, Lennox G, Goodwin-Austen LB (1991) Dementia associated with cortical Lewy bodies: proposed diagnostic criteria. Dementia 2: 283-284

14. Ballard CG, Mohan RNC, Patel A, Bannister C (1993) Idiopathic clouding of consciousness – do the patients have cortical Lewy body disease? Int J Geriatr Psychiatry 8: 571-576

15. Shergill S, Mullen E, D'Ath P, Katrona C (1994) What is the clinical prevalence of Lewy body dementia. Int J Geriatr Psychiatry 9: 907-912

16. Ballard CG, Saad K, Gahir M, Solis M, Coope B, Wilcock G (1995) The prevalence and phenomenology of psychotic symptoms in dementia sufferers. Int J Geriatr Psychiatry 10: 477-485

17. Dickson DW, Ruan D, Crystal H, Mark MH, Davies P, Kress Y, Yen SH (1991) Hippocampal degeneration differentiates diffuse Lewy body disease (DLBD) from Alzheimer's disease: light and electron microscopic immunocytochemistry of CA2-3 neurites specific to DLBD. Neurology 41: 1402-1409

18. Bancher C, Lassman H, Budka H, Jellinger K, Grundke-Iqbal I, Iqbal K, Wiche G, Seitelberger F, Wisniewski HM (1989) An antigenic profile of Lewy bodies: immunocytochemical indication for protein phosphorylation and ubiquitination. J Neuropathol Exp Neurol 48:81-93

19 Pollanen MS, Dickson DW, Bergeron C (1993) Pathology and biology of the Lewy body. J Neurophatol Exp Neurol 52: 183-191

20. Masliah E, Mallory M, DeTeresa R, Alford M, Hansen L (1993) Differing patterns of aberrant neuronal sprouting in Alzheimer's disease with and without Lewy bodies. Brain Res 617: 258-266

21. Ince PG, Irving D, MacArthur F, Perry RH (1991) Quantitative neuropathological study of Alzheimer-type pathology in the hippocampus: comparison of senile dementia of Alzheimer type, senile dementia of Lewy body type, Parkinson's disease and non-demented elderly control patients. J Neurol Sci 106: 142-152

22. Wu E, Lipton RG, Dickson DW (1992) Amyloid angiopathy in diffuse Lewy body disease. Neurology 42: 2131-2135

23. McKeith IG, Fairbairn AF, Bothwall RA, Moore PB, Ferrier IN, Thompson P, Perry RH

(1994) An evaluation of the predictor validity and inter-rater reliability of clinical diagnostic Criteria for SDLT. Neurology 44: 872-877

24. Galasko D, Katzman R, Salmon DR, Thal LJ, Hanson L (1996) Clinical and neuropathological findings in Lewy body dementias. Brain Cogn 31: 106-175

25. Klakta LA, Louis ED, Schiffer RB (1996) Psychiatric factors in diffuse Lewy body disease. Findings in 28 pathologically diagnosed cases. Neurology 47: 1148-1152

26. Cummings JL (1991) Behavioural complications of drug treatment of Parkinson's disease. J Am Geriatr Soc 39: 708-716

27. Howard R, Levy R (1994) Charles-Bonnet syndrome plus complex visual hallucinations of Charles Bonnet type in late paraphrenia. Int J Geriatr Psychiatry 9: 399-404

28. McShane R, Godling K, Reading M, McDonald B, Esiri MM, Hope T (1995) Prospective study of relations between cortical lewy bodies, poor eyesight and hallucinations in Alzheimer's disease. J Neurol Neurosurg Psychiatry 59: 185-188

29. Ballard CG, O'Brien J, Coope B, Fairbairn A, Abid F, Wilcock G (1997) A prospective study of psychotic symptoms in dementia sufferers. Int Psychogeriatr 9: 57-64

30. Cheng A, Ferrier IN, Morris CM, Jabeen S, Sahgal A, McKeith IG, Edwardson JA, Perry EK, Perry RH (1991) Cortical serotonin-S2 receptor binding in Lewy body dementia, Alzheimer's and Parkinson's disease. J Neurol Sci 106: 50-55

31. Langlais PJ, Thal L, Hansen L, Galasko D, Alford M, Masliah E (1993) Neurotransmitters in basal ganglia and cortex of Alzheimer's disease with and without Lewy bodies. Neurology 43: 1927-1934

32. Cohen RA, Kaplan RF, Meadows ME, Wilkinson H (1994) Habituation and sensitization of the orienting response following bilateral anterior cingulotomy. Neuropsychologia 32: 609-617

33. Degos JD, da Fonseca N, Gray F, Cesaro P (1993) Severe frontal syndrome associated with infarcts of the left anterior cingulate gyrus and the head of the right caudate nucleus. A clinico-pathological case. Brain 116: 1541-1548

34. McKeith IG, Fairbairn AF, Perry R, Thompson P, Perry EK (1992b) Neuroleptic sensitivity in patients with senile dementia of Lewy body type. BMJ 305: 653-678

35. Marshall EF, Perry EK, Perry RH, McKeith IG, Fairbairn AF, Thompson P (1993) Dopamine metabolism in postmortem caudate nucleus in neurdegenerative disorders. Neurosci Res Commun 14: 17-25

36. Piggott MA, Perry EK, Marshall EF, McKeith IG, Johnson M, Melrose HL, Court JA, Lloyd S, Fairbairn A, Brown A, Thompson P, Perry RH (1998) Nigrostriatal dopaminergic activities in dementia with Lewy bodies in relation to neuroleptic sensitivity: comparisons with Parkinson's disease. Biol Psychiatry 44: 765-777

37. Piggott MA, Marshall EF (1996) Neurochemical correlates of pathological and iatrogenic extrapyramidal symptoms. In: McKeith E, Perry E (eds) Recent advances in dementia with Lewy bodies. Cambridge University Press, Cambridge, pp 449-467

38. Förstl H, Burns A, Luthert P, Cairns N, Levy R (1993) The Lewy body variant of Alzheimer's disease: clinical and pathological findings. Br J Psychol 162: 385-392

39. Sahgal A, Galloway PH, McKeith IG, Edwardson JA, Lloyd S (1992) A comparative study of attentional deficits in senile dementia of alzheimer's and Lewy body types. Dementia 3: 350-354

40. Ballard C, Patel A, Oyebode F, Wilcock G (1996) Cognitive decline in patients with Alzheimer's disease, vascular dementia and Senile dementia of Lewy body type. Age Ageing 25: 209-213

41. Corder EH, Saunders AM, Strittmatter WJ, Schmechel DE, Gaskell PC, Small GW, Roses AD, Haines JL, Pericak-Vance MA (1993) Gene dose of apolipoprotein E type 4 allele and the risk of Alzheimer's disease in late onset families. Science 261: 921-923

42. Martins RN, Clarnette R, Fisher C, Broe GA, Brooks WS, Montgomery P, Gandy SE (1995) Apo E genotypes in Australia: roles in early and late onset Alzheimer's disease and Down's syndrome. Neuroreport 6: 1513-1516

43. Corder EH, Saunders AM, Strittmatter WJ, Schmechel DE, Gaskell PC, Rinnuler JB, Locke PA, Conneally PM, Schmader KE, Small GW, Roses AD, Haines JL, Pericak-Vance MA (1994) Protective effect of apolipoprotein E type 2 allele for late onset Alzheimer's disease. Nat Genet 7: 180-184

44. Castaño EM, Prelli F, Wisniewski T, Golabek A, Kumar RA, Soto C, Frangione B (1995) Fibrillogenesis in Alzheimer's disease of the amyloid β peptides and apolipoprotein E. Biochem J 306: 599-604

45. Harrington CR, Louwagie J, Rossau R, Vanmechelen E, Perry RH, Perry EK, Xuereb JH, Roth M, Wischik CM (1994) Influence of apolipoprotein E genotype on senile dementia of the Alzheimer and Lewy body types. Significance for etiological theories of Alzheimer's disease. Am J Pathol 145: 1472-1484

46. Morris CM, Massey HM, Benjamin R, Leake A, Broadbent C, Griffiths M, Lamb H, Brown A, Ince PG, Tyrer S, Thompson P, McKeith IG, Edwardson JA, Perry RH, Perry EK (1996) Molecular biology of APO E alleles in Alzheimer's and non-Alzheimer's dementias. J Neural Transm Suppl 47: 205-218

47. Koller WC, Glatt SL, Hubble JP, Paolo A, Tröster AI, Handler MS, Horvat, RT, Martin C, Schmidt K, Karst A, Wijsman EM, Yu C-E, Schellenberg GD (1995) Apolipoprotein E genotypes in Parkinson's disease with and without dementia. Ann Neurol 37: 242-245

48. Saitoh T, Xia Y, Chen X, Masliah E, Galasko D, Shults C, Thal LJ, Hansen LA, Katzman R (1995) The CYP2D6B allele is overrepresented in the Lewy body variant of Alzheimer's disease. Ann Neurol 37: 110-112

49. Brøsen K, Gram LF (1989) Clinical significance of the sparteine/debrisoquine oxidation polymorphism. Eur J Clin Pharmacol 36: 537-547

50. Gonzalez FJ, Skoda RC, Kimura S, Demierre A (1988) Characterization of the common genetic defect in humans deficient in debrisoquine metabolism. Nature 331: 442-446

51. Kurth MC, Kurth JH (1993) Variant cytochrome P450 CYP2D6 allelic frequencies in Parkinson's disease. Am J Med Genet 48: 166-168

52. Kamboh MI, Shanghera DK, Ferrel RE, DeKosky ST (1995) APO E4-associated Alzheimer's disease risk modified by αI-antichymotrypsin polymorphism. Nature Genet 10: 486-488

53. Wragg M, Hutton M, Talbot C (1996) Genetic association between intronic polymorphism in presenilin-1 gene and late-onset Alzheimer's disease. Alzheimer's Disease Collaborative Group. Lancet 347: 509-512

54. Wilcock GK (1990) Tetrohydro amino acridine and Alzheimer's disease. BMJ 300: 939

55. Levy R, Eagger S, Griffiths M (1994) Lewy bodies and response to tacrine. Lancet 343: 176

56. Schneider LS, Pollock VE, Lyness SA (1990) A meta analysis of controlled trials of neuroleptic treatment in dementia. J Am Geriatr Soc 38: 553-563

57. Auzou P, Cote L, Fahn S (1995) Clozapine Parkinson's disease. Lancet 345: 516-517

58. Lee H, Cooney JM, Lewlor BA (1994) Case report: the use of risperidone, an atypical neuroleptin Lewy body disease. Int J Geriatr Psychiatry 9: 415-417

59. McKeith IG, Ballard CG, Harrison RWS (1995) Neuroleptic sensitivity to risperidone in Lewy body dementia. Lancet 346: 699

60. Cummings JL, Gorman DG, Shapira J (1993) Physostigmine ameliorates the delusions of Alzheimer's disease. Biol Psychiatry 33: 536-541

Frontotemporal Dementia

L. GUSTAFSON and A. BRUN

Introduction

The principles for classification of organic dementia are based on our present knowledge of its etiology, clinical picture, and pathological characteristics such as the type and predominant location of the brain damage. This paper concerns the clinical and pathological findings in dementia cases with primary degenerative changes within the frontal and temporal lobes and the possibility to differentiate this frontotemporal dementia (FTD) from Alzheimer's disease (AD) and other dementias. The importance of localized cortical atrophy within the frontal and temporal lobes in dementia was pointed out by Arnold Pick more than 100 years ago [1]. The neuropathological characteristics of this condition were given by Alzheimer [2], who described the ballooned cells and argentophilic inclusion bodies. The clinicopathological entity of this "Pick's disease" or "lobar atrophy" was delineated by Onari and Spatz [3] and Schneider [4], and further elaborated in a series of important contributions [5–9].

Pure Pick's disease is rare, reported with a prevalence of 1%–2% in post mortem studies of dementia compared to the prevalence of 40%–50% of AD [10, 11]. The overall frequency of Pick's disease has been estimated to be 24–60 per 100,000 in Minnesota, USA, and in Switzerland [12, 13]. A Pick to Alzheimer ratio of 1:3 was reported in Minnesota and 1:11 in Finland [14]. Longitudinal dementia studies in Lund (Sweden), Manchester (UK) and some other places have drawn attention to a larger group of FTD, clinically similar to Pick's disease but without the typical pathology of Pick's disease or AD [10, 15–20]. This disease has been named frontal lobe degeneration of non-Alzheimer type (FLD) [10, 21] and constitutes, together with Pick's disease and motor neuron disease (MND) with dementia, the major part of FTD. The clinical and pathological criteria for FTD have been described in a consensus statement by research groups in Lund and Manchester [22] and are presented in Appendix 1.

There is little information on the incidence and prevalence of FLD. In the Lund study 9% of 400 consecutive cases examined post mortem suffered from FTD, with FLD in 7.5% and Pick's disease in 1.5% [11]. The female to male ratio in 30 FLD cases was 1:1. The mean age at onset of FLD was 54 ± 8 years and in Pick's disease ($n=6$) 51 ± 11 years. The mean duration was 7.6 ± 3.6 (range 3–17) years in FLD and 9 ± 5.3 (range 4–17) years in Pick's disease. The Manchester study showed similar age characteristics, and Neary suggests that FTD may be responsible for as much as 20% of early-onset dementia. Knopman et al. [17] reported "dementia lacking distinctive histology" in 10% and Pick's disease in 6.8% of brain bank patients before 70 years of age. Constantinides et al. [8] found lobar atrophy without neuronal swellings and argyrophilic inclusions in one-third of a "typical Pick material." Pasquier et al. [23] have recently reported 49 FTD cases (according to the Lund-Manchester criteria) and 725 probable or possible AD [24] cases in 1,015 consecutive cases examined at the Memory Clinic in Lille. The prevalence of dementia in MND has been estimated to be 5% [25, 26], with a mainly presenile clinical onset at 53.5 years in men and 58.8 years in women [27].

Pathological Findings

Frontal Lobe Degeneration of Non-Alzheimer Type

Grossly there is mild-to-moderate atrophy of the frontal gyri and in a third of the cases also of the anterior temporal cortex, though in many cases there is no noticeable atrophy. The atrophy is never of the severity and circumscription seen in Pick's disease. On sectioning, slight widening of the frontal ventricular horns is often noticed. Limbic and central gray structures are grossly unremarkable, whereas in some cases the substantia nigra displays a mild depigmentation.

Microscopical examination reveals a consistent picture of a degenerative nature in the frontal or frontotemporal lobes, including the anterior cingulate gyral cortex, with postcentral parietal involvement only in advanced cases. The sensory and motor gyri stand out unchanged.

The microscopical changes consist of microvacuolation of lamina two and three with neuronal atrophy or loss and gliosis, which is most evident in lamina one [19]. These superficial laminae also show a 40%–50% reduction of synapse density [28]. Deeper cortical laminae usually appear largely unchanged. The superficial white matter reveals astrocytic gliosis which is mild to moderate and without prominent loss of myelin [29].

In less than one-third of the cases the striatum and amygdala show mild-to-moderate changes with neuronal shrinkage or loss, gliosis, and microvacuolation. The substantia nigra is the site of a moderate loss of neurons, though never of the severity associated with Parkinson's disease. There are no neuronal inclusions, no tau-positive structures and no signs of occurrence of prions, as shown with prion antibodies or protease-resistant prion protein analysis [30]. Plaques, tangles, or amyloid is not found except in a few older cases and then only to an extent compatible with age, and there is thus no overlap with AD. Lewy bodies have not been identified. A few cases with spinal anterior horn neuronal loss have been encountered, but without pyramidal tract demyelination. The pathological findings are summarized in Table 1.

Table 1. Pathological changes in FLD

- Grossly mild frontal atrophy
- Microscopical changes in frontal or frontotemporal cortex, including anterior cingulate gyrus; rare parietal involvement (advanced stages); cortical laminae 1–3 show: microvacuolation, neuronal loss or atrophy, gliosis, synapse loss
- Sometimes also in amygdala and striatum
- Superficial white-matter gliosis
- Moderate nigral neuronal loss
- No inclusions or prions, plaques, tangles or amyloid

Other Disorders of the Frontotemporal Dementia Group

Closely allied are the other disorders belonging to the FTD group defined in the Lund-Manchester consensus [22] besides FLD, i.e., Pick's disease, MND with dementia, and most likely some progressive aphasic disorders. The "Pick type" of degeneration [31] may be regarded as a variety of FLD.

The *"Pick type" of degeneration* involves the same areas as FLD, though more intense with a tendency to circumscription, involvement of the striatum and with more frequent asymmetry. Microscopically, the process involves all cortical layers with intense astrocytosis and some of the cases show Pick bodies and/or inflated neurons [31]. Also, the Constantinidis et al. [8] form of Pick's disease without inclusions and with a distribution of changes different from that in classical Pick's disease may fit in this category. The "Pick type" mentioned above may be regarded as a link to Pick's disease.

Pick's disease involves largely the same cortical areas as FLD, but more consistently the striatum and hippocampus. The process is also much more intense, causing a circumscription of involved cortical areas where all cortical laminae are degenerated, deserving the name of lobar atrophy. The microscopical picture is marked by Pick bodies and inflated neurons. This definition of Pick's disease has lately been endorsed by Verity and Wechsler [32] and Baldwin and Förstl [33] and seems at present to be the only tenable definition.

MND with dementia shows the same type and distribution of changes as FLD. In addition, there are ubiquitin-positive inclusions in cortical neurons and also hippocampal dentate gyrus neurons. Also, there are the histopathological changes seen in MND [27]. A comparison of main pathological features in the FTD group and AD is presented in Table 2.

In progressive non-fluent aphasia, in the frontal and anterior temporal cortex, especially on the left side, one finds degeneration of large pyramidal cells in lamina 3 and 5, microvacuolation of the superficial cortical laminae, and mild reactive astrocytosis, but no inclusions or ballooned cells [31].

Semantic dementia is marked by frontal and temporal cortical microvacuolation of superficial laminae, together with reactive astrocytosis, which is also found in the deeper cortical lamina although without neuronal loss. Adjoining white matter

Table 2. Comparison of pathology

Feature	FLD	Pick's disease	AD	MND dementia
Circumscribed frontal atrophy	–	+	–	–
Principal cortical topography	Frontotemporal	Frontotemporal	Parietotemporal	Frontal
Laminar involvement	I-III	I-VI	I-VI	I-III
Plaques, tangles and amyloid	–	–	+	–
Neuronal inclusions	–	+	–	+

shows astrocytic gliosis. Again, there are no neuronal inclusion bodies. Among subcortical nuclei only the amygdala reveals a loss of neurons and astrocytosis [31].

Other Similar Disorders of Differential Diagnostic Interest

Some other groups of disorders are also of differential diagnostic interest from a clinical point of view. They either show similar changes, similar localization, or just similar functional consequences but due to alterations of a different type and localization (Table 3).

Corticobasal degeneration involves areas of the frontal cortex but also parietal and most notably the sensory and motor cortex at variance with the FTD diseases. The microscopical picture includes the so-called chromatolytic cells seen also in Pick's disease and the Pick type. Besides the substantia nigra, which shows marked degeneration and the striatum, the basal ganglia and some brain stem nuclei are involved, adding to the features diverging from FLD [34].

The *chromosome 17-linked disorders* are relatively rare and mostly familial, with a frontotemporal distribution, but often also subcortical structures are engaged with an emphasis on the substantia nigra, adding a parkinsonian trait [35]. Of particular interest in this context is that in this spectrum of disorders one variety appears to be identical to FLD [36], which is why FLD in the future may prove to belong in this group. The familial variety of progressive subcortical gliosis also shows the same linkage. Here the deep cortical strata and superficial white matter show gliosis as well as the thalamus and inferior olivae.

If these disorders thus show divergent features that should at least in the fully expressed stage allow a clinical differential diagnosis against FTD disorders and especially FLD, other more common disorders may pose a problem in this respect, above all *AD* and *Creutzfeldt-Jakob disease with frontal emphasis*, which may at an early stage mimic clinically the frontal disorders [37]. However, both display their respective structural characteristics and on closer scrutiny reveal also non-frontal involvement, e.g. the posterior temporoparietal changes characteristic of AD. In a rare case, however, AD and FTD may coincide. Since they lack the traits of a pre-

Table 3. Organic dementias with frontal features

Frontotemporal dementia (FTD):
    Frontal lobe degeneration of non-Alzheimer type (FLD)
    (Pick-type degeneration)
    Pick's disease
    MND with dementia
    Progressive aphasia[a]
    Semantic dementia[a]

Dementia lacking distinctive histology
Rare familial forms (Kim)

Chromosome 3-linked disorder
Chromosome 17-linked disorders
    Corticobasal degeneration
    Familial progressive subcortical gliosis
    Rare familial syndromes

Dementias with occasional frontal accent
    Creutzfeldt-Jakob disease with frontal emphasis
    AD with frontal emphasis

Cerebrovascular dementias
    Binswanger's disease
    Selective incomplete white-matter infarction
    Strategic infarct dementia

Projected/secondary frontal dementia
    Huntington's disease
    Progressive supranuclear palsy

[a] Included in fronto-temporal lobar degeneration [31]

dominantly cortical frontal disorder, AD of the Lewy body type and Lewy body disease are not discussed here.

Cerebrovascular disorders may also mimic FTD disorders clinically. This is less obvious for multi-infarct dementia with its sudden and large infarcts, but the similarities with FLD are more intriguing with *Binswanger's disease*, which shows a clear-cut tendency to predominate in the frontal white matter. Here the lacunes, surrounded by large areas of partial destruction, add up in small steps creating the impression of a progressive degenerative disorder to be confused with FLD through undermining of frontal cortical areas [38].

Selective incomplete white-matter infarction shows the same picture save for the complete, lacunar infarcts [39]. Structurally, however, Binswanger's disease in the long run tends to involve also the basal ganglia and the brain stem and is more often marked by hypertensive changes not seen in selective white-matter infarction, which instead has a hypotensive background.

Strategic infarct dementia, a form of multi-infarct dementia with less extensive but strategically placed smaller infarcts such as bilaterally in the thalamus or in the supramarginal angular gyri, may be placed among disorders with a projected effect on the frontal cortex, which in itself need not be structurally involved.

Principally the same mechanisms are at play in *Huntington's disease* with striatal degenerative changes and in *progressive supranuclear palsy* (PSP) where the pathology resides in the brain stem. In Huntington's disease the pathology may, however, be more widespread and involve also the frontal cortex.

Etiology and Pathogenesis

In view of a 60% frequency of a similar disorder in a first-degree relative FLD most likely has a genetic etiology. This is further underscored by linkage to chromosome 17, found in a family with the same histological picture as in FLD [36]. On the same grounds, chromosome 3 may be a candidate gene since a form similar to FLD has been mapped to that chromosome [40]. There is also strong evidence of genetic factors with a positive family history reported in about 50% of patients with Pick's disease [5, 6, 11, 15, 41–44]. A linkage to chromosome 17 has been found in a family with a clinical picture of FTD, parkinsonism and amyotrophy [45]. The spectrum of clinical manifestations has been described in families with Pick's disease or FLD [46, 47]. Conflicting results exist concerning a relationship in FLD to chromosome 19 since an abnormal ApoE allele pattern has been found [48–50] and denied [51]. A prion etiology has been excluded in FLD, as noted above, and also in frontotemporal dementias linked to chromsome 17 [52].

From a pathogenetic point of view one might speculate that the loss of synapses in FLD is a primary feature, paralleling or leading to neuritic-dendritic degeneration and secondary atrophy-death of neurons, this process eliciting a reaction with gliosis and microvacuolation. Castellani et al. [53] have presented evidence that free radical damage and oxidative stress might play a role in the etiology of neurodegeneration in Pick's disease and corticobasal degeneration.

With the simple type of pathological changes and a known tendency for hypotension in FLD, anoxic-ischemic causes should also be discussed. They seem, however, to be ruled out by the topography of the lesions, including the tendency to spare structures sensitive to anoxia ischemia, and on the clinical side by the steadily and smoothly progressive character of the disease. The lack of amyloid in the microscopical picture excludes FLD from the amyloidogenic group.

Clinical Features

The following description of FTD is based on clinical observations in dementia cases with the diagnosis of FLD or Pick's disease confirmed post mortem. The clinical onset is insidious, usually starting in the presenium and seldom after the age of 70 years. In the early stage the disease is already characterized by changes and deterioration of personality and behavior, emotional features and a progressive reduction of expressive speech. Memory failure and lack of concentration are almost always found, although memory for daily events, spatial orientation, and practical abilities may remain even at a late stage of the disease. The Lund-Manchester consensus on core features (Appendix 1) might be useful as a checklist for clinical items, although at present not as a diagnostic rating scale with a defined scoring procedure.

Emotional and behavioral changes are always present. They usually develop slowly, and therefore it may be difficult to decide when the first symptoms of the dementing disease appear, and the duration may be difficult to estimate. Signs of disinhibition, lack of judgment, emotional unconcern and lack of insight into the present condition and its consequences are early manifestations. Impaired control of emotions is seen such as tearfulness, inadequate smiling and spells of crying and laughing. There may be an initial period of disinhibition with restlessness, irritability, unrestrained sexuality, but even more often changes in behavior and speech towards stereotypy, lack of spontaneity and in a later stage apathy, mutism, and amimia.

The emotional changes are sometimes difficult to differentiate from affective disorders. Elated mood, in combination with unrestrained talking, confabulation and signs of disinhibition may be misinterpreted as a hypomanic or manic state. Many FTD patients display episodes of depressed mood, with suicidal ideation. Slowly developing apathy, social withdrawal, hypomimia, and reduced speech may easily be misdiagnosed as a major depression and several of our FLD and Pick cases received antidepressant medication at an early stage of the disease. Although FTD patients usually deny any awareness of ill health, many of them complain of anxiety and somatic symptoms such as pain combined with bizarre hypochondriacal ideas. Most patients developing FTD are judged to have a fairly normal premorbid personality. However, nervousness and anxiety were sometimes reported [21]. Lebert et al. [54], in a study of 19 patients with FTD, found no relationship between the emotional features and premorbid personality traits, but rather a metabolic pattern on single-photon emission computerized tomography (SPECT). The correlation between SPECT findings and emotional symptoms in FTD [55] is probably not disease-specific, but is also found in other types of organic dementia [54, 56].

Deceptions are often reported in Pick's disease, FLD, and AD [21, 57, 58]. The association of psychotic symptoms with changes of personality, restlessness, and stereotyped behavior and speech in FTD sometimes leads to the diagnosis of schizophrenia. In our longitudinal study hallucinations and illusions were observed in about 20% of FLD and early-onset AD, and in 50% of the late-onset AD group. The psychotic symptoms in FLD are often bizarre, badly controlled and show important clinical variability. Sensory distortions, especially hyperesthesia, were found in about 30% of the FLD group and in 10% of the AD cases. The psychotic symptoms in FLD often give the impression of functional psychosis, while in AD such symptoms are more immediately related to dysgnosia and visuospatial dysfunction [59]. The association between psychosis and frontal lobe damage has also been described in vascular dementia and in AD with frontal lobe involvement [18, 37].

Restlessness and an irresistible impulse to explore the environment are often observed. This "utilization behavior" described in patients with various frontal lobe lesions [60] shows important similarities to the "hypermetamorphosis" and distractibility of the Klüver-Bucy syndrome. Other components of this syndrome in humans such as hyperorality, craving for affection and sexual contact, and bulimia are also prevalent in FLD and Pick's disease [21, 61]. Other serious consequences of disinhibition and lack of judgment in FTD are traffic accidents and changes in drinking behavior. The latter, which may also be considered a change in oral/dietary behavior, is sometimes misdiagnosed as alcoholism.

FTD is characterized by progressive reduction of expressive speech. "*Sprachverödung*" [4] or "*Dissolution du langage*" [62, 63] may to some extent be overshadowed by verbal stereotypy with a limited number of words and phrases. Echolalia is prevalent [6, 11], and receptive speech may be preserved even at a late stage, as is also the patient's handwriting. The late mutism, and amimia may, however, make communication extremely difficult. The symptom constellation of palilia (stereotyped phrases), echolalia, mutism, and amimia (PEMA syndrome of Guiraud) is typical of FTD. The handwriting in FTD may change in various ways, but these disturbances are different from the dysgraphia in AD [21]. The speech disorder in FTD has been described in a consistent way by various research groups [15, 17, 64]. There are important similarities between FTD at an early stage and the clinical spectrum of progressive aphasia as described by Mesulam [65] and Neary et al. [20]. The progressive aphasia is dominated by language disturbances and a relative preservation of memory and practical abilities. Ultimately, many of these cases seem to develop global dementia [66, 67]. Progressive aphasia and FTD may not represent two distinct disorders, but rather the clinical manifestations of the same histological process with different topographic distribution. The syndrome of semantic dementia is also accompanied by behavior changes, although less severe than those of FTD, and the brain pathology is mainly restricted to the middle and inferior temporal gyri [31]. Dynamic aphasia similar to that in FTD has been described in PSP [68].

Physical Signs

There are few pathological somatic findings in patients with FTD. Primitive reflexes may be found early while akinesia, rigidity, and tremor are a late phenomena. Epileptic seizures are rare, and most patients with FTD show normal or only slightly pathological EEG. When dementia of the early frontal lobe type combines with fasciculation, muscular wasting, dysarthria and dysphagia, this probably indicates MND, although such neurological features may sometimes appear in patients with Pick's disease and FLD [69, 70]. The frontal lobe symptoms in MND may appear early and even precede the neurological features. The nosological classification of MND with dementia is unsettled, although it seems justified to classify a majority as FTD. This is further supported by the high frequency of normal EEG and frontotemporal pathology on brain imaging in these cases [26, 71].

Surprisingly often, FTD patients have a low and labile blood pressure with a high prevalence of orthostatic blood pressure drops and syncopal attacks [21]. Low and labile blood pressure was found in more than 50% of unmedicated FTD cases, with a similar prevalence to that reported in AD and vascular dementia at a similar age [72]. Urinary incontinence, a late feature in uncomplicated AD, is reported early in about 50% of our FTD cases. This is probably due to the frontal lobe involvement since incontinence is also a common feature in vascular dementia and in AD with lesions in frontobasal cortex and the anterior cingulate gyrus [73, 74]. Endocrine functions are mainly normal in FTD, although thyroid hormone abnormalities are rather common in FTD (38%) compared to AD (9%) [75].

Investigations

EEG Findings

The EEG findings may be normal or only slightly pathological, and this might be found even at a comparatively late stage of FTD. This has been shown not only in FLD [15, 21, 76], but also in Pick's disease [77], MND dementia, and other dementias of the frontal lobe type [15, 17, 71]. By contrast, the EEG is almost always pathological in AD, even at an early stage. The differential diagnosis between FTD, AD and vascular dementia (VaD) can be improved by sleep studies [78] and quantitative EEG analysis [79].

Brain Imaging

Cortical atrophy with and without focal accentuation has been shown in Pick's disease and FLD with computed tomography (CT) and magnetic resonance imaging (MRI). The anterior-posterior gradient of atrophic changes may contribute to differentiating it from AD, although the findings are often non-conclusive with an important overlap between demented and normal subjects [43, 80, 81]. The recognition of vascular lesions in frontal gray and white matter and subcortical strategic infarcts is important since such lesions may cause dementia with frontal lobe characteristics [18, 37, 82]. The differential diagnosis between FTD and Huntington's disease may be improved if brain imaging shows degeneration of the caudate nuclei [83].

The techniques for measurement of regional cerebral blood flow (rCBF) and metabolism in the brain have strongly improved the differential diagnosis of dementia even at an early stage of the disease. The frontotemporal rCBF abnormalities in FLD and Pick's disease, shown with the xenon clearance technique, contrast with the temporoparietal pathology in AD [84–86]. The flow pattern in FTD is not disease-specific, but is also found in vascular brain damage, Creutzfeldt-Jakob disease and in AD with marked frontal lobe involvement [37]. Neary et al. [15], Jagust et al. [16] Kitamura et al. [8], Pasquier et al. [88] and Miller et al. [89], using SPECT, have consistently shown frontal hypoperfusion in FTD. This has also been shown with positron emission tomography (PET) in Pick's disease [90, 91], in PSP [92] and in MND [93]. SPECT studies of progressive aphasia often show asymmetric pathology with predominant left hemisphere involvement [20].

Neuropsychological Testing

Cognitive symptoms may appear early in FTD, although they are often difficult to evaluate because of the patient's behavioral changes and speech dysfunction. The early test profile is characterized by slow verbal production and relatively intact reasoning and memory, while intellectual and motoric speed is reduced [15, 43, 94]. By contrast, early AD is usually characterized by relatively intact verbal ability and simultaneous impairment of reasoning ability, verbal and spatial memory dysfunction, dysphasia and dyspraxia [94]. Frisoni et al. [95] reported poorer verbal fluency in FTD than in AD, while Pasquier, et al. [96] did not find significant differences between FTD and AD patients in verbal letter fluency test and verbal category tests. Receptive dysphasia with difficulties in understanding instructions, which may appear early in AD, are found only in a minority of FTD cases, mainly in Pick's disease [94]. Reading and writing may be comparatively preserved even at a late stage of FTD, when the patient has become completely mute. Misspelling and dyscalculia may sometimes be reported as an early symptom in FTD [21].

Reduced recent memory is a common finding in FTD, and remote memory is also disturbed, although to a less extent than in AD. Johanson and Hagberg [94] reported test results from 20 post mortem verified FTD cases, all of which showed memory disturbance, although verbal and spatial memory were within the normal range in 6 cases. These cases showed the best preserved hippocampal structures at autopsy. Systematic evaluation of behavior qualities such as cooperation, self-criticism, distractibility, flight reactions and strategy in the test situation strongly contributed to the differentiation from AD [94, 97]. Confabulation is rather frequent in FTD, especially in Pick's disease.

Correlations between rCBF and various measures of cognitive impairment in FTD have been calculated. Miller et al. [18] found a significant relationship between the Mini-Mental State Examination score (MMSE, [98]) and rCBF in posterior parietal regions, as measured with SPECT. Elfgren and co-workers have shown significant correlations between rCBF and the level of cognitive impairment in FTD and between verbal fluency and left frontal lateral, frontal medial and left temporal anterior inferior rCBF measured with SPECT [99, 100].

In our experience neuropsychological testing can be used for early recognition of FTD and for differentiation against AD and other dementias, normal aging and non-organic mental disease. For diagnostic purposes it is important to rely not only on quantitative measures, but also on systematic observation and recording of the patient's behavior and emotional reactions in the test situation [101, 102].

The characteristics of the progressive dynamic speech disorder in FTD with stereotypy, echolalia and late mutism have already been touched upon. Patients with progressive non-fluent aphasia also show reading and writing impairment while memory and visuospatial functions are comparatively spared [65, 67, 103]. Eventually most cases with progressive aphasia develop the cognitive profile of FTD in agreement with the distribution of brain pathology.

Neurochemistry

There are few systematic neurochemical studies on FTD. Nigrostriatal dopamine decrease has been reported in Pick's disease [104] and FLD [105], and reduction of serotonin receptors [105] and substance P levels in substantia nigra and frontal cortex has been found in Pick's disease [104]. Choline acetyltransferase activities comparable with controls rather than AD values have been reported in Pick's disease and in FLD [17, 106, 107] and in MND with dementia of the frontal lobe type [108]. Reduced cerebrospinal fluid (CSF) somatostatin levels have been found in FTD as well as in AD [109, 110]. Moreover, the delta sleep-inducing peptide (DSIP) was significantly reduced in AD, but not in FTD while the corticotropin-releasing factor (CRF) was significantly reduced in FTD but not in AD. The differences between AD and FTD in peptide expression might be related to differences in localization of the disease process or the sequence in which different neurons are affected [110]. One recent study reports increased levels of tau and PHF-tau in the CSF in AD but low levels in FTD [111]. Zinc has been suggested as an etiological factor in FTD [13], but the findings in this respect are inconsistent.

Differential Diagnosis

Most guidelines for diagnosis of dementia such as the NINCDS-ADRDA criteria [24] focus upon AD with criteria that easily may lead to the inclusion of FLD and Pick's disease. DSM-III-R [112] presents Pick's disease without further diagnostic guidelines, while DSM-IV [113] describes Pick's disease as "one of the pathologically distinct etiologies associated with frontotemporal brain atrophy." The ICD-10 [114] describes "dementia in Pick's disease" (F02.0) as a slowly progressive dementia commencing in middle life with predominance of frontal lobe features and selective atrophy of the frontal and temporal lobes, but without the pathological changes of AD. In the early stages memory and parietal lobe functions are relatively preserved.

The differential diagnosis of FTD, especially at an early stage, may be difficult, not only against other organic brain diseases with frontal features, but also against functional mental diseases with similar changes in personality, mood, and behavior. The clinical differences between FTD and AD are most clear at an early stage of the disease. The initial stage of FTD is dominated by emotional and personality changes and severe dyspraxia, memory failure and spatial disorientation developed comparatively late as a result of the relative sparing of the temporoparietal occipital cortical areas. By contrast, early-onset AD is characterized by memory failure, dyspraxia, dysgnosia, and spatial disorientation in agreement with a consistent pattern of cortical involvement. Early changes in personality with emotional unconcern, lack of insight, restlessness and other signs of disinhibition and hyperorality, utilization behavior, and stereotypy are more prevalent in FTD, while habitual personality traits, social awareness, and insight into own illness may remain even at a late stage of AD. Progressive dynamic aphasia with stereotypy of speech, echolalia, late mutism and amimia, the PEMA syndrome, is typical for FTD while global dysphasia, logoclonia, increased muscular tension, grand mal and myoclonia are more common in AD [21]. In a minority of AD cases (about 5%) frontal lobe features may appear early in addition to the temporoparietal symptom pattern. This is due to early and more severe involvement of the frontal lobes [37]. The normal EEG and the predominant frontotemporal pathology on brain imaging may also contribute to the differential diagnosis against AD where EEG is almost always pathological and the brain imaging pathology shows post central accentuation.

Vascular brain damage may cause frontotemporal cortical dysfunction by different mechanisms. The frontal lobe symptoms are usually related to lesions in the frontal gray and white matter, the anterior cingulate gyrus and thalamic nuclei with important connection to the anterior cortex [73, 115]. The vascular etiology is often indicated by a history of vascular risk factors and the clinical course of the dementing process. A frontal lobe clinic has been described in multi-infarct dementia with frontal emphasis and in patients with strategic infarcts such as bilateral thalamic infarcts [116]. Selective incomplete white-matter infarctions (SIWI) often show psychiatric symptoms of frontal lobe type and a diagnosis of FTD may be considered [37, 117]. Sometimes frontal vascular damage develops gradually without dramatic onset or fluctuations. This may be seen in Binswanger's disease with a clinical picture dominated by emotional lability with periods of euphoria, apathy, or depression while disinhibition is less prevalent [118]. The fact that about one third of all AD cases show an important SIWI in addition to the Alzheimer encephalopathy may contribute to diagnostic uncertainty.

A differential diagnosis between FTD and Huntington's disease may be critical when personality changes and psychotic features dominate and when the neurological characteristics of Huntington's disease are less obvious or appear late in the course. CT, MRI, SPECT and PET may, however, reveal the striatal involvement and genetic analysis the specific mutation of Huntington's disease. Dementia of the frontal and subcortical-frontal type is also found in Creutzfeldt-Jakob disease and in human immunodeficiency virus (HIV), which has to be considered in the diagnostic process. Patients with progressive supranuclear palsy may also present with a frontal lobe symptom pattern, probably secondary to deafferentation from subcortical structures [92]. The differential diagnosis should be based on the neurological findings in PSP, such as ocular palsy, dysarthria, bradykinesia, and rigidity. The progressive subcortical gliosis described by Neumann [119] and Neumann and Cohn [120] may also present a progressive dementia, starting with emotional and behavioral symptoms of the frontal lobe type. This rare disease has become increasingly important since linkage studies suggest a relation to chromosome 17 [121].

The clinical recognition of FTD is often possible by systematic analysis of the patient's history and clinical picture supported by neuropsychological testing, brain imaging and other diagnostic techniques. The Lund-Manchester consensus is recommended for this purpose. In our daily clinical work we use a combination of diagnostic rating scales: one for recognition of AD, one scale for diagnosis of FTD (FTD score) [116, 122, 123] and the Hachinski ischemic score [124]. Diagnoses based on the scoring profile in these three rating scales have been validated against rCBF findings and neuropathological diagnosis [116, 125].

Treatment and Care

FTD is heterogeneous from the etiological point of view, although genetic factors seem to be crucial in about 50% of FLD cases. Early diagnosis is a prerequisite for adequate treatment and care of the patient and for information and support to the family and other caregivers involved. There is at present no specific pharmacological treatment for the underlying degenerative process, but symptomatic treatment is often tried against mood changes, restlessness, aggressiveness, hallucinations, and delusions. A problem is, however, that patients with FTD are often sensitive to psychotropic medication and easily develop disturbing side effects and paradoxical reactions. One important consequence of the diagnostic process is differential diagnosis against early-onset AD, for which pharmacological treatment is now available. FTD patients are often restless, stereotype, and stimulus-bound. They show a strong need for physical activity that should be channelled rather than restricted, as should also the patient's comparatively preserved practical and spatial abilities. An individually structured program for daily activities may be rewarding and often reduces the need for pharmacological treatment. The daily activities should be carried out together with someone aware of the patient's difficulties to plan, initiate, and control emotions and behavior. The interaction with the FTD patient has to rely on the experience and strategies of the caregiver and on firm and repetitive verbal guidance of the patient. The long duration of the disease has a strong impact on the family that needs information, training, and support through all these critical years. When psychotic features and unpredictable aggressiveness prevail, special psychogeriatric services are needed. Special units or resources for diagnosis, treatment, and care should be established to handle the type of care problems related to frontal lobe dysfunction in dementia.

Appendix 1

Clinical diagnostic features of FTD (consensus statement by the research groups in Lund and Manchester [22])

Core diagnostic features

Behavioural disorder
- Insidious onset and slow progression
- Early loss of personal awareness (neglect of personal hygiene and grooming)
- Early loss of social awareness (lack of social tact, misdemeanours such as shoplifting)
- Early signs of disinhibition (such as unrestrained sexuality, violent behaviour, inappropriate jocularity, restless pacing)

- Mental rigidity and inflexibility
- Hyperorality (oral/dietary changes, overeating, food fads, excessive smoking and alcohol consumption, oral exploration of objects)
- Stereotyped and perseverative behaviour (wandering, mannerisms such as clapping, singing, dancing, ritualistic preoccupation such as hoarding, toileting and dressing)
- Utilization behaviour (unrestrained exploration/use of objects in the environment)
- Distractibility, impulsivity, and impersistence
- Early loss of insight into the fact that the altered condition is due to a pathological change of own mental state

Affective symptoms
- Depression, anxiety, excessive sentimentality, suicidal and fixed ideation, delusion (early and evanescent)
- Hypochondriasis, bizarre somatic preoccupation (early and evanescent)
- Emotional unconcern (emotional indifference and remoteness, lack of empathy and sympathy, apathy)
- Amimia (inertia, aspontaneity)

Speech disorder
- Progressive reduction of speech (aspontaneity and economy of utterance)
- Stereotypy of speech (repetition of limited repertoire of words, phrases or themes)
- Echolalia and perseveration
- Late mutism

Preserved spatial skills in praxis
(intact ability to negotiate the environment)

Physical signs
- Early primitive reflexes
- Early incontinence
- Late akinesia, rigidity and tremor
- Low and labile blood pressure

Investigations
- Normal EEG despite clinically evident dementia
- Brain imaging (structural and/or functional): predominant frontal and/or anterior temporal abnormality
- Neuropsychology: profound failure on "frontal lobe" tests in the absence of severe amnesia, aphasia or perceptual spatial disorder

Supportive diagnostic features

- Onset before 65

- Positive family history of similar disorder in a first-degree relative
- Bulbar palsy, muscular weakness and wasting, fasciculations (motor neuron disease)

Diagnostic exclusion features

- Abrupt onset with ictal events
- Head trauma related to onset
- Early severe amnesia
- Early spatial disorientation, lost in surroundings, defective localisation of objects
- Early severe apraxia
- Logoclonic speech with rapid loss of train of thought
- Myoclonus
- Cortical bulbar and spinal deficits
- Cerebellar ataxia
- Choreo-athetosis
- Early severe pathological EEG
- Brain imaging predominant post-central structural or functional deficit. Multifocal cerebral lesions on CT or MRI
- Laboratory tests indicating brain involvement or inflammatory disorder (multiple sclerosis, syphilis, AIDS and herpes simplex encephalitis)

Relative diagnostic exclusion features

- Typical history of chronic alcoholism
- Sustained hypertension
- History of vascular disease (such as angina and claudication)

References

1. Pick A (1892) Über die Beziehungen der senilen Hirnatrophie zur Aphasie. Prag Med Wochenschr 17: 165-167
2. Alzheimer A (1911) Über eigenartige Krankheitsfälle des späteren Alters. Z Ges Neurol Psychiatry 4: 356-385
3. Onari K, Spatz H (1926) Anatomische Beiträge zur Lehre von der Pickschen umschriebene Grosshirnrinden–Atrophie (Picksche Krankheit). Z Ges Neurol Psychiatry 101: 470-511
4. Schneider C (1927) Über Picksche Krankheit. Monatsschr Psychiatry Neurol 65: 230-275
5. Sjögren T, Sjögren H, Lindgren AGH (1952) Morbus Alzheimer and morbus Pick. A genetic, clinical and patho-anatomical study. Munksgaard, Copenhagen
6. Van Mansvelt (1954) Pick's disease. A syndrome of lobar, cerebral atrophy; its clinico-anatomical and histopathological types. Thesis. Enschede, Utrecht
7. Delay J, Brion S (1962) Les démences tardives. Masson, Paris
8. Constantinidis J, Richard J, Tissot R (1974) Pick's disease. Histological and clinical correlations. Eur Neurol 11: 208-217

9. Munoz-Garcia D, Ludwin SK (1984) Classic and generalized variants of Pick's disease: a clinicopathological, ultrastructural and immunocytochemical comparative study. Ann Neurol 16: 467-480

10. Brun A (1987) Frontal lobe degeneration of non-Alzheimer type. I. Neuropathology. Arch Gerontol Geriatr 6: 193-208

11. Gustafson L (1993) Clinical picture of frontal lobe degeneration of non-Alzheimer type. Dementia 4: 143-148

12. Heston LL, Mastri AR (1982) Age at onset of Pick's disease and Alzheimer's dementia: implications for diagnosis and research. J Gerontol 37: 422-424

13. Constantinidis J, Richard J, Tissot R (1985) Pick dementia: anatomoclinical correlations and pathophysiological considerations. Interdisc Top Gerontol 19: 72-97

14. Sulkava R, Haltia M, Paetau A, Wikström J, Palo J (1983) Accuracy of clinical diagnosis in primary degenerative dementia: correlation with neuropathological findings. J Neurol Neurosurg Psychiatry 46: 9-13

15. Neary D, Snowden JS, Northen B, Goulding PJ (1988) Dementia of frontal lobe type. J Neurol Neurosurg Psychiatry 51: 353-361

16. Jagust WJ, Reed BR, Seab JP, Kramer JH, Budinger TF (1989) Clinical-physiologic correlations of Alzheimer's disease and frontal lobe dementia. Am J Physiol Imaging 4: 89-96

17. Knopman DS, Mastri AR, Frey WH, Sung JH, Rustan T (1990) Dementia lacking distinctive histologic features: a common non-Alzheimer degenerative dementia. Neurology 40: 251-256

18. Miller BL, Cummings JL, Villanueva-Meyer J, Boone K, Mehringer CM, Lesser IM, Mena I (1991) Frontal lobe degeneration: clinical, neuropsychological and SPECT characteristics. Neurology 42: 1374-1382

19. Brun A (1993) Frontal lobe degeneration of non-Alzheimer type revisited. Dementia 4: 126-131

20. Neary D, Snowden JS, Mann DMA (1993) The clinical pathological correlates of lobar atrophy. Dementia 4: 154-159

21. Gustafson L (1987) Frontal lobe degeneration of non-Alzheimer type. II. Clinical picture and differential diagnosis. Arch Gerontol Geriatr 6: 209-233

22. Brun A, Englund B, Gustafson L, Passant U, Mann DMA, Neary D, Snowden JS (1994) Clinical and neuropathological criteria for frontotemporal dementia. J Neurol Neurosurg Psychiatry 57: 416-418

23. Pasquier F, Lebert F, Amouyel P (1995) Épidémiologie. In: Pasquier F, Lebert F (eds) Les démences fronto-temporales, épidémiologie. Masson, Paris, pp 23-29

24. McKhann G, Drachman D, Folstein M, Katzman R, Price D, Stadlan EM (1984) Clinical diagnosis of Alzheimer's disease: report of the NINCDS-ADRDA work group under the auspices of the Department of Health and Human Services Task Force on Alzheimer's Disease. Neurology 34: 939-944

25. Hudson AJ (1981) Amyotrophic lateral sclerosis and its association with dementia, Parkinsonism and other neurological disorder: a review. Brain 104: 217-247

26. Neary D, Snowden JS, Mann DMA (1990) Frontal lobe dementia and motoneuron disease. J Neurol Neurosurg Psychiatry 53: 23-32

27. Mitsuyama Y (1993) Presenile dementia with motoneuron disease. Dementia 4: 137-142

28. Liu X, Brun A (1996) Regional and laminar synaptic pathology in frontal lobe degeneration of non-Alzheimer type. Int J Geriatr Psychiatry II: 47-55

29. Englund E, Brun A (1987) Frontal lobe degeneration of non-Alzheimer type II. White matter changes. Arch Gerontol Geriatr 6: 235-243

30. Collinge J, Hardy J, Brown J, Brun A (1994) Familial Pick's disease and dementia in frontal lobe degeneration of non-Alzheimer type are not variants of prion disease. J Neurol Neurosurg Psychiatry 57: 762

31. Snowden JS, Neary D, Mann DMA (1996) Fronto-temporal lobar degeneration: fronto-temporal dementia, progressive aphasia, semantic dementia. Churchill Livingstone, New York

32. Verity A, Wechsler FV (1987) Progressive subcortical gliosis of Neumann: a clinico-pathologic study of two cases with review. Arch Gerontol Geriatr 6: 189-195

33. Baldwin B, Förstl H (1993) Pick's disease – 101 years on still there but in need of reform. Br J Psychiatry 163: 100-104

34. Thomson PD, Marsden CD (1992) Corticobasal degeneration. In: Rossor MN (ed) Baillière's clinical neurology. Unusual dementias. Baillière Tindall, London, pp 677-686

35. Wilhelmsen K, Lynch T, Pavlou G, Higgins M, Nygaard T (1994) Localization of disin-hibition-dementia-parkinsonism-amyotrophy complex to chromosome 17 q21-22. Am J Hum Genet 55: 1159-1165

36. Basun H, Almqvist O, Axelman K, Brun A, Campbell TA, Collinge J, Forsell C, Froelich S, Wahlund L-O, Wetterberg L, Lannfelt L (1997) Clinical characteristics of a chromo-some 17-linked rapidly progressive familial frontotemporal dementia. Arch Neurol 54: 539-544

37. Brun A, Gustafson L (1991) Psychopathology and frontal lobe involvement in organic dementia. In: Iqbal K, McLachlan DRC, Winbald B, Wisnewski HM (eds) Alzheimer's disease: basic mechanisms, diagnosis and therapeutic strategies. Wiley, London, pp 27-33

38. Brun A, Fredriksson K, Gustafson L (1992) Pure subcortical arteriosclerotic en-cephalopathy (Binswanger's disease). A clinicopathologic study II. Pathological fea-tures. Cerebrovasc Dis 2: 87-92

39. Brun A, Englund E (1986) A white matter disorder in dementia of the Alzheimer type. A pathoanatomical study. Ann Neurol 19: 253-262

40. Brown V, Asworth A, Gydesen S, Soranden A, Rossor M, Hardy D, Colinge J (1995) Familial non-specific dementia maps to chromsome 3. Hum Mol Genet 4: 1625-1628

41. Heston LL (1978) The clinical genetics of Pick's disease. Acta Psychiatr Scand 57: 202-206

42. Groen JJ, Endtz LJ (1982) Hereditary Pick's disease: second re-examination of a large family and discussion of other hereditary cases, with particular references to electroencephalography and computerized tomography. Brain 105: 443-459

43. Knopman DS, Christensen KJ, Schut LJ, Harbaugh RE, Reeder T, Ngo T, Frey W (1989) The spectrum of imaging and neuropsychological findings in Pick's disease. Neurolo-gy 39: 362-368.

44. Schmitt HP, Yang Y, Förstl H (1995) Frontal lobe degeneration of non-Alzheimer type in Pick's atrophy: lumping or splitting? Eur Arch Psychiatry Clin Neurosci 245: 299-305

45. Lynch T, Sano M, Marder KS, Bell KL, Foster NI, Defendini RF, Sima AAF, Keohane C, Nygaard TG, Fahn S, Mayeux R, Rowland LP, Wilhelmsen KC (1994) Clinical charac-teristics of a family with chromosome 17-linked disinhibition-dementia-parkinson-ism-amyotrophy-complex. Neurology 44: 1878-1884

46. Schenk VWD (1959) Re-examination of a family with Pick's disease. Ann Hum Genet 23: 325-333

47. Passant U, Gustafson L, Brun A (1993) Spectrum of frontal lobe dementia in a Swedish family. Dementia 4: 160-162

48. Frisoni GB, Calabresi L, Geroldi C, Bianchetti A, D'Acquarica AL, Govoni S, Sirtori CR, Trabucchi M, Franceschini G (1994) Apolipoprotein E 4 allele in Alzheimer's disease and vascular dementia. Dementia 5: 240-242

49. Farrer LA, Abraham CR, Volicer L, Foley EJ, Kowall NW, McKee AC, Wells JM (1995) Allele epsilon 4 of apolipoprotein E shows a dose effect on age at onset of Pick disease. Exp Neurol 136: 162-170

50. Gustafson L, Abrahamsson M, Grubb A, Nilsson K, Fex G (1997) Apolipoprotein E genotyping in Alzheimer's disease and frontotemporal dementia. Dementia Geriatr Cogn Disord 8: 240-243

51. Pickering-Brown SM, Siddons M, Mann DMA, Owen E, Neary D, Snowden JS (1995) Apolipoprotein E allelic frequencies in patients with lobar atrophy. Neurosci Lett 188: 205-207

52. Foster NL, Wilhelmsen K, Sima AAF, Jones MZ, D'Amato C, Gildman S, and consensus participants (1998) Frontotemporal dementia and parkinsonism linked to chromosome 17: a consensus conference. Ann Neurol 41: 706-715

53. Castellani R, Smith MA, Richey PL, Kalaria R, Gambetti P, Perry G (1995) Evidence for oxidative stress in Pick disease and corticobasal degeneration. Brain Res 696: 268-271

54. Lebert F, Pasquier F, Petit H (1995) Personality traits and frontal lobe dementia. Int J Geriatr Psychiatry 10: 1047-1049

55. Luauté JP, Favel P, Rémy C, Sanabria E, Bidault E (1994) Troubles de l'humeur et démence de type frontal. Hypothèse d'un rapport pathologénique. Encéphale 20: 27-36

56. Miller BL, Chang L, Mena I, Boone K, Lesser IM (1993) Progressive right frontotemporal degeneration: clinical, neuropsychological and SPECT characteristics. Dementia 4: 204-213

57. Eiden H-F, Lechner H (1950) Über pscyhotische Zustandsbilder bei der Pickschen und Alzheimerschen Krankheit. Arch Psychiatr Ze Neurol 1984: 393-412

58. Burns A, Jacoby R, Levy R (1990) Psychiatric phenomena in Alzheimer's disease. I. Disorders of thought content. Br J Psychiatry 157: 72-76

59. Gustafson L, Risberg J (1992) Deceptions and delusions in Alzheimer's disease and frontal lobe dementia. In: Katona C, Levy R (eds) Delusions and hallucinations in old age. Gaskell, London, pp 218-229

60. Lhermitte F, Pillon B, Serdaru M (1986) Human autonomy and the frontal lobes. I. Imitation and utilization behaviour: a neuropsychological study of 75 patients. Ann Neurol 19: 326-334

61. Cummings JL, Duchen LW (1981) Klüver-Bucy syndrome in Pick's disease: clinical and pathological correlations. Neurology 31: 1415-1422

62. Delay J, Neveu P, Desclaux P (1944) Les dissolutions du langage dans la maladie de Pick. Rev Neurol 76: 37-38

63. Escourolle R (1958) La maladie de Pick. Étude critique d'ensamle et synthèse anatomoclinique. Foulon, Paris

64. Snowden JS, Neary D (1993) Progressive language dysfunction and lobar atrophy. Dementia 4: 226-231

65. Mesulam MM (1982) Slowly progressive aphasia without generalized dementia. Ann Neurol 11: 592-598

66. Green J, Morris JC, Sandson J, McKeel DW, Miller JW (1990) Progressive aphasia: a precursor of global dementia? Neurology 40: 423-429

67. Snowden JS, Neary D, Mann DMA, Goulding PJ, Testa HJ (1992) Progressive language disorder due to lobar atrophy. Ann of Neurol 31: 174-183

68. Esmonde T, Giles E, Xuereb J, Hodges J (1996) Progressive supranuclear palsy presenting with dynamic aphasia. J Neurol Neurosurg Psychiatry 60: 403-410

69. Wikström J, Paetau A, Palo J, Sulkava R, Haltia M (1982) Classic amyotrophic lateral sclerosis with dementia. Arch Neurol 39: 681-683

70. MitsuyamaY, Kogoh H, Ata K (1985) Progressive dementia with motor neuron disease. An additional case report and neuropathological review of 20 cases in Japan. Eur Arch Psychiatr Neurol Sci 235: 1-8

71. Talbot PR, Goulding PJ, Lloyd JJ, Snowden JS, Near D, Testa HJ (1995) The interrelationship between "classical" motor neurone disease and frontotemporal dementia: a neuropsychological and single photon emission tomographic study. J Neurol Neurosurg Psychiatry 58: 541-547

72. Passant U, Warkentin S Karlson S, Nilsson K, Edvinsson L, Gustafson L (1996) Orthostatic hypotension in organic dementia: relationship between blood pressure, cortical blood flow and symptoms. Clin Auton Res 6: 29-36

73. Andrew J, Nathan PW (1964) Lesions of the anterior frontal lobes and disturbances of micturition and defecation. Brain 87: 232-262

74. Wilson DH, Chang AE (1974) Bilateral anterior cingulectomy for the relief of intractable pain (report of 28 patients). Confin Neurol 36: 61-68

75. Fäldt R, Passant U, Nilsson K, Wattmo C, Gustafson L (1996) Prevalence of thyroid hormone abnormalities in elderly patients with symptoms of organic brain disease. Aging Clin Exp Res 8: 347-353

76. Johannesson G, Brun A, Gustafson L, Ingvar DH (1977) EEG in presenile dementia related to cerebral blood flow and autopsy findings. Acta Neurol Scand 56: 89-103

77. Delay J, Brion S, Escourolle R (1957) L'opposition anatomo-clinique des maladies de Pick et d'Alzheimer. Étude de 38 cas. Presse Med 65: 1495-1497

78. Blois R, Gaillard J-M, Richard J (1989) Clinical and sleep EEG findings. In: Hovaguimian T, Henderson S, Khachaturian Z, Orley J (eds) Classification and diagnosis of Alzheimer disease – an international perspective. Hogrefe and Huber, Toronto Lewinston New York Bern Stuttgart Göttingen, pp 145-151

79. Rosén I, Gustafson L, Risberg J (1993) Multichannel EEG frequency analysis and somatosensory-evoked potentials in patients with different types of organic dementia. Dementia 4: 43-49

80. Gustafson L, Brun A, Cronqvist S, Dalfelt G, Risberg J, Riesenfeld V, Rosén I (1989) Regional cerebral blood flow, MRI, and BEAM in Alzheimer's disease. J Cerebr Blood Flow Metab 9[Suppl 7]: 513

81. Förstl H, Hentschel F, Besthorn C, Geiger-Kabisch C, Sattel H, Schreiter-Gasser U, Bayerl JR, Schmitz F, Schmitt HP (1994) Frontal und temporal beginnende Hirnatrophie. Nervenarzt 65: 611-618

82. Frisoni GB, Beltramello A, Geroldi C, Weiss C, Bianchetti A, Trabucchi M (1996) Brain atrophy in frontotemporal dementia. J Neurol Neurosurg Psychiatry 61: 157-165

83. Mazziotta JC (1989) Huntington's disease: studies with structural imaging techniques and positron emission tomography. Semin Neurol 9: 360-369

84. Gustafson L, Brun A, Holmkvist-Franck A, Risberg J (1985) Regional cerebral blood flow in degenerative frontal lobe dementia of non-Alzheimer type. J Cerebr Blood Flow Metabol 5: 141-142

85. Risberg J (1987) Frontal lobe degeneration of non-Alzheimer type. III. Regional cerebral blood flow. Arch Gerontol Geriatr 6: 225-233

86. Risberg J, Passant U, Warkentin S, Gustafson L (1993) Regional cerebral blood flow in frontal lobe dementia of non-Alzheimer type. Dementia 4: 186-187

87. Kitamura S, Araki T, Sakamotot S, Ilio M, Terashi A (1990) Cerebral blood flow and cerebral oxygen metabolism in patients with dementia of frontal lobe type. Rinsho Shinkeigaku 30: 1171-1175

88. Pasquier F, Lebert F, Lavenu I, Jacob B, Steinling M, Petit H (1997) The use of SPECT in a multidisciplinary memory clinic. Dementia Geriatr Cogn Disord 8: 85-91

89. Miller BL, Itti L, Li J, Darby AI, Booth R, Chang L, Mena I (1995) Atrophy-corrected cerebral blood flow in fronto-temporal dementia. Facts Res in Gerontol [Suppl 1]: 93-103

90. Kamo H, McGeer PL, Harrop R, McGeer EG, Calne DB, Martin WR, Pate BD (1987) Positron emission tomography and histopathology in Pick's disease. Neurology 37: 439-445

91. Salmon E, Franck G (1989) Positron emission tomographic study in Alzheimer's disease and Pick's disease. Arch Gerontol Geriatr Suppl 1: 241-247

92. D'Antona R, Baron JC, Samson Y, Serdaru M, Viader F, Agid Y, Cambier J (1985) Subcortical dementia. Brain 108: 785-799

93. Ludolph AC, Langen KJ, Regard M, Herzog H, Kemper B, Kuwert T, Böttger IG, Feinendegen L (1993) Frontal lobe function in amyotrophic lateral sclerosis: a neuropsychologic and positron emission tomography study. Acta Neurol Scand 85: 81-89

94. Johanson A, Hagberg B (1989) Psychometric characteristics in patients with frontal lobe degeneration of non-Alzheimer type. Arch Gerontol Geriatr 8: 129-137

95. Frisoni GB, Pizzolato G, Geroldi C, Rossato A, Bianchetti A, Trabucchi M (1995) Dementia of the frontal type: neuropsychological and (99Tc)-HMPAO SPET features. J Geriatr Psychiatry Neurol 8: 42-48

96. Pasquier F, Lebert F, Grymonprez L, Petit H (1995) Verbal fluency in dementia of frontal lobe type and dementia of Alzheimer type. J Neurol Neurosurg Psychiatry 58: 81-84

97. Pachana NA, Brauer-Boone K, Miller BL, Cummings JL, Berman N (1996) Comparison of neuropsychological functioning in Alzheimer's disease and frontotemporal dementia. J Int Neuropsychol Soc 2: 505-510

98. Folstein MF, Folstein SE, McHugh PR (1975) "Mini-Mental State." A practical method for grading the cognitive state of patients for the clinician. J Psychiatr Res 12: 189-198

99. Elfgren C, Passan U, Risberg J (1993) Neuropsychological findings in frontal lobe dementia. Dementia 4: 214-219

100. Elfgren C, Ryding E, Passant U (1996) Performance on neuropsychological tests related to single photon emission computerised tomography findings in frontotemporal dementia. Br J Psychiatry 169: 416-422

101. Hagberg B (1987) Behaviour correlates to frontal lobe dysfunction. Arch Gerontol Geriatr 6: 311-321

102. Johanson A, Gustafson L, Smith GJW, Risberg J, Hagberg B, Nilsson B (1990) Adaptation in different types of dementia and in normal elderly subjects. Dementia 1: 95-101

103. Tyrrell PJ, Warrington EK, Frackowiak RSJ, Rossor MN (1990) Heterogeneity in progressive aphasia due to focal cortical atrophy. A clinical and PET study. Brain 113: 1321-1336

104. Kanazawa I, Kwak S, Sasaki H, Muramoto O, Mizutani T, Hori A, Nukina N (1988) Studies on neurotransmitter markers of the basal ganglia in Pick's disease, with special reference to dopamine reduction. J Neurol Sci 83: 63-74

105. Gilbert JJ, Kish SJ, Chan L-J, Morito C, Shammak K, Hornykiewicz O (1988) Dementia, parkinsonism and motor neuron disease: neurochemical and neuropathological correlates. Neurology 24: 688-691

106. Francis PT, Holmes C, Webster M-T, Stratmann GC, Procter AW, Bowen DM (1993) Preliminary neurochemical findings in non-Alzheimer dementia due to lobar atrophy. Dementia 4: 172-177

107. Hansen LA, Deteresa R, Tobias H, Alford M, Terry RD (1988) Neocortical morphometry and cholinergic neurochemistry in Pick's disease. Am J Pathol 131: 507-528
108. Clark AW, White CL III, Manz JH, Parhad II, Curry B, Whitehouse PJ, Lehman L, Cole JT (1986) Primary degenerative dementia without Alzheimer pathology. Can J Neurol Sci 13: 462-470
109. Minthon L, Edvinsson L, Ekman R, Gustafson L (1990) Neuropeptide levels in Alzheimer's disease and dementia with frontotemporal degeneration. J Neurol Transm 30: 57-67
110. Edvinsson L, Minthon L, Ekman R, Gustafson L (1993) Neuropeptides in cerebrospinal fluid of patients with Alzheimer's disease and dementia with frontotemporal lobe degeneration. Dementia 4: 167-171
111. Blennow K, Wallin A, Ågren H, Spenger C, Sigfrid J, Vanmechelen E (1996) Tau protein in cerebrospinal fluid: a biochemical marker for axonal degeneration in Alzheimer's disease? Mol Chem Neuropathol 26: 231-245
112. American Psychiatric Association (APA) (1987) Diagnostic and statistical manual of mental disorders, (DSM-III-R), 3rd edition, revised. APA, Washington DC
113. American Psychiatric Association (APA) (1994) Diagnostic and statistical manual of mental disorders (DSM-IV). APA, Washington DC
114. World Health Organization (WHO) (1992) The ICD-10 classification of mental and behavioural disorders. Clinical descriptions and diagnostic guidelines. World Health Organization, Geneva
115. Ishii N, Nishihara Y, Imamura T (1986) Why do frontal lobe symptoms predominate in vascular dementia with lacunes? Neurology 36: 340-345
116. Brun A, Gustafson L (1988) Zerebrovaskuläre erkrankungen. In: Kisker KP, Lauter H, Meyer J-E, Muller C, Strömgren E (eds) Psychiatrie der Gegenwart. Band 6. Organische Psychosen. Springer-Verlag, Berlin Heidelberg New York, pp 253-295
117. Englund E, Brun A, Gustafson L (1989) A white-matter disease in dementia of Alzheimer's type - clinical and neuropathological correlates. Int J Geriatr Psychiatry 4: 87-102
118. Fredriksson K, Brun A, Gustafson L (1992) Pure subcortical arterioslecortic encephalopathy (Binswanger's disease): a clinicpathologic study. Part 1. Clinical features. Cerebrovasc Dis 2: 82-86
119. Neumann MA (1949) Pick's disease. J Neuropathol Exp Neurol 8: 255-282
120. Neumann MA, Cohn R (1967) Progressive subcortical gliosis: a rare form of presenile dementia. Brain 90: 405-418
121. Petersen RB, Tabaton M, Chen SG, Monari L, Richardson SL, Lynch T, Menetto V, Lanska D, Markesbery WR, et al (1995) Familial progressive subcortical gliosis. Presence of prions and linkage to chromosome 17. Neurology 45: 1062-1067
122. Gustafson L, Nilsson L (1982) Differential diagnosis of presenile dementia on clinical grounds. Acta Psychiatr Scand 65: 194-209
123. Brun A, Gustafson L (1993) I. The Lund longitudinal dementia study: a 25-year perspective on neuropathology, differential diagnosis and treatment. In: Corain B, Nicolini M, Winblad B, Wisniewski H, Zatta P (eds) Alzheimer's disease. Advances in clinical and basic research. Wiley, London, pp 4-18
124. Hachinski VC, Iliff LD, Zilhka E, du Boulay GH, McAlliste V, Marhsall J, Ross Russel RW, Symon L (1975) Cerebral blood flow in dementia. Arch Neurol 32: 632-637
125. Risberg J, Gustafson L (1988)Regional cerebral blood flow in psychiatric disorders. In: Knezevic S, Maximilian VA, Mubrin Z, Prohovnik I, Wade J (eds) Handbook of regional cerebral blood flow. Erlbaum, Hillsdale, pp 219-240

Pseudodementia

D. De Leo and W. Padoani

Introduction

Memory, orientation, and intelligence are the keys to contact with the outside world and the formation of personal experience. In some circumstances, however, these very functions may be solely a cause for suffering. On perception of a definitive gulf between environmental demands and the capacity to cope with events, some individuals seem to operate, as their last remaining defence mechanism, by completely refusing to respond to increasing outside pressure, which may originate in their personal sociorelationship scenario, or be the consequence of intimate psychic suffering or, ultimately, the result of a complex interaction between the two. In such circumstances, an essential component appears to come amiss that forces individuals to use their cognitive tools to the full in an endeavor to gain complete control of their own experiences, or rather their very reason for living.

In late life, the substrate altered by biological aging, exclusion from decisional and productive processes, the culturally accepted interpretation of the elderly person's social role, are all situations which, in association with severe psychic suffering, may precipitate such a radical refusal to relate to the outside world. Many syndromes of this kind, commonly referred to as "pseudodementias," may thus be included among those forms of self-injury, defined elsewhere as "suicidal erosion" [1, 2], such as refusing food or medical treatment, in this case taking the shape of a desperate "cognitive exile" from the world.

Since its introduction, most likely by Kiloh in 1961 [3], the term "pseudodementia" has frequently been used in the literature to indicate various clinical conditions sharing three essential criteria: a dementing type of cognitive deficit, and the absence of a causal organic substrate and reversibility.

According to the apparent definition by Wells [4], pseudodementia is "a syndrome in which dementia is mimicked or caricatured by functional psychiatric illness." Distinctive elements include a past history of psychiatric illness, rapidly progressing symptomatology, complaints of cognitive impairment, behavioral disorders not consistent with the apparent degree of deficit, "don't know" answers, and variability of performance on neuropsychological testing.

The work by Caine [5] provides a useful reference insofar as it establishes several univocal clinical criteria and proposes four distinctive elements for identifying pseudodementia: cognitive impairment associated with a primary psychiatric disorder; neuropsychological alterations that recall neuropathologically induced cognitive deficits; reversible intellectual disorders; no organic substrate supporting the disorder. Although this definition is generally accepted, it has received much criticism.

As regards the first criterion, clinical application of the term "pseudodementia" still remains rather broad, hindering comparison of the case studies reported in the literature, which still lack homogeneity. Although the term is most frequently used in cases of reversible cognitive deficits in depressed subjects, the same definition includes all cases of cognitive impairment present in patients with bipolar disorder in the manic stage, personality disorders, hysterical conversion reactions, Ganser syndrome, anxiety states, and depersonalization or schizophrenic states [6–12].

The second criterion, based solely on subjective clinical assessment, seems excessively vague and of little heuristic use.

The reversibility of the disorder, indicated in the third criterion, is crucial in determining the clinical usefulness of the term, since it attaches particular attention to the different prognostic implications between dementia and pseudodementia. Nonetheless, while there are undoubtedly numerous descriptions of a positive outcome to this disorder, several authors stress how depression, even in the absence of organic impairment, may constitute the premonitory symptoms of a subsequent development of dementia [10, 13–23].

In an 8-year follow-up, Kral and Emery showed how 89% of a group of patients, initially identified as suffering from a reversible dementialike disorder, proceeded to develop an irreversible form of dementia [24]. Furthermore, Alexopoulos and Chester reported how recent follow-ups have revealed that between 9% and 25% of cases per year of patients with initial, reversible dementia subsequently presented irreversible dementia [25–29].

On the basis of these data, Alexopoulos and Chester concluded that in the elderly, depression associated with reversible cognitive impairment has an annual development rate in dementia that is 2.5 to 6 times higher than in the general geriatric population, thereby constituting a significant risk factor [29].

Lastly, various criticisms have been directed at the exclusion of an organic substrate in pseudodementia. Numerous similarities, for example, have been identified between the clinical presentation of pseudodementia and subcortical dementia, which is chiefly characterized by a slowdown in cognitive, ideational and psychomotor functions, amnesia, and alterations in affectivity and personality [5, 30–32]. Various authors, by contrast, have suggested considering pseudodementia as dementia related to a specific, reversible neurophysiological or neurochemical-type organic substrate [31]. Rabins, moreover, hypothesized the presence of small anatomic anomalies in the brain, which appear to predict subsequent pseudodementia in the presence of a comorbid depressive picture [17]. Mitchell and Dening [23] recently attributed a central role, in the genesis of cognitive dysfunctioning associated with depression, to the activity of the hypothalamic-pituitary-adrenal axis.

Exclusion of an organic cause (or at least an immediately identifiable one) does, however, help more clearly mark the clinical boundaries of pseudodementia and should prevent syndromes such as reversible secondary dementias and delirium being indicated by this term, although this does occasionally occur. The term pseudodementia is not, however, clinically justified in the former disorders, since they constitute dementia to all intents and purposes, albeit reversible once the cause has been removed (endocrine and metabolic disorders, hereditary metabolic disorders, infectious and inflammatory diseases of the CNS, states of deficiency, endocranial expansive processes, cranial trauma, paraneoplastic syndromes, cardiovascular or respiratory diseases, action of toxic substances or drugs, etc.).

While some authors consider delirium to characterize maybe the most common form of pseudodementia [33], this assimilation does not fit very well with Caine's definition, since cognitive deficits are not associated with a primary psychiatric disorder and since an organic substrate is identifiable or very presumably present, meeting both DSM-IV and ICD-10 criteria (neurological disorders, cerebrovascular diseases, brain trauma, infectious or inflammatory states of the CNS, endocranial neoformations, metabolic encephalopathies, toxoinfectious states, cardiovascular diseases, withdrawal from drugs or other substances, etc.).

Nevertheless, the reversibility of the disorder and pre-eminence of cognitive impairment do remain two important areas of overlap between delirium and pseudodementia. The essential presence, in the former, of attention and consciousness deficits, the extreme brevity of onset of the disorder (hours or a few days), the clearly circadian fluctuation in the symptomatology, with symptoms worsening in the evening and in conditions of sensorial deprivation, the marked frequency of disorders of perception, combined with the fact that symptoms are of somatic origin rather than accountable to a primary psychiatric disorder, should suffice to clinically distinguish between delirium, dementia, and pseudodementia. This distinction, however, is not always easy (e.g., the frequent overlap of episodes of delirium in underlying dementia and, in the case of depressive pseudodementia, the possible onset of delirium following the start of antidepressant therapy).

Lastly, in some cases of pseudodementia, as will be illustrated in the remaining discussion, there are episodes that seem to fake the characteristics of delirium, as in the case of "manic delirium."

The difficulty of identifying a consensus of clinical criteria is due not only to the extreme diversity of depressive symptomatology in the elderly, but also to the limited uniformity in the studies conducted in the area (especially as regards type of diagnostic criteria to employ), the smallness of many of the case studies reported to date in the literature, the varying quantitative assessment criteria for the symptomatology presented, the limited number of diachronic assessments of the disorder, and the various therapeutic approaches adopted for the cases in question [34]. The need for a more rigorous nosological and clinical definition of pseudodementia is further highlighted by the many and varied interpretations of the syndrome, even by the very psychiatrists working in the field [35].

As concerns the incidence and prevalence of pseudodementia, as early as 1952, Madden and associates found that 10% of elderly psychotic patients presented symptoms characteristic of dementia, which regressed following appropriate therapy [36]. A recent review of the literature also showed how 13% of patients with diagnoses of dementia were, instead, affected by a reversible disease which, in 5% of the subjects analyzed, was depression [37].

The bulk of data on the phenomenon, however, only refers to depressive pseudodementia. Its incidence among institutionalized older adults stands at around 13%, compared to 30% of patients affected by various somatic pathologies attending a geriatric unit [38]. According to some reports, between 10% and 20% of depressed elderly individuals do in fact present significant cognitive impairment [39–42], though other case studies indicate findings in the order of 70% [43].

In relation to the problems of differentiating between dementia and pseudodementia, Marsden and Harrison diagnosed a functional-type cognitive disorder in 10% of patients who had been hospitalized to investigate suspected dementia. Several authors also suggested that a percentage of between 8% and 15% of subjects initially considered to be demented were subsequently diagnosed as depressed, with all the consequences of the subsequent therapeutic sequelae [4, 30, 44–49].

The discussion that follows first considers the problem of depressive pseudodementia, which represents the most commonly found syndrome, with the greatest clinical impact. This is followed by a presentation of other psychiatric pathologies that may present important cognitive deficits: bipolar disorder in the manic stage, schizophrenia, hysterical and regressive pseudodementia, a note on Ganser syndrome and AIDS-related pseudodementia.

Depressive Pseudodementia

The depressive form is the most frequently encountered clinical condition coming under the umbrella of pseudodementia. As indicated above, cognitive dysfunctions can be found in 10%-20% and, according to some authors, in as many as 70% of depressed elderly people [39–43]. The reasons why this cognitive manifestation of depression is particularly marked among the elderly rather than other age groups are not yet quite clear, although they seem to strengthen the hypothesis that pseudodementia brings to the fore organic brain disease or mild neurochemical imbalance, which only become manifest in co-occurrence with a truly depressive clinical picture.

The cognitive symptomatology accompanying depression in younger adults is nebulous and seems above all to reflect a motivational deficit [50]. Such motivational deficit does, however, also appear to be present in older adults, which some authors associate with a generalized alteration of the catecholaminergic system [51]. Hence it is reflected in a quantitative decline in processed information rather than in a qualitative modification in cognitive and decisional processes, which seem on the whole to be slowed down.

Interesting correlations, which aid comprehension of the various forms of pseudodementia, may be made between the cognitive alterations accompanying elderly depression and the type manifested in depressed individuals affected by AIDS, in the absence of demonstrated organic brain damage (see below). It is also clear that a certain degree of cognitive impairment is almost physiological in the elderly, a fact that may accentuate subjective perception of such decline in the case of severe depression. In any event, many observations appear to exclude any real difference between elderly and nonelderly subjects in depression-related cognitive impairment [52].

Sociorelational dynamics undoubtedly also play a role in this respect. Young people, on the one hand, tend to focus depressive feelings chiefly on their own self-image and social acceptance. Older adults, by contrast, tend to alternate their focus between somatic symptomatology and cognitive deficits (memory disorders, difficulties in attention and concentration, changes in learning and intellectual capacities). This consideration is reinforced by evidence that subjective perception of cognitive deficits and complaints about the same are more frequent among depressed individuals with pseudodementia than among those with dementing cognitive impairment [53–55].

The difficulties encountered, especially in elderly people, in distinguishing a depressive picture with cognitive impairment from depressive aspects associated with dementia, particularly in the early stages, have led to endeavors to more clearly characterize the various possible clinical presentations. Feinberg and Goodman, for example, proposed four main associations between dementia and depression, namely, depression presenting as dementia ("depressive pseudodementia"), dementia presenting as depression ("pseudodepression"), depression with secondary depression ("dementia syndrome of depression") and dementia with secondary dementia ("depressive syndrome of dementia") [35, 40, 56].

Emery and Oxman [10], instead, suggested that there were a number of ways to reach the common endpoint of the phenotype of dementia, and depressive dementia shows a certain vulnerability for such final presentation. The authors proposed that cognitive impairment in depression should be viewed as a continuum with dual polarity: presence-absence of organic degenerative damage and reversibility-irreversibility of the disorder. On the basis of this, they postulated the existence of five different groups of subjects, defined by eminently clinical criteria: major depression without depressive dementia, depressive dementia, degenerative dementia without depression, depression in degenerative dementia, independent co-occurence of degenerative dementia and depression.

Clinical Characteristics

Depressive pseudodementia is essentially an important cognitive deficit secondary to the presence of a depressive disorder. In general, in contrast to dementia syndromes, patients' relatives are well aware of the presence of a memory deficit and are able to indicate its onset quite precisely. Patients explicitly complain of their memory loss or other cognitive deficits. On testing, they tend to provide "don't know" type answers, and often demonstrate a certain irritability when encouraged or urged on by the tester. Patients with pseudodementia often communicate considerable despair, although not generally in a verbal fashion. The history of the disorder is rather short, and a personal or familial history of depressive disorder is often present. Characteristic behavioral traits include insomnia, mood changes during the day, and occasional indications of pessimistic ideation and ruin. Such patients often prove to be perplexed, taciturn, and unwilling to talk [4, 57]. Both short- and long-term memory are impaired, probably because of attention deficits, with difficulties in concentration and distraction, which compromise their overall learning ability. Little effort is made in neuropsychological tests and, unlike demented patients, there is apparently no evidence of performance anxiety, or the desire to conceal cognitive deficits, which, on the contrary, are often emphasized. Confabulation, agnosia and aphasia are invariably absent [58–62].

Rather than indicating progressive decline, cognitive test results tend to vary and change over time. Spatiotemporal orientation and calculation abilities are also better preserved [63, 64]. Mood is always depressed, even where this does not appear to be the prevalent aspect [65]. High suicide risk is present and may be associated with dysphoria, anxiety, and emotive lability with vegetative correlates typical of depression.

The diagnostic distinction between depressive pseudodementia and mild-or-moderate dementia remains an essential clinical question. Besides the above-mentioned clinical criteria [4, 62, 66], other specific tests may prove useful, although none may be considered sufficient on its own. Appropriate testing, starting from Folstein's Mini-Mental State Examination (MMSE) [67], may indicate seriously impaired performance, but very rarely comparable to levels associated with severe dementia. As indicated above, less impairment tends to be present in spatio-temporal orientation, gnostic and praxic functions, visuospatial and ideoperceptual capacities and language [18, 68, 69]. As different from the early stage of dementia, recent and remote memory are probably simultaneously impaired. During early differential diagnosis, some authors encourage the combined use of MMSE with other neuropsychological tests such as Wechsler's Logical Memory, Wechsler's Visual Reproduction and Kendrick's Object Learning tests [70].

Specific dementia assessment scales such as the Clinical Dementia Rating Scale [71], the Mattis Dementia Rating Scale [72], the CAMDEX Scale [73], the Alzheimer Dementia Assessment Scale [74], the Global Deterioration Scale [75] may undoubtedly be of use, but can only be considered determining in the case of severe dementia, i.e., where the clinical picture is self-explanatory [76, 77].

Depression assessment scales such as the Hamilton Scale [78], the Geriatric Depression Scale [79], and Self-Rating Depression Scale [80] are seriously limited by the fact that they are relatively unreliable in the presence of severe cognitive impairment and are unable to distinguish between primary depressive disorders and those related to early cognitive decline of organic origin.

Another most useful tool for discriminating between dementia and pseudodementia, based on case history, behavioral and cognitive criteria, is Wells' Check-

List Differentiating Pseudodementia from Dementia Scale [6]. More specifically, according to results published by Reynolds and associates, the integration of individual items from three different assessment scales (the MMSE orientation to time item; the inability to find one's way about familiar streets or about indoors and difficulty with dressing items from the Blessed Scale; delayed insomnia, psychological anxiety and loss of libido from the Hamilton Scale) distinguished pseudodementia from dementia in 90.5% of patients [18].

The dexamethasone suppression and TRH tests have instead proved to be of little use. EEG, CT and RMN tests tend to be negative in patients with pseudodementia, although this may also be the case in the early stages of organic decline. Moreover, in some cases of primary depression, brain CT has revealed signs of cerebral atrophy [10] and profound ischemic foci in the subcortical white matter [9]. Studies indicating a significantly lower density of 3H-imipramine binding sites in subjects affected by major depression compared to patients with Alzheimer's disease [81, 82] and research reporting significant differences between individuals with pseudodementia and dementia yielded by sleep EEGs, particularly in relation to continuity, architecture, and REM sleep time distribution [18, 83], do not seem to furnish really feasible prospects for differential diagnosis. By contrast, imaging with PET, which indicates a smaller anteroposterior gradient in glucose metabolism in depressed subjects only [84] and SPECT techniques [85, 86], seem to be more promising. Blennow and associates have [87] reported another interesting finding in this respect, which requires closer examination: they suggest testing CSF concentrations of an abnormally phosphorylated form of tau protein (associated with microtubules), which is only significantly present in the demented.

Lastly, attention should be paid to the link between hypercortisolemia and mnemonic decline. According to studies on rats, this reduction may be explained by the action of corticosteroid hormones on the granular cells of the hippocampal dentate gyrus. Since this very effect appears to be cancelled by narcotic antagonists, it has been suggested that naltrexone may distinguish subjects affected by pseudodementia (where the hippocampal system is intact and any mnemonic deficit appears related to depression-related hypercortisolemia) from those with dementia, in whom there is no response to naltrexone because of hippocampal degeneration [88–91].

Nonetheless, in view of the considerable difficulties inherent in distinguishing between these two syndromes, it is essential to identify the presence of any depressive symptoms and treat them appropriately (resolute, prolonged treatment is often required to bring about improvement). This is not only of undoubted help to the patient but may also provide an interesting *ex iuvantibus* diagnostic criterion [92, 93].

The prognosis and therapy to adopt are clearly the same as for other depressive disorders. Generally, therefore, appropriate treatment of pseudodementia brings about a positive outcome. As stated above, this opinion is not, however, shared by some authors who report on patients with cognitive disorders that are initially secondary to depressive syndromes, but subsequently develop into dementia [10, 13–15, 19–22, 24–29, 94].

To conclude, stress should once again be laid on the need to recognize and adequately treat depressive manifestations in elderly people with suspected dementia, possibly implementing a prolonged follow-up study and using various antidepressants, whenever a response is late in arriving [95, 96].

Manic Pseudodementia

Another dementialike clinical picture responsive to specific therapy is the type associated with late-onset bipolar disorder. While differential diagnosis is relatively easy for subjects with a past history of manic episodes, the problem becomes more complex where onset occurs in late life. According to Cowdry and Goodwin, between 1% and 4% of bipolar patients encounter their first manic episode after 60 years of age; various other authors describe onset even after 80 years of age [90–93].

Dementia is probably mistakenly diagnosed by reason of the peculiar nature of manic episodes in advanced age, namely, the tendency to present irritable, dysphoric rather than elated mood, episodes occurring in rapid succession, suggesting a chronic rather than episodic disorder [101], disorientation and confusion, which may conceal manipulative tendencies, severe insomnia, flight of ideas, hypersexuality and disinhibited behavior [8].

A particularly surreptitious picture of mania is "manic delirium" [102]. Certain elderly bipolar patients in the manic phase do in fact appear to develop a markedly disorganized picture, with symptoms suggesting delirium. In this context, cognitive functions are very difficult to test and may result in erroneous diagnoses. Catatonic like aspects, negative symptoms and perseveration may also emerge. Excitement and disinhibited behavior are common to both manic pseudodementia and organic pathologies, with prevalent frontal lobe atrophy (in particular, Pick's disease) [12].

The prognosis for manic pseudodementia is generally good, considering the very positive response in these cases to therapy with mood regulators, such as lithium or carbamazapine, employing lower doses than in the younger population. Particular attention should nonetheless be paid to the use of these drugs in older adults, with regular controls being made on serum levels of lithium or carbamazepine and electrolytes, blood composition, kidney, heart, and thyroid functioning, together with an initial hospital admission to assess the optimum dose of the drugs in question [8].

Schizophrenia

The natural course of schizophrenia and the chronic use of neuroleptics lead to progressive intellectual and cognitive impairment, particularly exacerbated by the frequent, extensive periods of institutionalization that characterize the lives of schizophrenics. In such situations, it is not uncommon to identify a late stage of the disorder, characterized by behavioral deficits and marked cognitive impairment [103].

In acute manifestations of schizophrenia, however, dementialike clinical pictures may appear and are characterized by impairment of language, thought, and intellectual and motor functioning [9]. Poor performance on tests of mnestic functioning, intelligence and psychomotor abilities may also be exhibited in the period between one acute episode and the next [9, 92]. Cognitive impairment is generally associated with signs of atrophy on brain CT testing, a prevalence of negative symptoms and, consequently, poor response to neuroleptics. These findings confine pictures of pseudodementia to the type II schizophrenia proposed by Crow [104–106].

Various elements suggest that the cognitive impairment accompanying psychosis is caused by an alteration in control functions such as attention and motivation. Hence, rather than a deficit in specific cognitive functions, which prove instead to be integral, there appears to be an impairment of "willed intention" [107].

The evident characteristics of schizophrenia and the very history of schizophrenic individuals make nonrecognition of the diagnosis almost impossible, although stress should be laid on the risk of interpreting potentially reversible regressive symptoms as definitive impairment, with consequent negative implications for the therapeutic sequelae and rehabilitative dynamics of these patients.

To emphasize the risks of hurried diagnoses further, which are still unfortunately all too frequent in elderly patients, it is worth briefly illustrating the encounter (albeit rather rare) with major depressive or bipolar disorders presenting progressive deterioration common to schizophrenia during intercritical periods (resulting in frequent hospital admissions), which are evident from the very onset of the disease [108]. Right from the studies by Kraepelin, this finding has been complicating one of the reference points of differential diagnosis between major mood disorders and schizophrenia, illustrating the difficulties, in some borderline cases, of distinguishing between these clinical pictures. This problem is further complicated in the case of elderly individuals, in whom symptoms of schizophrenia are, on the one hand, generally attenuated, while the impairment associated with development of the depressive disorder is, on the other, at its peak.

Hysterical Pseudodementia

Hysteria has often been considered a phenomenon typical of the young. Consequently, many consider that in present society this nosological definition has much less room than in the past, given the continued increase in the elderly population and the considerable change in diagnostic criteria for psychiatric disorders. Some authors even suggest that hysterical conversion has "disappeared" [109]. Leff, by contrast, has advanced the hypothesis that rather than disappearing, hysterical symptoms may have changed from an eminently somatic to a psychological manifestation, and hysterical pseudodementia may be a paradigm of such development [11, 110].

Several studies have reported many cases of hysterical or conversion pseudodementia [11, 33, 48, 111–113]. Neuropsychological testing in these cases reveals impaired performance, and dementia-related behavioral disorders are absent [9]. Cognitive dysfunction does, however, exhibit marked fluctuations and is clearly subject to environmental changes. It is often associated with histrionic behavior, ostentation, accentuation of symptoms, and a past history of conversion or somatization disorders [114].

Hysterical conversion pseudodementia is, essentially, one of the many adaptation strategies adopted by the elderly to cope with the relational, social, and environmental changes that characterize late life. A surplus concentration of stressors, in excess of personal adaptive capacities, forces the elderly to come to terms with their own psychophysical decline and with exclusion from society's decision-making and productive processes, of which they were a part in earlier life. This may lead either to depressive isolation, potentially precipitating serious suicidal behavior, which generally expresses a marked lack of confidence in the value of their own existence, or alternatively, to the use of their own psychic or physical deficiencies for communicating with the rest of the world [115, 116]. This tendency is reinforced by the secondary advantages it often yields [9], and it is not rare, in families with patients with hysterical pseudodementia, to find relatives with a severe cognitive deficit, which may have provided a precise model to imitate [11].

Kirby and Harper [115, 116], basing their results on their own observations of such cases, described a positive outcome following specific psychotherapy, or more simply, after targeted intervention in the individual's sociorelational setting. They also emphasized the importance of adopting a multidisciplinary approach to elders to avoid running the risk of considering curable situations to be irreparable and irreversible and eventually opting for institutionalization. There is indeed a very marked contrast between the diagnostic commitment required by these patients and the relative ease with which a favorable outcome can be brought about by simply changing some aspects of the patient's milieu. Lastly, it is interesting to note how, according to the literature, subjects with hysterical pseudodementia are relatively younger than individuals with dementing disorders [11, 112, 113].

Regressive Pseudodementia Syndrome

A condition very similar to hysterical pseudodementia, characterized by integral cognitive performance, sharply contrasting with marked behavioral changes and an apparently less favorable prognosis [12, 92, 117], is dependence-related pseudodementia, also defined here as regressive pseudodementia.

Elderly persons, in response to environmental and sociorelational stressors, may develop a form of adaptation that, rather than emerging through hysterical conversion, exacerbates a tendency towards regressive behavior with pathological dependency [92].

Regression in elderly patients is characterized by progressive introversion, combined with infantile, help-seeking behavior and abandonment of all interests enjoyed prior to the disorder; it seems to be precipitated by perception of loss of former role within the social and family context [118]. Conditions of this type are also described by Spar and La Rue, who observe a progressive decline in social relations and physical activities, neglect of personal hygiene and household cleanliness, with possible cognitive deficits. Although the depressive symptomatology is in these cases rather generic, antidepressant therapy may lead to marked improvement [93]. A dependency-related dementing syndrome with alterations in behavior, progressive loss of autonomy in activities of daily living, rapid mood swings, decline in social relations and poor self-care, is reported by Howells and Beats [117].

Three cases of regressive pseudodementia, with important environmental dependency by the patient and marked motor impairment associated with more or less integral cognitive performance, are described by De Leo and Predieri [12]. These subjects, despite progressively worsening regressive behavior, achieved scores of 30 on the MMSE [67]. One of these cases is illustrated below.

A.A., a 62-year-old married man, father of two children, was referred to the psychogeriatric service during hospitalization in a neurological ward, where early Alzheimer's disease had been diagnosed, despite negative CT and EEG tests. The patient had been referred for severe depression, partly in reaction to a series of family problems. The patient appeared markedly depressed, anxious, worried and complained of severe asthenia. His behavior showed marked dependency and regression and he exhibited rigidity, amimia, walked with short steps and was tearful. This attitude, characteristic of severe dementia, suggested a diagnosis of "depressive pseudodementia," despite achieving the highest score on the MMSE. Treatment with mianserin (90 mg/d.i.e.) and oxazepam (45 mg/d.i.e.) brought about a marked improvement in the patient's conditions and reduced his anxiety level.

After discharge, the patient was examined periodically at the psychogeriatric service. During this period, important changes emerged in his personality, with progressive abandonment of social activities and all interests. He reached the point of definitively shutting himself indoors, where he tyrannized his wife, at times aggressively, with vexing requests for attention and exclusive care which, moreover, he only accepted from her. Two years after the first aforementioned contact, the patient was admitted to an internal medicine ward for pneumonia, during which time CT and SPECT were again performed, both producing negative results.

In the period which followed, the patient's behaviour exhibited further decline. He had practically ceased walking and passed from armchair to bed, complaining of severe asthenia. During a home visit, requested by the family, the patient appeared depressed, markedly anxious, worried and suffered from insomnia. He complained of "not being able to go on," was no longer able to eat by himself, was occasionally incontinent and presented severe weight loss. Restorative and antidepressive therapy were prescribed, based on trazodone (300 mg/d.i.e.) and mianserin (60 mg/d.i.e.), associated with anxiolytic therapy with oxazepam and triazolam. Despite improvements in mood and anxiety, the regressive behavior continued, coupled with almost total immobility. Another admission to the

Department of Neurology followed, where instrumental tests on the organic substrate remained negative and where early Alzheimer's disease was newly diagnosed, despite the complete integrity of cognitive functioning. In the same year, during another outpatient examination, four years after the first consultation, the patient appeared indifferent and detached, reacting angrily and swearing that he "couldn't go on," when his wife accused him of feigning to be ill and behaving like a child. Memory functioning remained good and his MMSE score was 30. On that occasion, the antidepressant dosage was reduced. In a telephone conversation one-and-a-half year later, his wife referred to an essentially unvaried situation.

In the above case, as in two other cases described elsewhere by the authors, the subjects tended to adapt to their circumstances through increasing dependency and, through their demented behavior symptoms, appeared to find comfort in being relieved of responsibilities and being able to attract attention that would otherwise be difficult to obtain. Collusion with this type of psychological and cognitive attitude both by members of the family and, more usually, in institutionalized settings only serves to encourage the elderly persons to definitively accept their "demented" role, with all the prognostic implications implicit in this type of situation, which are not generally very comforting.

Other Pictures of Pseudodementia

Other pictures of pseudodementia of considerable clinical interest, particularly owing to their relatively early onset, which sets them apart from the above-mentioned conditions typical of older adults, are Ganser syndrome [119] and cognitive deficits related to acquired immunodeficiency syndrome, where clinical and instrumental assessment excludes organic brain damage ("AIDS dementia complex"). In the latter case, the cognitive deficit is probably related to depressive pseudodementia, though the peculiar aspects of the pathological picture, the young age of the patients, the usefulness of comparing these data with similar syndromes in elderly patients (common psychophysical decline, presence of conditions theoretically compatible with a diagnosis of dementia, possible co-occurrence of minimal organic damage with an eminently functional pathology, sense of exclusion from society's participatory, decision-making and productive dynamics, are some of the elements common to both syndromes) warrant closer examination of the diagnostic, prognostic and therapeutic characteristics of these forms [120–126].

Discussion

On examining the problem of pseudodementia, it immediately becomes clear how the boundaries of this research area should be precisely marked out in order to render studies on the subject more homogeneous. The most commonly used international classification systems are also of little help in this respect: neither

DSM-IV [127] nor ICD-10 [128] provides a specific definition of either pseudo-dementia or cognitive impairment associated with depression. They simply indicate the possible presence of a certain decline in psychomotor functioning and general difficulties in attention and concentration among the diagnostic criteria for major depression. The difficulties in reaching a precise operational definition have already been outlined [18], but it is clear that the route to a heuristic diagnostic definition first and foremost requires the consensus of opinion of the scientific community. The option at present seems to be between whether it is appropriate to consider pseudodementia from a dichotomic standpoint based on a rigid definition of reversibility or irreversibility and presence or absence of organic dysfunction, concepts already inherent in the nineteenth century definition of "vesanic dementia" [33], or along a continuum between two extremes, represented, for example, by simple major depression and dementia [10, 35, 40, 56].

The above-mentioned definition by Caine [5] seems to provide a useful source of theoretical reference, although it lays itself open to criticism, especially in terms of the reversibility of the disorder and the absence of a clearly demonstrable organic basis [5, 10, 13–15, 17, 19–31]. The present state of knowledge therefore dictates a longitudinal view of pseudodementia, and hence the need to increase the number of follow-up studies on this type of patient.

Despite the variability of some traits, the syndromes that may be included in the definition of pseudodementia we have painstakingly attempted to reconstruct are not only of the depressive type, even though this is the most significant area, but also within the field of bipolar disorders, schizophrenia, certain hysterical manifestations, Ganser syndrome, cognitive maifestations of AIDS and regressive pseudodementia. If we dwell on the most common manifestation of pseudodementia, i.e., depressive dementia, the question as to its diagnostic definition may in fact be extended to encompass the more general need to consider elderly depression more specifically. Depression at this age presents numerous peculiar features compared to other age groups and inherent implications (above all the high percentage of suicides typical of old age, which are frequently interrelated with the appearance or complication of organic pathologies) and direct and indirect social costs resulting from the fact that this diagnosis is frequently mistaken in the elderly, a fact which we will examine more closely in the discussion. The urgent need to specify diagnostic guidelines more clearly is confirmed by the limited number of structured diagnostic instruments that can be employed in elderly people with suspected pseudodementia. These subjects are known frequently to manifest their depressive experience through complaints of vegetative symptoms that are clearly not pathognomonic of depression, since they are frequently found in dementia syndromes or normal elderly subjects suffering from objectively demonstrable organic pathologies. The Hamilton Depression Rating Scale [78], comprising items on working efficiency and activities of daily living or psychomotor performance, which are often impaired in the elderly population irrespective of depression, tend, for example, to overestimate the latter diagnosis in subjects affected by dementia. DSM-IV diagnostic criteria which, by contrast, envisage the obligatory identification of frankly depressed mood or loss of interest or ability to feel pleasure, tend on the other hand to underestimate the diagnosis of depression in elderly subjects with predominantly impaired vegetative or cognitive symptoms [127, 129]. An important research field which some authors have undertaken is the development of new specific diagnostic intruments for distinguishing cognitive symptoms of depressive origin from elderly dementia [6, 39, 68, 130, 131].

As discussed above, not even biochemical and instrumental assessments are determining in the discrimination between pseudodementia and dementing syndromes [18, 81, 83, 88–91]. The clinical characteristics of depressive pseudodementia have already been extensively illustrated above, but stress should be laid on the need to avoid underestimating depressive symptomatology in the elderly and to treat all important clinical cases, even where their association with dementia is uncertain. Besides being of significant help for the psychic well-being and an important *ex iuvantibus* diagnostic criterion of the patient in question, such clinical orientation is also an invaluable instrument for controlling the social costs of elderly health care. It has in fact been demonstrated that there is an interaction between depression and disability related to organic pathologies, in the sense that depression favors not only exacerbated perception of somatic deficits of pathologies already present, but also furthers the very onset of strictly medical disorders in the elderly population, which is by its very nature more vulnerable. In turn, depression linked to prevalently organic diseases tends to aggravate the overall clinical picture and to stress the resulting disability [95, 132–139]. This finding becomes even more important if we consider that any cognitive dysfunction associated with depression may itself cause disability or disproportionately increase the low spirits caused by the depressed mood alone. Cognitive deficits, moreover, are among the potential predictors of depression-related disability [139].

On the basis of a survey carried out by the British National Health Service, approximately 190,000 people aged over 65 receiving home care services in England and Wales presented potentially treatable depression [140]. According to a study on the topic, only 16% of these were receiving antidepressant treatment and only 9% were in contact with psychiatric facilities. The same study also reported how applications for home care services were only partially ascribable to actual physical disability, but were, instead presumably due to the depression itself [141]. If we then add that elderly subjects with cognitive impairment, including depression-related syndromes, are frequently admitted to institutions, or other residential facilities for chronic patients, there are clear economic implications for missed diagnoses of depressive dementia and potentially reversible cognitive deficits in many elderly subjects.

These findings further emphasize the need to establish appropriate prevention strategies for elderly depression and all related manifestations, including depressive pseudodementia. Suitable treatment of depressed elderly subjects not only ensures marked improvement in their quality of life, but also reduces the need for support from public health services, thereby releasing precious resources for other rehabilitation or health care facilities. For this purpose, primary health care facilities globally and general practitioners in particular, play an extremely important role. Family doctors are often, however, better trained to recognize dementing syndromes rather than the depressive origin of some symptomatolog-ical manifestations of late life [142]. The need to increase knowledge on pseudo-dementia is also further emphasized by the fact that there is considerable varia-tion in its interpretation even among psychiatrists working in this area. As we have seen, of the various clinical manifestations of pseudodementia alongside depressive dementia, there are a number of syndromes that warrant further investigation, in various directions.

Hysterical pseudodementia, for example, is interesting both by reason of the appearance in late life of hysterical manifestations, once deemed more character-istic of the younger population, and considering Western society's perception of the role of the elderly, a scenario which appears to encourage manifestation of their true psychological needs, not recognized by the common cultural context, in the form of indirect demands for help. Patients with hysterical pseudodementia tend to be younger than the demented in general, and it is interesting to note the gulf between the diagnostic commitment often required in these cases and the rel-ative ease with which a favorable outcome can be brought about by intervention specifically targeted at the socio-relational entourage of which they are a part.

A scenario relatively similar to the one described above is regressive pseudo-dementia. What is striking in these cases is the discrepancy between the clinical picture, characterized by marked behavioral changes, substantial disability in activities of daily living and a rather unfavorable prognosis, and integral cogni-tive performance [12, 92, 117].

The interest aroused by cognitive deficits diagnosed in AIDS syndromes in the absence of organic brain damage caused by AIDS dementia complex is, instead, linked to the marked affinity between this clinical condition and depressive pseu-dodementia in the elderly, but presenting in such circumstances at a much younger age.

Lastly, the therapeutic approach is inevitably linked to the specific etiology of each individual form of pseudodementia and should take account of the general-ly advanced age and inevitable metabolical modifications of treated subjects. More specifically, in the case of depressive pseudodementia, consideration must be made of the frequent need to employ more incisive, diversified antidepressant treatment, where response to treatment is not immediately satisfactory, as is often the case. In this respect, future research should also take account of the implica-tions of adopting certain types of drug, such as substances with anticholinergic effects, on the possible future development of organic brain or neurotransmitter alterations in the elderly subjects' brains [10].

Conclusions and Future Research Guidelines

This paper highlights the need to determine an operational definition not only for pseudodementia, but also for depressive symptomatology in the elderly that would receive the consensus of opinion of the scientific community and render studies on the subject more uniform. Such studies should also address endeavors towards increasing clinical observations with a view to providing a more extensive, uniform, and hence statistically more significant case series, as well as establishing long-term follow-ups, which is of crucial importance in clarifying the relationship between pseudodementia and degenerative brain syndromes of known etiology.

Another line of work is the development of psychodiagnostic assessment instruments and the identification of laboratory and instrumental techniques designed to distinguish effectively between pseudodementia and true dementing disorders. This work should nonetheless be part of a more global prevention scheme for late-life psychiatric disorders, requiring the development of both highly sensitive, low-cost instruments to be used in an initial, generalized screening program, and more specific (and generally more costly) diagnostic equipment to be used in subsequent stages only.

The invaluable role of primary health care prevention schemes and the need to provide health professionals, especially (but not only) doctors involved in this delicate field, with adequate refresher training remains undisputed. Older adults go to see their doctors most frequently; hence the latter's competence and ability to listen might save these subjects from dangerous feelings of uselessness and therapeutic nihilism and at the same time assure them of an adequate quality of life, thereby helping to avoid or postpone social exile in institutions or other public health care facilities. Stress, too, has already been laid on how this may influence savings of public resources allocated to health care.

A further moot point relates to the association between pseudodementia and the age of the subjects affected. The identification of early-onset pseudodementia syndromes, such as Ganser syndrome and depressive AIDS-related pseudodementia, appears to exclude confinement of this syndrome to old age and to make possible interesting speculative implications.

Lastly, the identification of types of pseudodementia connected with hysterical conversion reactions or regressive pseudodementia opens new prospects in the study of the various pathological adaptation strategies adopted by elderly individuals to cope with what are often seen as insuperable demands from the outside world with respect to what they perceive as an unavoidable psychophysical decline, or more simply, to the unacceptability of their own solitude.

References

1. De Leo D, Diekstra RFW (1990) Depression and suicide in late life. Hogrefe and Huber, Toronto Lewinston New York Bern Stuttgart Göttingen
2. Diekstra RF (1993) The epidemiology of suicide and parasuicide. Acta Psychiatr Scand Suppl 371: 9-20
3. Kiloh LG (1961) Pseudo-dementia. Acta Psychiatr Scand 37: 336-361
4. Wells CE (1979) Pseudodementia. Am J Psychiatry 136: 895-900
5. Caine ED (1981) Pseudodementia: current concepts and future directions. Arch Gen Psychiatry 38: 1359-1364
6. Wells CE (1979) Diagnosis of dementia. Psychosomatics 20: 517-522
7. Brun M, Aspea A (1984) Les fausses démences du sujet agé. In: Simeone I, Abraham G (eds) Introduction à la psychogériatrie. Simep, Lyon

8. Casey DA, Fitzgerald BA (1988) Mania and pseudodementia. J Clin Psychiatry 49: 73-74
9. Cummings JL, Benson DF (1992) Dementia: a clinical approach, 2nd edn. Butterworth-Heinemann, Boston
10. Emery VO, Oxman TE (1992) Update on the dementia spectrum of depression. Am J Psychiatry 149: 305-317
11. Liberini P, Faglia L, Salvi F, Grant RP (1993) What is the incidence of conversion pseudodementia. Br J Psychiatry 162: 124-126 (letter)
12. De Leo D, Predieri M (1994) Pseudodemenze. In: De Leo D, Stella A (eds) Manuale di psichiatria dell'anziano. Piccin Nuova Libraria, Padova, pp 373-392
13. Busse EW (1975) Aging and psychiatric disease of late life. In: Arieti S (ed) American handbook of psychiatry. Basic Books, New York
14. Goldstein K (1975) Functional disturbances in brain damage. In: Arieti S (ed) American handbook of psychiatry. Basic Books, New York
15. Wells CE (1977) Dementia: definition and description. In: Wells CE (ed) Dementia, 2nd edn. Davis, Philadelphia
16. McAllister TW (1983) Overview: pseudodementia. Am J Psychiatry 140: 528-533
17. Rabins PV, Merchant A, Nestadt G (1984) Criteria for diagnosing reversible dementia caused by depression: validation by 2-year follow-up. Br J Psychiatry 144: 488-492
18. Reynolds CF, Hoch CC, Kupfer DJ, et al (1988) Bedside differentiation of depressive pseudodementia from dementia. Am J Psychiatry 145: 1099-1103
19. Cunha UG (1990) An investigation of dementia among elderly outpatients. Acta Psychiatr Scand 82: 261-263
20. Mattos P (1995) Citalopram, depression and pseudodementia: a neuropsychological case study. Arq Neuropsiquiatr 53: 841-848
21. Speck CE, Kukull WA, Brenner DE, et al (1995) History of depression as a risk factor for Alzheimer's disease. Epidemiology 6: 366-369
22. Buntinx F, Kester A, Berges J, Knottnerus JA (1996) Is depression in elderly people followed by dementia? A retrospective cohort study based in general practice. Age Ageing 25: 231-233
23. Mitchell AJ, Dening TR (1996) Depression-related cognitive impairment: possibilities for its pharmacological treatment. J Affect Disord 36: 79-87
24. Kral VA, Emery OB (1989) Long-term follow-up of depressive pseudodementia of the aged. Can J Psychiatry 34: 445-446
25. Reding M, Haycox J, Blass J (1985) Depression in patients referred to a dementia clinic: a 3-year prospective study. Arch Neurol 42: 894-896
26. Reynolds CF, Kupfer DJ, Hock CC, et al (1986) Two-year follow-up of elderly patients with mixed depression and dementia. Clinical and electroencephalographic sleep findings. J Am Soc 34: 793-819
27. Alexopoulos GS (1990) Clinical and biological findings in late-onset depression. Am Psychiatr Press Rev Psychiatry 9: 249-262
28. Alexopoulos GS, Abrams RC (1991) Depression in Alzheimer's disease. Psychiatr Clin North Am 14: 327-340
29. Alexopoulos GS, Chester JG (1992) Outcomes of geriatric depression. Clin Geriatr Med 8: 363-376
30. McAllister TW, Price TRP (1982) Severe depressive pseudodementia with and without dementia. Am J Psychiatry 139: 626-629
31. Mahendra B (1985) Depression and dementia: the multi-faceted relationship. Psychol Med 15: 227-236
32. Sellal F (1996) Les démences sous-corticales. Rev Med Intern 17: 419-424

33. Bulbena A, Berrios G (1986) Pseudodementia: facts and figures. Br J Psychiatry 148: 87-94
34. Reynolds CF, Kupfer DJ, Houck PR, et al (1988) Reliable discrimination of elderly depressed and demented patients by electroencephalographics sleep data. Arch Gen Psychiatry 45: 258-264
35. O'Shea B, Rahill M, MC Collam C (1986) Pseudodementia: facts and figures. Br J Psychiatry 148: 611-612 (letter)
36. Madden JJ, Luhan JA, Kaplan LA, et al (1952) Nondementing psychoses in older persons. JAMA 150: 1567-1570
37. Clarfield AM (1988) The reversible dementias: do they reverse? Ann Int Med 109: 476-486
38. Gurland BJ, Golden R, Dean L (1980) Depression and dementia in the elderly of New York City. In: Planning for the elderly in New York City. New York, Community Council of Greater New York
39. Roth M (1976) The psychiatric disorders of late life. Psychiatr Ann 6: 417-444
40. Folstein MF, MC Hugh PR (1978) Dementia syndrome of depression. In: Katzman R, Terry RD, Beck KL (eds) Aging, vol 7. Alzheimer's disease, senile dementia and related disorders. Raven, New York, pp 87-93
41. Rabins PV (1981) The prevalence of reversible dementia in a psychiatric hospital. Hosp Community Psychiatry 32: 490-492
42. Reifler BV, Larson E, Hanley R (1982) Coexistence of cognitive impairment and depression in geriatric outpatients. Am J Psychiatry 139: 623-626
43. Abas MA, Sahakian BJ, Levy R (1990) Neuropsychological deficits and CT scan changes in elderly depressives. Psychol Med 20: 507-520
44. Marsden CD, Harrison MJ (1972) Outcome of investigation of patients with presenile dementia. BMJ 2: 249-252
45. Duckworth GS, Ross H (1975) Diagnostic differences in psychogeriatric patients in Toronto, New York and London, England. Can Med Assoc J 112: 847-851
46. Nott PN, Fleminger JJ (1975) Presenile dementia: the difficulties of early diagnosis. Acta Psychiatr Scand 51: 210-217
47. Ron MA, Toone BK, Garralda ME, et al (1979) Diagnostic accuracy in presenile dementia. Br J Psychiatry 134: 161-168
48. Good MI (1981) Pseudodementia and physical findings masking significant psychopathology. Am J Psychiatry 138: 811-814
49. Shamoian CA (1985) Assessing depression in elderly patients. Hosp Community Psychiatry 36: 338-339
50. Crook T (1979) Psychometric assessment in the elderly. In: Raskin A, Jarvik LF (eds) Psychiatric symptoms and cognitive loss in the elderly. Hemisphere, New York
51. Hart RP, Kwentus JA (1987) Psychomotor slowing and subcortical-type disfunction in depression. J Neurol Neurosurg Psychiatry 50: 1263-1266
52. Tarbuck AF, Paykel ES (1995) Effects of major depression on the cognitive function of younger and older subjects. Psychol Med 25: 285-295
53. Kahn RL, Zarit SH, Hilbert NM, Niederehe G (1975) Memory complaint and impairment in the aged: the effect of depression and altered brain function. Arch Gen Psychiatry 32: 1569-1573
54. Wells CE (1980) The differential diagnosis of psychiatric disorders in the elderly. In: Cole J, Barrett J (eds) Psychopathology in the aged. Raven, New York, p 28
55. Magni E, Frisoni GB, Rozzini R, De Leo D, Trabucchi M (1996) Depression and somatic symptoms in the elderly: the role of cognitive function. Int J Geriatr Psychiatry 11: 517-522

56. Feinberg T, Goodman B (1984) Affective illness, dementia, and pseudodementia. J Clin Psychiatry 45: 99-103

57. Post F (1982) Affective disorders in old age. In: Paykel ES (ed) Handbook of affective disorders. Guilford, New York

58. Gallagher D, Thompson L (1983) Depression. In: Lewinsohn P, Teri L (eds) Clinical neuropsychology. Pergamon Elmosford, New York

59. Klerman GL, Davidson JM (1984) Memory loss and affective disorders. Psychosomatics 25: 29-32

60. Straker M (1984) Demenza o depressione? Med Paziente 10: 1427-1435

61. Giberti F (1985) La depressione tra psichiatria e psicoanalisi. In: Giberti F (ed) L'altra depressione. Apporti psicoanalitici alla psichiatria. Piccin, Padova

62. Pitt B (1986) Characteristics of depression in the elderly. In: Murphy E (ed) Affective disorders in the elderly. Churchill Livingstone, Edinburgh

63. Alexopoulos GS, Young RC, Mattis S (1989) Cognitive disturbances in geriatric depression with reversible dementia. Presented at the annual meeting of the American Association for Geriatric Psychiatry, Orlando

64. Albert M (1984) Assessment of cognitive function in the elderly. Psychosomatics 25: 310-317

65. Salzman, Schatzberg AF, Liptzin B, Satlin LA, Cole JO (1984) Diagnosis of affective disorders in the elderly. Psychosomatics 25: 126-131

66. Post F (1976) Diagnosis of depression in geriatric patients and treatment modalities appropriate for the population. In: Gallant DM (ed) Depression. Spectrum, New York

67. Folstein MF, Folstein SE, MC Hugh PR (1975) Mini-mental state. A practical method for grading the cognitive state of patients for the clinician. J Psychiatr Res 12: 189-198

68. Wells CE, Buchanan BC (1977) The clinical use of psychological testing in evaluation in dementia. In: Wells CE (ed) Dementia, 2nd edn. Davis, Philadelphia

69. Padovani A (1994) La valutazione clinica e neuropsicologica. In: De Leo D, Stella A (eds) Manuale di psichiatria dell'anziano. Piccin Nuova Libraria, Padova, pp 183-198

70. Des Rosiers G, Hodges JR, Berrios G (1995) The neuropsychological differentiation of patients with very mild Alzheimer's disease and/or major depression. J Am Geriatr Soc 43: 1256-1263

71. Hughes CP, Berg L, Danziger WL, et al (1982) A new clinical scale for the stating of dementia. Br J Psychiatry 140: 566-572

72. Mattis S (1976) Mental status examination for organic mental syndromes in the elderly patients. In: Bellak L, Karasu TB (eds) Geriatric psychiatry: a handbook for psychiatrists and primary care physicians. Grune and Stratton, New York, pp 77-121

73. Roth M, Tym E, Mountjoy CQ, et al (1968) CAMDEX: a standardized instrument for the diagnosis of mental disorder in the elderly with special reference to the early detection of dementia. Br J Psychiatry 149: 698-709

74. Rosen WG, Mohs RC, Davis KL (1984) A new rating scale for Alzheimer's disease. Am J Psychiatry 141: 1356-1364

75. Reisberg B, Ferris SH, De Leon MJ, et al (1982) The global deterioration scale for the assessment of primary degenerative dementia. Am J Psychiatry 139: 1136

76. Jarvik LF, Neshkens RE (1985) Dementia and delirium in old age. In: Brocklehurst JC (ed) Textbook of geriatric medicine and gerontology. Churchill Livingstone, Edinburgh, pp 309-327

77. Fuld PA (1982) Psychological testing in the differential dementias. In: Katzman R, Terry RD, Bick KL (eds) Alzheimer's disease: senile dementia and related disorders. Raven, New York

78. Hamilton M (1960) A rating scale for depression. J Neurol Neurosurg Psychiatry 23: 56-61
79. Yesavage JA, Brinj TL, Rose TL, et al (1983) The development and validation of geriatric depression screening scale: a preliminary report. J Psychiatr Res 17: 37-49
80. Zung WK (1965) A self-rating depression scale. Arch Gen Psychiatry 12: 63-70
81. Surany-Cadote BE, Gauthier S, Lafaille F, et al (1985) Platelet 3H-imipramine binding distinguishes depression from Alzheimer's dementia. Life Sci 37: 2305-2311
82. De Leo D, Schifano F, Magni G (1988) Results of dexamethasone suppression test in early Alzheimer dementia. Eur Arch Psychiatry Neurol Sci 238: 19
83. Buysse DJ, Reynolds CF, Kupfer DJ, et al (1988) Electroencephalographic sleep in depressive pseudodementia. Arch Gen Psychiatry 45: 568-575
84. Buchsbaum MS, De Lisi LE, Holcomb HH, et al (1984) Antero-posterior gradients in cerebral glucose use in schizophrenia and affective disorders. Arch Gen Psychiatry 41: 1159-1166
85. Parker G, Austin MP (1995) A clinical perspective on SPECT. Aust N Z J Psychiatry 29: 38-47
86. Golan H, Kremer J, Freedman M, Ichise M (1996) Usefulness of follow-up regional cerebral blood flow measurements by single photon emission computed tomography (SPECT) in the differential diagnosis of dementia. J Neuroimaging 6: 23-28
87. Blennow K, Wallin A, Agren H, et al (1995) Tau protein in cerebrospinal fluid: a biochemical marker for axonal degeneration in Alzheimer's disease? Mol Chem Neuropathol 26: 231-245
88. Collier TJ, Quirk GJ, Routtenberg AR (1987) Separable roles of hippocampal granule cells in forgetting and pyramidal cells in remembering spatial information. Brain Res 409: 316-328
89. Sloveters RS, Valiquette G, Abrams GA, et al (1989) Selective loss of hippocampal granule cells in mature rat brain after adrenalectomy. Science 243: 535-538
90. Wolkowitz OM, Reus VI, Weingartner H, et al (1990) Cognitive effects of corticosteroids. Am J Psychiatry 147: 1297-1303
91. Lipschitz A (1991) Naltrexone challenge for hypercortisolemic pseudodementia. Am J Psychiatry 148: 953 (letter)
92. Kaplan HJ, Sadock BJ (1985) Comprehensive textbook of psychiatry, 4th edn. Williams and Wilkins, Baltimore
93. Spar JE, La Rue A (1990) Geriatric Psychiatry. American Psychiatric Press, Washington DC
94. Sahakian BJ (1991) Depressive pseudodementia in the elderly. Int J Geriatr Psychiatry 6: 453-458
95. Butler RN, Lewis MI (1995) Late-life depression: when and how to intervene. Geriatrics 50: 44-57
96. Fisman M (1988) Intractable depression and pseudodementia: a report of two cases. Can J Psychiatry 33: 628-630
97. Shulman K, Post F (1980) Bipolar affective disorder in old age. Br J Psychiatry 136: 26
98. Summers WK (1983) Mania with onset in the eighth decade: two cases and a review. J Clin Psychiatry 44: 141
99. Walter-Ryan WG (1983) Mania with onset in the ninth decade. J Clin Psychiatry 44: 430 (letter)
100. Kellner MB, Neher F (1991) A first episode of mania after age 80. Can J Psychiatry 36: 607-608
101. Zis AP, Groff P, Webster M, et al (1980) Prediction of relapse in recurrent affective disorder. Psychopharmacol Bull 16: 47-49

102. Shulman KI (1986) Mania in old age. In: Murphy E (ed) Affective disorders in the elderly. Churchill Livingstone, Edinburgh
103. Arieti S (1978) Interpretazione della schizofrenia. Feltrinelli, Milano
104. Crow TJ (1980) Molecular pathology of schizophrenia: more than one disease process? Br J Psychiatry 280: 66-68
105. Crow TJ (1982) The biology of schizophrenia. Experimentia 38: 1275-1282
106. Andreasen NJC, Olsen SA, Dennert JW, et al (1982) Ventricular enlargement in schizophrenia: relationship to positive and negative symptoms. Am J Psychiatry 139: 297-302
107. Della Sala S (1990) Le demenze sottocorticali. In: Dene G, Pizzamiglio L (eds) Manuale di Neuropsicologia. Zanichelli, Bologna
108. Vocisano C, Klein DN, Keefe RSE, Dienst ER, Kincaid MM (1996) Demographics, family history, premorbid functioning, developmental characteristics, and course of patients with deteriorated affective disorder. Am J Psychiatry 153: 248-255
109. Brody J (1985) Prospects for an ageing population. Nature 315: 463-466
110. Leff J (1988) Psychiatry around the globe. Gaskell, London
111. Cavenar JO, Maltbie AA, Austin L (1979) Depression simulating organic brain disease. Am J Psychiatry 136: 521-523
112. McEvoy JP, Wells CE (1979) Conversion pseudodementia. J Clin Psychiatry 40: 447-449
113. Friedman MJ, Liposwki ZJ (1981) Pseudodementia in a young PhD. Am J Psychiatry 138: 381-382
114. Seva A (1991) The European handbook of psychiatry and mental health. Prensas Universitarias de Zaragoza, Zaragoza
115. Kirby HB, Harper RG (1987) Team assessment of geriatric mental patients: the care of functional dementia produced by histerical behavior. Gerontologist 27: 573-576
116. Kirby HB, Harper RG (1988) Team assessment of geriatric mental patients (II): behavioral dynamics and psychometric testing in the diagnosis of functional dementia due to hysterical behavior. Gerontologist 28: 260-262
117. Howells R, Beats B (1989) A dementia syndrome of dependency? Br J Psychiatry 154: 872-876
118. Donini G (1982) Psichiatria geriatrica. In: Reda GC (ed) Trattato di psichiatria. Uses, Firenze
119. Adler R, Touyz S (1989) Ganser syndrome in a 10 year old boy: an 8 year follow-up. Aust N Z J Psychiatry 23: 124-126
120. Krikorian R, Wrobel AJ (1991) Cognitive impairment in HIV infection. AIDS 5:1501-1507
121. Wilkins JW, Robertson KR, Snyder CR, et al (1991) Implications of self-reported cognitive and motor dysfunctions in HIV-positive patients. Am J Psychiatry 148: 641-643
122. Freedman JB, O'Dowd MA, Wyszinski B, et al (1994) Depression, HIV dementia, delirium, posttraumatic stress disorder. Gen Hosp Psychiatry 16: 426-434
123. Perry SW (1994) HIV related depression. Res Publ Assoc Res Nerv Ment Dis 72: 223-238
124. Desi M, Seibel N, Korezlioglu J, et al (1995) Troubles cognitives au cours de l'infection par le VIH. Validation d'une batterie courte d'evaluation neuropsychologique. Encephale 21: 289-294
125. Seilhean D, Duyckaerts C, Hauw JJ (1995) Demence et VIH: neuropathologie. J Neuroradiol 22: 161-162
126. Silvestre D, Linard F, Desi M, et al (1995) Anxious depressive state and cognitive deficit in HIV infection. Encephale 21: 285-288
127. American Psychiatric Association (1994) Diagnostic and statistical manual of mental disorders, 4th edn. American Psychiatric Association, Washington DC

128. World Health Organization (1993) The ICD-10 classification of mental and behavioural disorders. Diagnostic criteria for research. World Health Organization, Geneva

129. Lazarus LW, Newton N, Cohler B, Lessor J, Schweon C (1987) Frequency and presentation of depressive symptoms in patients with primary degenerative dementia. Am J Psychiatry 144: 41-45

130. Sunderland T, Alterman IS, Yount D, et al (1988) A new scale for the assessment of depressed mood in demented patients. Am J Psychiatry 145: 955-959

131. Burke WJ, Houston MJ, Boust SJ, et al (1979) Use of Geriatric Depression Scale in dementia of the Alzheimer type. J Am Geriatr Soc 37: 856-860

132. Craig TJ, Van Natta PA (1983) Disability and depressive symptoms in two communities. Am J Psychiatry 140: 598-601

133. Griffith RA, Good WR, Watson NP, et al (1987) Depression, dementia and disability in the elderly. Br J Psychiatry 150: 482-493

134. Wells KB, Stewart A, Hays RD, et al (1989) The functioning and well-being of depressed patients: results from the Medical Outcomes Study. JAMA 262: 914-919

135. Kennedy GJ, Kelman HR, Thomas C (1990) The emergence of depressive symptoms in late life: the importance of declining health and increasing disability. J Commun Health 15: 93-104

136. Laukkanen P, Kauppinen M, Era P, Heikkinen E (1993) Factors related to coping with physical and instrumental activities of daily living among people born in 1904-1923. Int J Geriatr Psychiatry 8: 287-296

137. Bruce ML, Seeman TE, Merril SS, Blazer DG (1994) The impact of depressive symptomatology on physical disability: MacArthur studies of successful aging. Am J Public Health 84: 1796-1799

138. Forsell Y, Jorm AF, Winblad B (1994) Association of age, sex, cognitive dysfunction, and disability with major depressive symptoms in an elderly sample. Am J Psychiatry 151: 1600-1604

139. Alexopoulos GS, Vrontou C, Kakuma T, et al (1996) Disability in geriatric depression. Am J Psychiatry 153: 877-885

140. OPCS (1993) OPCS population estimates unit. OPCS monitor PP1 93/2. HMSO, London

141. Banerjee S, MacDonald A (1996) Mental disorder in an elderly home care population: associations with health and social service use. Br J Psychiatry 168: 750-756

142. O'Connor DW, Pollitt PA, Hyde JB, et al (1988) Do general practitioners miss dementia in elderly patients? BMJ 297: 1107-1110

Neuropsychiatric Symptoms in Dementia Patients

S.T. Chen and J.L. Cummings

Introduction

Neuropsychiatric symptoms and alterations are common and diverse in scope in the dementias. Alzheimer's disease (AD) is associated with agitation, psychosis, personality changes, and depression [1–6]. Vascular dementia (VaD) is accompanied by apathy, depression, and lability [7–9]. Frontotemporal dementias (FTD) produce apathy, euphoria, irritability, and disinhibition [7–10]. Depression and anxiety are associated with Parkinson's disease (PD) patients [11, 12]. Mood changes, personality changes, and psychosis are observed in Huntington's disease (HD) [13].

Neuropsychiatric symptoms and alterations of dementias are significant in their diagnostic, management, and prognostic implications. Neuropsychiatric features may be the presenting manifestations of dementias, appearing before any evidence of cognitive impairment [14–16], and change through the course of illness, requiring reassessment of problems and interventions [17]. Neuropsychiatric disturbances are a major source of caregiver burden and a common reason for institutionalization [18–20]. Neuropsychiatric aspects of dementias are often responsive to pharmacotherapy and other interventions [7, 21, 22]. Specific neuropsychiatric symptoms, as in the case of delusions, are associated with more rapid cognitive decline in AD [23–26].

This chapter will discuss dementia-related neuropsychiatric symptoms and alterations, including psychosis, depression, anxiety, agitation, and personality changes, and their phenomenology, assessment, pathogenesis, and pharmacologic management. Aggression, wandering, and the Klüver-Bucy (KBS) syndrome are also addressed. (Nonpharmacologic interventions are discussed in the chapter by Zanetti and Trabucchi, this volume).

Assessment

The assessment and measurement of neuropsychiatric symptoms in demented patients present unique challenges. Evaluation of cognitive aspects of dementia may include well-established neuropsychological tests and instruments such as the Mini-Mental State Examination [27] based on items with correct or incorrect answers. However, assessment of neuropsychiatric aspects of dementia is based on clinical observation or subjective report by patient or caregivers and may be more subject to bias. In addition, neuropsychiatric symptoms such as apathy or anxiety occur on a continuum with normal behavior.

There are several instruments with established reliability and validity used in the assessment of neuropsychiatric status in dementia. Instruments commonly used in assessing multiple behaviors include the BEHAVE-AD [6], the Neurobehavioral Rating Scale (NRS) [28] and the Neuropsychiatric Inventory (NPI) [29]. (see Table 1). Other instruments that focus on specific symptoms and behaviors are discussed in the sections below.

The BEHAVE-AD was developed specifically for assessment of AD-related neuropsychiatric disturbances, including psychosis, affective disturbance, aggressiveness, and anxiety. It lacks other behaviors commonly accompanying AD and other dementias, including apathy, disinhibition, appetite changes, and irritability. The BEHAVE-AD has 25 items, each rated on a 4-point scale based on severity by an observer [6].

The NRS was developed for assessment of cognitive and behavioral sequelae of head trauma. Content and convergent validity of the NRS were subsequently demonstrated in AD and VaD patients [31]. The NRS has 28 items (18 of which address noncognitive alterations), each rated on a seven-point scale of severity

Table 1. Instruments for assessment of neuropsychiatric status in dementia and categories of behavioral disturbance assessed

BEHAVE-AD (adapted from [30])	NRS[a] (adapted from [28])	NPI (adapted from [29])
Paranoid and delusional ideation	Unusual thought content	Delusions
Suspiciousness	Hallucinations	Hallucinations
Hallucinations	Agitation	Agitation
Activity disturbances	Fatigability	Disinhibition
Aggressiveness	Decreased motivation	Apathy/indifference
Affective disturbance	Motor retardation	Aberrant motor behavior
Anxieties and phobias	Excitement	Depression/dysphoria
Diurnal rhythm disturbance	Disinhibition	Euphoria/elation
	Hostility	Irritability/lability
	Depressed mood	Anxiety
	Blunted affect	Night-time behavior
	Mood lability	Appetite/eating change
	Emotional withdrawal	
	Guilt	
	Somatic concern	
	Anxiety	
	Tension	

[a] NRS includes ten additional items that measure cognition/insight and verbal output disturbance

rated by a clinician observer in conjunction with a semistructured interview. The NRS has been used to show differing behavioral profiles between AD and VaD patients [32].

The NPI was developed specifically for the assessment of neuropsychiatric disturbances occurring in dementia patients and has demonstrated validity and reliability [29]. The NPI has 12 items, each rated for frequency and severity by a regular caregiver after a set of questions is read aloud by an examiner. A screening question for each item minimizes administration and scoring time; if a screening question elicits an absence of a behavioral disturbance, the item is scored zero and the next item is assessed. The NPI has been shown to differentiate patients with AD from patients with FTD [33] and to distinguish AD from progressive supranuclear palsy [34].

The source of information in an assessment can have a significant impact on the neuropsychiatric profile of a demented patient: there is often little agreement between a physician's examination and an interview with an informant on symptoms of anxiety, depression and psychosis [35].

Psychosis

Delusions

A delusion is a fixed, false belief that is without basis in reality and not part of a cultural tradition. Delusions are a common neuropsychiatric symptom of dementia. Reported prevalence rates of delusions in AD range from 10% [36] to 73% [37], with most studies reporting between 30% and 38% [38]. Studies comparing AD and VaD patients suggested similar frequencies of delusions between groups [8, 32]. One study comparing AD and FTD patients found somewhat higher rates among AD patients, though not to a statistically significant degree [33].

There are several types of delusions observed in dementia. Perhaps the most common type of delusion in AD patients involves paranoid themes – beliefs that one is the victim of theft, spousal infidelity (Othello's delusion), or other harm. Other types of delusions present in AD and other dementias include the beliefs that family members or other familiar persons have been replaced by impostors (Capgras syndrome), a stranger is living in the home (phantom boarder), characters from television are in the home (picture sign), and their house is not their home [5, 6].

Studies have examined the relationship between delusions and other aspects of dementia, including cognition, other neuropsychiatric disturbances, prognosis, neuroimaging, and neuropathology/neurochemistry. There is little consensus on the relationship between delusions and cognitive impairment in dementia. Studies have reported no association [39, 40], a positive correlation [41], weak positive associations [42, 43], a trend toward a negative association [8], or a mixed relationship [1–4, 30, 44]. One recent large-scale, prospective study that compared AD patients with and without delusions reported that the delusional group had greater cognitive impairment, especially disorders suggesting frontal/temporal dysfunction [45].

Demented patients with delusions exhibit significantly greater aggression than those without delusions [43, 46]. PD patients with dementia and psychosis have significantly more insomnia, agitation, personality changes, and self-care problems and are significantly more unmanageable at home than their nonpsychotic counterparts [47].

Investigators have studied delusions as a possible prognostic factor in terms of both progression of cognitive impairment and functional decline (e.g., dressing, feeding, shopping, and personal finances) in dementia. With the exception of one study [48], prospective longitudinal studies reported that the presence of delusions in dementia predicts a faster progression of cognitive decline [23–26]. Only one of those studies found that delusions predicted more rapid functional decline [24].

Demented patients with psychosis have cerebral perfusion or metabolic patterns different from those without psychosis. The characteristic pattern observed in AD is bilateral parietal and temporal hypoperfusion [49, 50] or hypometabolism [51]. Positron emission tomography (PET) studies demonstrated a significant positive correlation between psychosis and frontal hypometabolism in AD patients [52, 53]. Studies using single-photon emission computed tomography (SPECT) showed relative left (vs right) frontal hypoperfusion [54] and bilateral temporal hypoperfusion [55] in delusional AD patients when compared to non-delusional AD patients.

There have been few studies examining neuropathologic and neurochemical correlates of psychosis in AD. Delusions are associated with significantly increased densities of neuritic plaques in the prosubiculum and of neurofibrillary tangles in the middle frontal cortex, along with a relative preservation of norepinephrine in the substantia nigra and a significant reduction of serotonin in the prosubiculum [56]. The authors posited that preservation of noradrenergic functioning in AD patients might contribute to the development of psychosis. Physostigmine, a cholinergic agent, ameliorates delusions in AD patients, suggesting that the cholinergic deficit in AD patients might contribute to the development of psychosis [57]. The pathologic correlates of psychosis in PD are unknown [58].

Psychosis is associated with other dementias, including Lewy body dementia (LBD) [59], HD [13], normal pressure hydrocephalus [60], and Fahr's disease [7].

Hallucinations

A hallucination is a sensory perception that occurs in the absence of an external stimulus. Hallucinations are a less common neuropsychiatric symptom than delusions in patients with dementia. In a review Wragg and Jeste [38] reported a median occurrence rate of 28% in AD patients; visual hallucinations were most common (median prevalence 22%), followed by auditory hallucinations (median prevalence 13%) alone or in combination with visual hallucinations. Hallucinations are more common in VaD than in AD [8]. Characteristic visual hallucinations include persons from the past, animals, intruders, complex scenes, or inanimate objects [5, 61]. Auditory hallucinations are often persecutory, argumentative or commanding and are usually accompanied by delusions [62]. Hallucinations in AD are associated with parietal hypometabolism, a different cortical region from that associated with delusions [54], suggesting possible distinct pathophysiological pathways for the two symptoms.

In a study of 14 LBD patients, visual hallucinations were associated with an imbalance between monoaminergic and cholinergic activities in the temporal cortex [63].

Hallucinations associated with PD are largely medication-related. Approximately 30% of treated patients experience visual hallucinations, while other types of hallucinations and delusions are less frequent [64]. Drug-induced hallucinations are more common among demented than nondemented PD patients. Any antiparkinsonian agent can result in visual hallucinations. However, hallucinations associated with anticholinergic medication are characterized as more threatening, less well-formed, combined with auditory and tactile experiences, and accompanied by delirium when compared those associated with dopaminergic agents [65].

Treatment

Although neurochemical studies of psychosis in dementia implicate neurotransmitter systems other than dopamine, neuroleptics, which block postsynaptic dopamine receptors, are the primary class of pharmacologic agents used to treat this condition. Studies of the use of neuroleptics in dementia have largely focused on patients with agitated behaviors (discussed below), many of whom also suffered from psychosis. There are few treatment studies that examined psychosis separately. In a small but relevant study, eight AD outpatients, all of whom had psychosis, seven of whom had other behavioral disturbances, were treated with haloperidol and placebo in a crossover design. Haloperidol was superior to placebo in reduction of psychosis and overall behavioral disturbances in five of the eight patients [66]. These results compare favorably to the small effect size reported in a meta-analysis of neuroleptic treatment in dementia-related noncognitive disturbances, which showed that 18 of 100 patients benefited from neuroleptic treatment beyond that of placebo [21]. Though the data are limited, some have posited that psychosis in dementia may be more responsive to neuroleptics than are other behavioral disturbances [67].

A number of neuroleptic agents have been studied for the treatment of dementia-related psychosis. (A list of neuroleptics and suggested dose ranges is presented in Table 2). Controlled studies suggest that there are no significant differences in efficacy between neuroleptic agents [68–70]. There are, however, noteworthy differences in adverse effects among the neuroleptics. High-potency dopamine-2 (D2) antagonist agents such as haloperidol are generally associated with more extrapyramidal symptoms (EPS) and akathisia than low-potency agents such as chlorpromazine and thioridazine. Low-potency agents, however,

Table 2. Medications used in the treatment of psychotic symptoms in demented patients

Generic name	Trade name[a]	Suggested daily oral dose range
Haloperidol	Haldol	0.25–2 mg
Fluphenazine	Prolixin	0.25–2 mg
Thiothixene	Navane	0.5–5 mg
Trifluoperazine	Stelazine	1–8 mg
Thioridazine	Mellaril	12.5–100 mg
Chlorpromazine	Thorazine	12.5–100 mg
Risperidone	Risperdal	0.5–2 mg
Clozapine	Clozaril	12.5–100 mg
Olanzapine	Zyprexa	5–10 mg

[a] Examples of trade names (USA)

have greater anticholinergic activity, which, in addition to causing adverse peripheral effects such as urinary retention and constipation, may worsen cognition, particularly in patients with cholinergic-deficit states such as AD. Whether the level of cognitive impairment in demented patients associated with one agent is greater than that with another has not been studied. One study reported that the high-potency agent haloperidol tended to worsen cognition in AD patients [66]. Sedation and orthostatic hypotension, mediated through antihistaminic and antiadrenergic activity, respectively, are also common adverse effects of neuroleptics and are more often associated with low-potency agents.

Clozapine, olanzapine, and risperidone, believed to produce antipsychotic activity through multiple receptor types, are associated with less EPS in young adults with schizophrenia [71, 72]. Published controlled studies of these agents in the elderly demented population are lacking. Clozapine and olanzapine, though, may be particularly useful in managing PD patients with drug-induced psychosis [73, 74].

Depression

Depression is found in all types of dementias. The degree to which depression is a disturbance in dementia, particularly in AD, is difficult to determine. Frequency, characteristics, and assessment of depression in dementia are controversial. Reported prevalence rates of depression in AD range from zero [75] to 87% [44], with a median rate of 41% [38]. Methodological variables that might explain this broad range include assessment instrument, source of information, patient population, and definitions of dementia, AD, and depression.

There are several instruments used to assess depression in dementia, most of which were not developed for specific use with demented or elderly patients. Widely used tools such as the Hamilton Depression Rating Scale [76] and the structured clinical interview for the third edition of the *Diagnostic and Statistical*

Manual of Mental Disorders (DSM-III) [77, 78] contain several items that do not distinguish between depression and dementia, such as sleep and appetite disturbances, psychomotor retardation and agitation, and somatic symptoms, which may result in overestimation of rates of depression in dementia [79]. Self-rating scales such as the Geriatric Depression Scale (GDS) [80], Beck Depression Inventory [81], and Zung Rating Scale [82] may not be appropriate for patients with limited memory and insight; the GDS was shown not to maintain validity in large numbers of AD patients [83] or in AD patients who deny cognitive deficits [84]. Instruments designed specifically for dementia patients, such as the Cornell Scale for Depression in Dementia (CSDD) [85], may prove to be more useful in this population; the CSDD was the most sensitive among five depression rating scales in detecting depression in AD patients [86]. The BEHAVE-AD, NRS, and NPI have depression subscales useful in assessing mood changes.

Other factors that may affect the rate of depression in AD are the source of information and the patient population. Caregiver reports yield a substantially higher frequency of depression than clinician observation [87]; this finding may be related to the reported high rate of depression among AD patient caregivers [88]. Patient self-report tends to yield lower degrees of depression than either caregiver report or clinician evaluation [89]. In a 1989 review, Wragg and Jeste [38] found that higher frequencies of depression were reported in populations drawn from inpatient psychiatric units and clinics associated with acute care settings, while lower frequencies were reported in dementia clinic outpatients and research subjects. One study of over 1, 300 AD patients from research databases reported only a 1.5% prevalence and 1.3% 2-year incidence of major depression [90]. AD patients, however, commonly have depressive symptoms without meeting criteria for major depression [62].

There is greater concurrence on the frequency of depression in other dementias. Depression is more frequent in VaD than in AD: at least 25% of VaD patients have major depression, while up to 60% have significant depressive symptomatology [8, 91, 92]. Depression is the most common neuropsychiatric feature of PD, affecting approximately 40% of patients [11], occurring with equal frequency in those with (30% of patients) and without dementia [93, 94]. Depression occurs in HD patients with a mean frequency of 23% [13] and is observed in patients with progressive supranuclear palsy [95] and with normal pressure hydrocephalus [96].

Researchers have attempted to describe possible clinical features that distinguish depression in dementia and primary depressive disorders. Depression in AD may be less pervasive and more modifiable by the environment than in primary depressive disorders [44]. Depression in PD may include more dysphoria, pessimism, irritability, suicidal ideation, and anxiety, but less guilt, self-blame, and psychosis, when compared to primary depressive disorders [97–99]. Studies comparing depressed and nondepressed PD patients found more bradykinesia, rigidity, and gait disturbance and less tremor in the depressed group [100, 101].

Studies have examined the relationship between depression and other aspects of dementia, including cognition, functional ability, and neurobiologic markers.

There are mixed data on the relationship between depression and cognition in dementia. Some researchers have found associations between depression and milder cognitive impairment in AD [102, 103], while others have found no association or greater cognitive impairment with depression [104, 105]. No association between depression and cognitive impairment was found in VaD patients [102]. Most studies reported that depression in AD and PD was associated with greater functional difficulties [103, 106, 107].

Neuropathologic studies implicate dysfunction of serotonergic and noradrenergic systems in depression in dementia [108–111]. Administration of cholinergic agents may precipitate depression in some AD patients [112]. Cummings et al. [8] hypothesized that the cholinergic deficit in AD may be protective against the occurrence of major depression.

Studies using PET found lower metabolism in the caudate nuclei and orbitofrontal cortex in depressed PD and HD patients when compared to their nondepressed counterparts [113, 114]. In AD patients, depression is associated with the absence rather than with the presence of frontal hypometabolism [52] and is correlated with parietal hypometabolism [53]. In VaD patients, depression and overall severity of neuropsychiatric symptoms are associated with the extent of subcortical white-matter ischemia [115].

Treatment

Difficulties with the assessment of depression in dementia may explain the limited number and conclusions of well-controlled pharmacotherapy studies. Three double-blind, placebo-controlled studies involving imipramine [116], maprotiline [117], and citalopram [118] have been performed. None of these studies found an overall significant difference in efficacy between drug and placebo. Nyth and Gottfries [118], however, found improvement in emotional bluntness, confusion, irritability, anxiety, fear/panic, depressed mood and restlessness only in the citalopram-treated group. In a double-blind study of nondepressed demented patients, Olafsson and colleagues [119] reported a trend towards improved mood, restlessness, anxiety, and irritability with fluvoxamine over placebo. Results from uncontrolled studies, though requiring greater caution in their interpretation, are more encouraging. Selective serotonin-reuptake inhibitors (SSRIs) such as fluoxetine and sertraline were reported to be useful in managing depression associated with dementia [120] and to specifically improve affect and reduce food refusal [121]. A significant advantage of most of the SSRIs over the tricyclic agents such as imipramine and maprotiline in demented patients is the absence of anticholinergic activity, which may have deleterious effects on already compromised cognition. In earlier studies, the monoamine oxidase inhibitors (MAOIs) selegiline and moclobemide were also shown to improve depression in demented patients [122, 123].

Few conclusions can be drawn from the existing data on the treatment of depression in dementia. The small number of controlled studies that exist provide no evidence that antidepressants are more efficacious than placebo. Uncontrolled

Table 3. Medications used in the treatment of depressive symptoms in demented patients

Class/generic name	Trade name[a]	Suggested daily oral dose range
Serotonin-reuptake inhibitors		
Fluoxetine	Prozac	10–20 mg
Paroxetine	Paxil	10–20 mg
Sertraline	Zoloft	25–100 mg
Fluvoxamine	Luvox	50–200 mg
Nefazodone	Serzone	100–400 mg
Trazodone	Desyrel	50–300 mg
Tricyclics		
Desipramine	Norpramin	25–150 mg
Nortriptyline	Pamelor	10–100 mg
Monoamine oxidase inhibitors		
Selegiline/l-deprenyl	Eldepryl	5–20 mg
Moclobemide	Aurorix, Manerix	100–400 mg
Other antidepressants		
Bupropion	Wellbutrin	75–300 mg
Venlafaxine	Effexor	75–225 mg
Psychostimulants		
Dextroamphetamine	Dexedrine	5–20 mg
Methylphenidate	Ritalin	5–20 mg

[a] Examples of trade names (USA)

studies suggest that SSRIs and MAOIs may be helpful and should be studied further. Some of the studies reported improvement in emotional factors other than mood, which should also be studied more rigorously. Additional drugs such as bupropion, nefazodone, and venlafaxine, which have relatively favorable side effect profiles, deserve attention as well. A list of agents potentially useful in treating mood changes in dementia and their recommended daily doses is presented on Table 3.

Anxiety

Relatively little is known about anxiety in demented patients. There are difficulties in studying anxiety in this population, some of which are inherent to anxiety and some of which are inherent to the population. Anxiety symptoms lie on a continuum with symptoms experienced by normal individuals; agreement on a case definition of anxiety is lacking [124]. Recognition of anxiety is largely dependent on subjective report, which may not be accurately elicited due to cognitive impairment and the tendency for the elderly to focus on somatic rather than psychological symptoms [125]. Comorbid depression [126, 127] and medical illnesses, which can produce physiological symptoms of anxiety, also make accurate assessment difficult.

Among the general elderly population, the 1-month prevalence of anxiety disorders, which include DSM-III panic disorder, phobic disorders, and obsessive-compulsive disorder [77], is 5.5% [128]. Studies that examined a broad spectrum of neuropsychiatric features in demented patients using caregiver report or clinician observation found higher prevalence rates of anxiety, from 27% to 48%; there were no significant differences between types of dementia or association with dementia severity in these studies [31, 33, 62]. One study that specifically focused on anxiety and depression in dementia, using a self-rated scale, reported a 38% prevalence rate of anxiety among mild dementia patients, significantly higher than the 9% in the control group; depression was also significantly higher in the patient group [129]. There was no significant correlation between anxiety and cognitive function. The relationship between depression and anxiety was not examined; therefore, it is not known whether the higher rate of anxiety may be attributable to depression rather than dementia.

Treatment

There are no published data on the pharmacologic treatment of anxiety in dementia to guide the clinician. Several classes of agents, including benzodiazepines, may be used (see Table 4). Oxazepam, lorazepam, and temazepam, because of their shorter elimination half-lives and lack of active metabolites, are preferred over other benzodiazepines for use in the elderly. All benzodiazepines should be used with caution, however, because of the risk of delirium, amnesia, disinhibition, and withdrawal reactions, particularly in this population [130]. Buspirone may be an effective alternative to benzodiazepines as an anxiolytic.

Table 4. Medications used in the treatment of anxiety symptoms in demented patients

Class/generic name	Trade name[a]	Suggested daily oral dose range
Serotonin reuptake inhibitors	See Table 3	See Table 3
Tricyclics	See Table 3	See Table 3
Benzodiazepines		
Clonazepam	Klonopin	0.25–1.0 mg
Lorazepam	Ativan	0.5–4.0 mg
Oxazepam	Serax	10–60 mg
Temazepam	Restoril	15–60 mg
Beta-blockers		
Atenolol	Tenormin	25–100 mg
Propranolol	Inderal	30–90 mg
Pindolol	Visken	10–60 mg
Buspirone	Buspar	15–60 mg

[a] Examples of trade names (USA)

Agitation

Agitation is defined by Cohen-Mansfield and Billig [131] as "inappropriate verbal, vocal, or motor activity that is not explained by needs or confusion [which are] not apparent"; manifestations include anxiety, irritability, uncooperativeness, self-injurious behaviors, physical or verbal aggressiveness, repetitive motor and vocal activity, sleep disturbance, and wandering. However, studies on agitation do not consistently adhere to this definition or explicitly define the term. Some investigations address behaviors such as wandering and aggressiveness as separate problems. Here agitation is discussed as a heterogeneous cluster of behaviors similar to the above as well as individual behaviors where appropriate, with the understanding that there exists variability in the use of the term "agitation." Table 5 presents commonly used assessment instruments for agitation.

Agitation occurs frequently in dementia, affecting between 38% and 65% of patients cross-sectionally [31, 62, 135, 136]. Agitated behaviors are associated with a number of comorbid conditions and poor outcomes, including psychosis and depression [137, 138], greater cognitive impairment [136, 139], greater functional impairment [135], faster cognitive decline [24, 140], increased caregiver burden and institutionalization [20, 141], and increased physical abuse directed at the patients [142]. Identification of comorbid psychosis and depression and differentiation of agitation from neuroleptic-induced akathisia have important implications for treatment.

Aggression

Aggressive behaviors have been studied separately from other agitated behaviors because of their direct impact on care, safety, and milieu. Indeed, aggressive behavior was one of three syndromes of agitation determined by factor analysis of agitation symptoms, the others being physically nonaggressive behavior and verbally agitated behavior [132]; this syndrome included hitting, spitting, cursing, intentionally falling, and sexual advances. Aggression is extremely common in special populations. A study of cognitively impaired nursing home patients found that 86.3% exhibited some form of aggression during a 7-day period [134]. However, in this study, recordings were made by nursing staff, who are among the most frequent victims of aggression [143, 144], which may have biased the data. These studies found that most aggressive behavior occurred in the context of caregiving and assistance with activities of daily living [134, 143, 144].

Wandering

Wandering is defined as moving about without a fixed course, aim or goal, or towards an inappropriate, impossible goal [145]. Very little is known about wandering, despite its inherent potential for danger, caregiver burden, and tendency to precipitate institutionalization. Reported prevalence rates are as high as 70% among community-dwelling patients [146]. Dementia patients who wander have

Table 5. Instruments for assessment of agitation in dementia and their respective items

Cohen-Mansfield Agitation Inventory (adapted from [132])	Overt Aggression Scale (adapted from [133])	Ryden Aggression Scale (adapted from [134])
Wandering	Loud noises, angry shouting	Biting
Inappropriate dress/disrobing	Yelling insults	Elbowing
Spitting	Cursing/moderate verbal threats	Hitting/punching
Cursing/verbal aggression	Clear verbal threats of violence toward others	Kicking
Unwarranted request for help	Slamming door, scattering clothing, making mess	Pinching/squeezing
Repetitive sentences/questions	Throwing objects down, kicking furniture, marking walls	Pushing/shoving
Hitting	Breaking objects, smashing windows	Pulling hair
Kicking	Setting fires, throwing objects dangerously	Scratching
Grabbing others	Scratching self, pulling own hair	Slapping
Pushing	Banging head, throwing self on floor, hitting objects with fists	Spitting
Throwing things	Inflicting small cuts, bruises, minor burns on self or others	Tackling
Strange noises	Inflicting deep cuts, bites that bleed, loss of consciousness, internal injury, fractures on self or others	Making threatening gestures
Screaming	Making threatening physical gestures, swinging, grabbing at others	Throwing an object
Biting	Striking, kicking, pushing others, pulling others' hair	Striking a person with an object
Scratching		Brandishing a weapon
Trying to get to different place		Using a weapon
Intentional falling		Property damage
Complaining		Cursing at a person
Eating/drinking inappropriate substances		Hostile language
Hurting self or others		Making verbal threats
Handling things inappropriately		Name calling
Hiding things		
Hoarding things		
Destroying property		
Repetitious mannerisms		
Verbal sexual advances		
Physical sexual advances		
General restlessness		

been reported to exhibit greater graphomotor perseveration [147], disruption of circadian rhythm [148], cognitive impairment [40], and other agitated behaviors [149], as well as to sustain more fractures [150], than nonwandering demented patients. Anecdotal strategies reported to successfully manage wandering in a structured setting include placing a two-dimensional grid pattern on the floor which demented patients perceive as a barrier [151, 152] and scheduling opportunities for structured movement and sensory stimulation in a nonrestrictive environment [153].

Treatment

Pharmacologic treatment of agitation in dementia is generally discussed in terms of two broad categories: neuroleptic and nonneuroleptic treatment. Refer to Table 2 and the section on psychosis for a list of agents, recommended daily doses, and adverse effects. Neuroleptics are the most commonly prescribed psychotropic medication in demented patients [154, 155]. Despite their widespread use, there are relatively few well-controlled studies of neuroleptics in the treatment of dementia-related behavioral disturbances. Devanand and Levy [67] identified only three random-assignment, double-blind, placebo-controlled studies published over the past two decades; the medications tested were thioridazine and loxapine [156], thiothixene [68], and haloperidol and loxapine [157]. Each of these studies demonstrated (marginal, in the first two studies) superiority of the neuroleptic over placebo, but suffered from at least one of the following methodological flaws: the inclusion of different types of dementia in subject samples, the use of concomitant psychotropic medications, a high placebo response, and global rather than specific outcome measures.

A meta-analysis of 17 placebo-controlled studies published between 1955 and 1985 was conducted by Schneider and colleagues [21]. The medications studied included chlorpromazine, thioridazine, haloperidol, loxapine, trifluoperazine, acetophenazine, thiothixine, penfluridol, and milenperone. Doses used in these studies ranged from 66 to 267 mg/day chlorpromazine equivalents. Trial duration ranged from 3 to 18 weeks. Sample sizes ranged from 18 to 71. No individual study demonstrated neuroleptic superiority over placebo according to the reviewers' calculations. Overall results showed that neuroleptic treatment changed the improvement rate from 0.41 to 0.59, or that 18 of 100 patients benefited from neuroleptic treatment beyond that of placebo, a relatively small effect size. One may interpret these results at least two different ways: withholding neuroleptics may deny improvement to 18% of patients; conversely, the high placebo rate and medication nonresponse rate suggest that a significant number of patients are receiving neuroleptics unnecessarily. However, there are considerable caveats to any interpretation of these results. Only nine studies contained at least a large predominance of dementia patients (> 78% of subjects), who may respond differently to treatment than patients with other disorders accompanied by agitation. All 17 studies were of inpatients, which might suggest that dementia severity in the study population was greater than in the general dementia population. Not all of the individual studies considered factors such as clinical worsening, noncompleters, and the impact of treatment-emergent effects, which, if reported, could have substantially affected the overall effect size in the meta-analysis. Together, the results and limitations of these past studies suggest that the role of neuroleptics in the treatment of agitation in dementia remains to be determined.

A variety of nonneuroleptic medications has been studied, some in comparison to placebo or to neuroleptics, in the treatment of agitation in demented patients. Table 6 provides a list of agents and their recommended daily doses. Benzodiazepines, along with neuroleptics, are among the most commonly used class of medications in this setting. However, there are no compelling data to support their use. Three placebo-controlled studies found benzodiazepines to be helpful in elderly patients with behavioral problems. Chesrow and colleagues [158] found that oxazepam was effective in 87% of patients, compared to 51% with chlordiazepoxide and 0% with placebo. Chlordiazepoxide was associated with more drowsiness, dysarthria, ataxia, and dizziness than oxazepam or placebo. In another study, diazepam reduced anxiety, hyperactivity, and pacing in comparison to placebo, but the 7.5 mg dose had to be reduced due to drowsiness in a majority of subjects [159]. Beber [160] reported improvement in anxiety, tension, irritability, insomnia, and agitation with oxazepam over placebo (68% vs 18%). A more recent, non-placebo-controlled comparison trial of haloperidol, oxazepam, and diphenhydramine demonstrated modest efficacy with all three agents and no statistically significant differences among them [161].

The anticonvulsants, carbamazepine and valproate, may have a role for treatment of agitation in demented patients. One double-blind, placebo-controlled study found significant benefits of carbamazepine (mean dose 300 mg/day, mean level 5.7 μg/ml) over placebo [162], while another study did not [163]. Open stud-

Table 6. Medications used in the treatment of agitated behaviors in demented patients

Class/generic name	Trade name[a]	Suggested daily oral dose range
Neuroleptics	See Table 2	See Table 2
Benzodiazepines	See Table 4	See Table 4
Anticonvulsants		
Carbamazepine	Tegretol	100–800 mg
Valproic acid	Depakote, Depakene	125–1000 mg
Serotonin-reuptake inhibitors	See Table 3	See Table 3
Selegiline/l-deprenyl	Eldepryl	5–20 mg
Buspirone	Buspar	15–60 mg
Lithium		150–900 mg
Beta-blockers	See Table 4	See Table 4
Tacrine	Cognex	80–160 mg
Donepezil	Aricept	5–10 mg

[a] Examples of trade names (USA)

ies reported up to a 50% reduction in aggressive behaviors [164–166]. The largest prospective study of valproate (375–750 mg/day) enrolled ten subjects, eight of whom showed a 50% or greater reduction in agitation frequency [167].

Serotonergic agents have been studied in agitated demented patients. Rationale for their use includes evidence of a relationship between serotonergic dysfunction and aggression in nondemented subjects [168, 169], and of reduced agitation and restlessness following administration of a selective serotonin receptor agonist, metachlorophenylpiperazine (m-CPP), in AD patients [170]. Recent small, open studies of trazodone have reported decreased irritability, restlessness, and agitation [171, 172]. In a double-blind comparison of trazodone and haloperidol, equal improvement in overall psychopathology was found in both groups, with the suggestion that trazodone (mean dose 218 mg/day) was more effective for repetitive behaviors, verbal aggression, negativism, and resistance to care [173]. A large, well-controlled study of citalopram (20–30 mg/day) showed decreased irritability and restlessness [118]. Double-blind, placebo-controlled studies of alaproclate [174] and fluvoxamine [119], and an open trial of fluoxetine [175] reported no significant benefits for treating agitation in dementia.

Open trials of selegiline suggest possible benefits for agitation, tension, excitement and other neuropsychiatric disturbances at 10 mg/day [123, 176, 177]. A double-blind study of mild-to-moderate AD found no difference between selegiline 10 mg/day and placebo at 2 months, but at 15 months, noted a modest improvement in overall psychopathology in the selegiline-treated group [178].

Additional medications that have been reported (in small, controlled studies) to have some benefits in treating agitated behaviors in demented patients include buspirone [179, 180], lithium [181], and beta-receptor-blocking agents [182, 183]. Of interest also is one open trial of tacrine, a cholinesterase inhibitor initially promoted to improve cognition in AD, which demonstrated improvement in anxiety, apathy, hallucinations, aberrant motor behaviors, and disinhibition in AD patients after treatment [184].

Personality Changes

Dementia affects a broad range of personality traits. In comparison to nondemented retirees of similar age, AD patients are more out of touch, more unreasonable, less self-reliant, less enthusiastic, less mature, more lifeless, less affectionate, less kind, and less generous [15]. As with agitation, personality changes refer to a heterogeneous constellation of disturbances; however, there exist studies on more common specific personality changes such as apathy and disinhibition, which accompany dementia.

Studies suggest that the assessment of personality alterations may help distinguish between different types of dementias. Apathy and disinhibition are significantly greater in severity and frequency in FTD patients than in AD patients [33]. AD and VaD patients are largely similar in their personality alterations, with the exception of loss of maturity, which is significantly greater in AD patients [185].

Apathy is significantly greater in degree and frequency in patients with progressive supranuclear palsy than in patients with AD [34]. The prevalence of apathy in HD and AD patients is similar, but the degree of apathy is greater in HD patients [186]. Personality changes also occur with PD [187] and normal-pressure hydrocephalus [96].

Data on the relationship between personality changes and dementia severity in AD are mixed. Some cross-sectional studies reported no relationship [185, 188]. Reichman and colleagues [189] found a positive correlation. Litvan and colleagues [34] found a significant association between apathy and executive dysfunction. Longitudinal studies found that the percentage of AD patients with personality changes increases substantially over time [190], but that such changes at the onset of disease may follow a stable course in individual AD patients [17].

Apathy

Apathy is defined as diminished motivation not attributable to a decreased level of consciousness, cognitive impairment, or emotional distress [191]. Individuals exhibit social and occupational withdrawal and reduced interpersonal engagement. Apathy is a syndrome distinct from depression, which has features related to diminished motivation, such as anhedonia, anergy, and psychomotor retardation [192].

Neuroanatomical and neurochemical correlates of apathy and related disturbances have been described. Motivation is mediated via a medial frontal-subcortical circuit originating in the anterior cingulate gyrus [193], and disease processes such as FTD affecting this circuit produce apathy. Levy and colleagues [33] reported that 21 of 22 FTD patients exhibited apathy, the most frequent neuropsychiatric disturbance in the group. Apathy is sometimes the earliest symptom in FTD [10]. Apathy is also common in VaD [194], PD [187], and HD [186], all disease processes that can affect the medial frontal-subcortical circuit. Passivity and indifference are related to dysfunction of the parietal lobes, medial temporal cortex, and convexity of the frontal lobes, structures affected in AD [17, 195]. Placidity can result from deficits in cholinergic input from the limbic system to the cortex due to atrophy of the nucleus basalis of Meynert in AD [196, 197]. The high prevalence of apathy in AD, between 61 and 80%, supports these models [33, 62, 188].

Disinhibition

Disinhibition and related behaviors in dementia include impulsivity, intrusiveness, impatience, irritability, mood lability, and inappropriate verbalizations and gestures. Disruption of orbitofrontal-subcortical circuitry results in such behaviors; dementing illnesses that may affect parts of this circuit include FTD, VaD, HD, PD, progressive supranuclear palsy, and Creutzfeldt-Jakob disease [195]. Disinhibition is perhaps most prevalent in FTD, occurring in 68% of patients [33], and not uncommonly as the earliest symptom [10]. Disinhibition is also common in AD, occurring in 23 to 36% of patients [31, 33, 62]. Sexual disinhibition occurs at a much lower frequency, occurring in 2.9% to 7% of AD patients [4, 135].

Klüver-Bucy Syndrome

The KBS was originally described in monkeys following bitemporal lobectomy, which resulted in these changes: visual agnosia, hyperorality, hypermetamorphosis (excessive attention and reaction to visual stimuli), hypersexuality, and placidity [198]. KBS has been described in humans following bitemporal lobectomy [199], in AD [200], and in Pick's disease [201].

Subsequent studies have demonstrated that AD and FTD patients commonly have features of KBS, but rarely the complete syndrome [4, 10]. Burns and colleagues [4] found that over 80% of AD patients had at least one feature, most commonly visual agnosia, but only one of the 178 subjects had the full KBS. In that study patients with at least one feature of KBS were more likely to have temporal atrophy on CT; however, individual features such as hyperorality and hypersexuality were associated with other areas of neuroanatomical change, while aggression, a behavior not part of KBS, was associated with temporal atrophy. Neuropathologic correlates of KBS features reported in AD were decreased number of neurons in the parahippocampal gyrus and parietal neocortex [202]. In individual FTD patients, temporal hypoperfusion and atrophy were not necessarily associated with or requisite for KBS behaviors [10]. There was one case report of KBS in HD, a subcortical dementia, which responded to treatment with haloperidol [203].

Treatment

There are case reports of improvement in personality changes after pharmacologic treatment. One study reported seven cases of apathy in a variety of neuropsychiatric disorders, including dementia, each of which improved with one or more of the following: amantadine, levodopa, bromocriptine, amphetamine, methylphenidate, bupropion, and selegiline [204]. Dementia-related disinhibition has been reported to respond to clomipramine [205] and buspirone [206].

Summary

Neuropsychiatric symptoms and alterations are important aspects of dementia. Prevalence studies support the common occurrence of these disturbances in demented patients. Accurate clinical assessment, which may be aided by available rating scales, provides valuable diagnostic information, as suggested by studies reporting different behavioral profiles among groups based on dementia type. Some neuropsychiatric symptoms are associated with specific neurobiological changes and may represent separate processes with meaningful prognostic and treatment implications. Recognition enables the family, caregiver, and clinician to better meet present and anticipate future needs of the demented patient, such as appropriate level of care, support for the caregiver, and pharmacologic intervention. Data suggest that dementia-related neuropsychiatric symptoms are responsive to pharmacologic treatment, though the paucity of well-controlled studies indicates a need for additional investigations.

References

1. Burns A, Jacoby R, Levy R (1990) Psychiatric phenomena in Alzheimer's disease. I. Disorders of thought content. Br J Psychiatry 157: 72-76
2. Burns A, Jacoby R, Levy R (1990) Psychiatric phenomena in Alzheimer's disease. II. Disorders of perception. Br J Psychiatry 157: 76-81
3. Burns A, Jacoby R, Levy R (1990) Psychiatric phenomena in Alzheimer's disease. III. Disorders of mood. Br J Psychiatry 157: 81-86
4. Burns A, Jacoby R, Levy R (1990) Psychiatric phenomena in Alzheimer's disease. IV. Disorders of behaviour. Br J Psychiatry 157: 86-94
5. Cummings JL, Victoroff JI (1990) Noncognitive neuropsychiatric syndromes in Alzheimer's disease. Neuropsychiatry Neuropsychol Behav Neurol 3: 140-158
6. Reisberg B, Borenstein J, Salob SP, Ferris SH, Franssen E, Georgotas A (1987) Behavioral symptoms in Alzheimer's disease: phenomenology and treatment. J Clin Psychiatry 48[Suppl 5]: 9-15
7. Cummings JL, Benson DF (1992) Dementia: a clinical approach, 2nd edn. Butterworth-Heinemann, Boston
8. Cummings JL, Miller B, Hill MA, Neshkes R (1987) Neuropsychiatric aspects of multi-infarct dementia and dementia of the Alzheimer type. Arch Neurol 44: 389-393
9. Dian L, Cummings JL, Petry S, Hill MA (1990) Personality alterations in multi-infarct dementia. Psychosomatics 31: 415-419
10. Miller BL, Cummings JL, Villanueva-Meyer J, Boone K, Mehringer CM, Lesser IM, Mena I (1991) Frontal lobe degeneration: clinical, neuropsychological, and SPECT characteristics. Neurology 41: 1374-1382
11. Cummings JL (1992) Depression and Parkinson's disease: a review. Am J Psychiatry 149: 443–453
12. Stein MB, Heuser IJ, Juncos JL, Uhde TW (1990) Anxiety disorders in patients with Parkinson's disease. Am J Psychiatry 147: 217-220
13. Mendez MF (1994) Huntington's disease: update and review of neuropsychiatric aspects. Int J Psychiatry Med 24: 189-208
14. Lesser IM, Miller BL, Boone KB, Hill-Gutierrez E, Mena I (1989) Psychosis as the first manifestation of degenerative dementia. Bull Clin Neurosc 54: 59-63
15. Petry S, Cummings JL, Hill MA, Shapira J (1988) Personality alterations in dementia of the Alzheimer type. Arch Neurol 45: 1187-1190
16. Rubin EH, Kinscherf DA (1989) Psychopathology of very mild dementia of the Alzheimer type. Am J Psychiatry 146: 1017-1021
17. Petry S, Cummings JL, Hill M A, Shapira J (1989) Personality alterations in dementia of the Alzheimer type: a three-year follow-up study. J Geriatr Psychiatry Neurol 4: 203-207
18. Deutsch LH, Bylsma FW, Rovner BW, Steele C, Folstein MF (1991) Psychosis and physical aggression in probable Alzheimer's disease. Am J Psychiatry 148: 1159-1163
19. Morriss RK, Rovner BW, Folstein MF, German PS (1990) Delusions in newly admitted residents of nursing homes. Am J Psychiatry 147: 299-302

20. Steele C, Rovner B, Chase GA, Folstein M (1990) Psychiatric symptoms and nursing home placement of patients with Alzheimer's disease. Am J Psychiatry 147: 1049-1051
21. Schneider LS, Pollock EE, Lyness SA (1990) A metaanalysis of controlled trials of neuroleptic treatment in dementia. J Am Geriatr Soc 38: 553-563
22. Tariot PN, Schneider LS, Katz IR (1995) Anticonvulsant and other non-neuroleptic treatment of agitation in dementia. J Geriatr Psychiatry Neurol 8[Suppl 1]: S28-S39
23. Drevets WC, Rubin EH (1989) Psychotic symptoms and the longitudinal course of senile dementia of the Alzheimer type. Biol Psychiatry 25: 39-48
24. Mortimer JA, Ebbitt B, Jun S-P, Finch MD (1992) Predictors of cognitive and functional progression in patients with probable Alzheimer's disease. Neurology 42: 1689-1696
25. Rosen J, Zubenko GS (1991) Emergence of psychosis and depression in the longitudinal evaluation of Alzheimer's disease. Biol Psychiatry 29: 224-232
26. Stern Y, Mayeux R, Sano M, Hauser WA, Bush T (1987) Predictors of disease course in patients with probable Alzheimer's disease. Neurology 37: 1649-1653
27. Folstein M, Folstein S, McHugh P (1975) Mini-Mental State: a practical method for grading the cognitive state of patients for the clinician. J Psychiatr Res 12: 189-198
28. Levin HS, High WM, Goethe KE, Sisson RA, Overall JE, Rhoades HM, Eisenberg HM, Kalisky Z, Gary HE (1987) The neurobehavior rating scale: assessment of the sequelae of head injury by the clinician. J Neurol Neurosurg Psychiatry 50: 183-193
29. Cummings JL, Mega M, Gray K, Rosenberg-Thompson S, Carusi DA, Gornbein J (1994) The neuropsychiatric inventory: comprehensive assessment of psychopathology in dementia. Neurology 44: 2308-2314
30. Reisberg B, Franssen E, Sclan S, Kluger A, Ferris SH (1989) Stage specific incidence of potentially remediable behavioral symptoms in aging and Alzheimer disease: a study of 120 patients using the BEHAVE-AD. Bull Clin Neurosci 54: 95-112
31. Sultzer DL, Levin HS, Mahler ME, High WM, Cummings JL (1992) Assessment of cognitive, psychiatric, and behavioral disturbances in patients with dementia: the Neurobehavioral Rating Scale. J Am Geriatr Soc 40: 549-555
32. Sultzer DL, Levin HS, Mahler ME, High WM, Cummings JL (1993) A comparison of psychiatric symptoms in vascular dementia and Alzheimer's disease. Am J Psychiatry 150: 1806-1812
33. Levy ML, Miller BL, Cummings JL, Fairbanks LA, Craig A (1996) Alzheimer disease and frontotemporal dementias: behavioral distinctions. Arch Neurol 53: 687-690
34. Litvan I, Mega MS, Cummings JL, Fairbanks L (1996) Neuropsychiatric aspects of progressive supranuclear palsy. Neurology 47: 1184-1189
35. Forsell Y, Jorm AF, Winblad B (1993) Variation in psychiatric and behavioural symptoms at different stages of dementia: data from physicians' examinations and informants' reports. Dementia 4: 282-286
36. Birkett DP (1972) The psychiatric differentiation of senility and arteriosclerosis. Br J Psychiatry 120: 321-325
37. Leuchter AF, Spar JE (1985) The late-onset psychoses. J New Ment Disord 173: 488-494
38. Wragg RE, Jeste DV (1989) Overview of depression and psychosis in Alzheimer's disease. Am Psychiatry 146: 577-587
39. Berrios GE, Brook P (1985) Delusions and the psychopathology of the elderly with dementia. Acta Psychiatr Scand 72: 296-301
40. Teri L, Larson EB, Reifler BV (1988) Behavioral disturbance in dementia of the Alzheimer's type. J Am Geriatr Soc 36: 1-6
41. Swearer JM, Drachman DA, O'Donnell BF, Mitchell AL (1988) Troublesome and disruptive behaviors in dementia. J Am Geriatr Soc 36: 784-790

42. Cooper JK, Mungas D, Weiler PG (1990) Relation of cognitive status and abnormal behaviors in Alzheimer's disease. J Am Geriatr Soc 38: 867-870
43. Flynn FG, Cummings JL, Gornbein J (1991) Delusions in dementia syndromes: investigation of behavioral and neuropsychological correlates. J Neuropsychiatry Clin Neurosci 3: 364-370
44. Merriam AE, Aronson MD, Gaston P, Wey SL, Katz I (1988) The psychiatric symptoms of Alzheimer's disease. J Am Geriatr Soc 36: 7-12
45. Jeste DV, Wragg RE, Salmon DP, Harris MJ, Thal LJ (1992) Cognitive deficits of patients with Alzheimer's disease with and without delusions. Am J Psychiatry 149: 184-189
46. Doody RS, Massman P, Mahurin R, Law S (1995) Positive and negative neuropsychiatric features in Alzheimer's disease. J Neuropsychiatry Clin Neurosci 7: 54-60
47. Naimark D, Jackson E, Rockwell E, Jeste DV (1996) Psychotic symptoms in Parkinson's disease patients with dementia. J Am Geriatr Soc 44: 296-299
48. Drachman DA, O'Donnell BF, Lew RA, Swearer JM (1990) The prognosis in Alzheimer's disease. "How far" rather than "how fast" best predicts the course. Arch Neurol 47: 851-856
49. Jagust WJ, Budinger TF, Reed BR (1987) The diagnosis of dementia with single photon emission computed tomography. Arch Neurol 44: 258-262
50. Woods SW, Pearsall HR, Seibyl JP, Hoffer PB (1991) The Quinn Essay: single-photon emission computed tomography in neuropsychiatric disorders. In: Hoffer PB (ed) Yearbook of nuclear medicine. Mosby Year Book, St Louis
51. Smith GS, de Leon MJ, George AE, Kluger A, Volkow ND, McRae T, Golomb J, Ferris SH, Reisberg B, Ciaravino J (1992) Topography of cross-sectional and longitudinal glucose metabolic deficits in Alzheimer's disease: pathophysiologic implications. Arch Neurol 49: 1142-1150
52. Grady CL, Haxby JV, Schapiro MB, Gonzalez-Aviles A, Kumar A, Ball MJ, Heston L, Rapoport SI (1990) Subgroups in dementia of the Alzheimer type identified using positron emission tomography. J Neuropsychiatry Clin Neurosci 2: 373-384
53. Sultzer DL, Mahler ME, Mandelkern MA, Cummings JL, Van Gorp WG, Hinkin CH, Berisford MA (1995) The relationship between psychiatric symptoms and regional cortical metabolism in Alzheimer's disease. J Neuropsychiatry Clin Neurosci 7: 476-484
54. Kotrla KJ, Chacko RC, Harper G, Jhingran S, Doody R (1995) SPECT findings on psychosis in Alzheimer's disease. Am J Psychiatry 152: 1470-1475
55. Starkstein SE, Vazquez S, Petracca G, Sabe L, Migliorelli R, Teson A, Leiguarda R (1994) A SPECT study of delusions in Alzheimer's disease. Neurology 44: 2055-2059
56. Zubenko GS, Moossy J, Martinez AJ, Rao G, Claassen D, Rosen J, Kopp U (1991) Neuropathologic and neurochemical correlates of psychosis in primary dementia. Arch Neurol 48: 619-624
57. Cummings JL, Gorman DG, Shapira J (1993) Physostigmine ameliorates the delusions of Alzheimer's disease. Biol Psychiatry 33: 536-541
58. Paulus W, Jellinger K (1991) The neuropathologic basis of different clinical subgroups of Parkinson's disease. J Neuropathol Exp Neurol 50: 743-755
59. McKeith IG, Perry RH, Fairbairn SJ, Jabeen S, Perry EK (1992) Operational criteria for senile dementia of the Lewy body type (SDLT). Psychol Med 22: 911-922
60. Roberts JK, Trimble MR, Robertson M (1983) Schizophrenic psychosis associated with aqueduct stenosis in adults. J Neurol Neurosurg Psychiatry 46: 892-898
61. Mendez MF, Martin RJ, Smyth KA, Whitehouse PJ (1990) Psychiatric symptoms associated with Alzheimer's disease. J Neuropsychiatry Clin Neurosci 2: 28-33
62. Mega MS, Cummings JL, Fiorello T, Gornbein J (1996) The spectrum of behavioral changes in Alzheimer's disease. Neurology 46: 130-135

63. Perry EK, Marshall E, Kerwin J, Smith CJ, Jabeen S, Cheng AV, Perry RH (1990) Evidence of a monoaminergic-cholinergic imbalance related to visual hallucinations in Lewy body dementia. J Neurochem 55: 1454-1456
64. Cummings JL (1991) Behavioral complications of drug treatment of Parkinson's disease. J Am Geriatr Soc 39: 708-716
65. Goetz CG, Tanner CM, Klawans HL (1982) Pharmacology of hallucinations induced by long-term drug therapy. Am J Psychiatry 139: 494-497
66. Devanand DP, Sackeim HA, Brown RP, Mayeux R (1989) A pilot study of haloperidol treatment of psychosis and behavioral disturbance in Alzheimer's disease. Arch Neurology 46: 854-857
67. Devanand DP, Levy SR (1995) Neuroleptic treatment of agitation and psychosis in dementia. J Geriatr Psychiatry Neurol 8[Suppl 1]: S18-S27
68. Barnes R, Veith R, Okimoto J, Raskind M, Gumbrecht G (1982) Efficacy of antipsychotic medications in behaviorally disturbed dementia patients. Am J Psychiatry 139: 1170-1174
69. Carlyle W, Ancill RJ, Sheldon L (1993) Aggression in the demented patient: a double-blind study of loxapine versus haloperidol. Int Clin Psychopharmacol 8: 103-108
70. Cowley LM, Glen RS (1979) Double-blind study of thioridazine and haloperidol in geriatric patients with a psychosis associated witih organic brain syndrome. J Clin Psychiatry 40: 411-419
71. Chouinard G, Jones B, Remington G, Bloom D, Addington D, MacEwan GW, Labelle A, Beauclair L, Arnott W (1993) A Canadian multicenter placebo-controlled study of fixed doses of risperidone and haloperidol in the treatment of chronic schizophrenic patients. J Clin Psychopharmacol 13: 25-40
72. Kurz M, Hummer M, Oberbauer H, Fleischhacker WW (1995) Extrapyramidal side effects of clozapine and haloperidol. Psychopharmacology 118: 52-56
73. Musser WS, Akil M (1996) Clozapine as a treatment for psychosis in Parkinson's disease: a review. J Neuropsychiatry Clin Neurosci 8: 1-9
74. Wolters EC, Jansen ENH, Tuynman-Qua HG, Bergmans PLM (1996) Olanzapine in the treatment of dopaminomimetic psychosis in patients with Parkinson's disease. Neurology 47: 1085-1087
75. Knesevich JW, Martin RL, Berg L, Danziger W (1983) Preliminary report on affective symptoms in the early stages of senile dementia of the Alzheimer type. Am J Psychiatry 140: 233-235
76. Hamilton M (1960) A rating scale for depression. J Neurol Neurosurg Psychiatry 23: 56-61
77. American Psychiatric Association (1980) Diagnostic and statistical manual of mental disorders, 3rd edn. American Psychiatric Press, Washington DC
78. Spitzer RL, Williams JBW, Gibbon M, First MB (1988) Structured clinical interview guide for DSM-III-R. New York State Psychiatric Institute, Biometrics Research Department, New York
79. Burke WJ, Rubin EH, Morris JC, Berg L (1988) Symptoms of "depression" in dementia of the Alzheimer type. Alzheimer Dis Assoc Disord 2: 356-362
80. Yesavage JA, Brink TL, Rose TL, Rose TL, Lum O, Huang V, Adey M, Leirer VO (1983) Development and validation of a geriatric depression screening scale: a preliminary report. J Psychiatr Res 17: 37-49
81. Beck AT, Ward CH, Mendelson M, Mock J, Erbaugh J (1961) An inventory for measuring depression. Arch Gen Psychiatry 4: 561-571
82. Zung WWK (1965) A self-rating depression scale. Arch Gen Psychiatry 12: 56-65

83. Burke WJ, Houston MJ, Boust SJ, Roccaforte WH (1989) Use of the Geriatric Depression Scale in dementia of the Alzheimer type. J Am Geriatr Soc 37: 856-860
84. Feher EP, Larrabee GJ, Crook TH (1992) Factors attenuating the validity of the Geriatric Depression Scale in a dementia population. J Am Geriatr Soc 40: 906-909
85. Alexopoulos GS, Abrams RC, Young RC, Shamoian CA (1988) Cornell scale for depression in dementia. Biol Psychiatry 23: 271-284
86. Logsdon RG, Teri L (1995) Depression in Alzheimer's disease patients: caregivers as surrogate reporters. J Am Geriatr Soc 43: 150-155
87. Mackenzie TB, Robiner WN, Knopman DS (1989) Differences between patient and family assessments of depression in Alzheimer's disease. Am J Psychiatry 146: 1174-1178
88. Cohen D, Eisdorfer C (1988) Depression in famiy members caring for a relative with Alzheimer's disease. J Am Geriatr Soc 36: 885-889
89. Teri L, Wagner AW (1991) Assessment of depreion in patients with Alzheimer's disease: concordance among informants. Psychol Aging 6: 280-285
90. Weiner MF, Edland SD, Luszczynska H (1994) Prevalence and incidence of major depression in Alzheimer's disease. Am J Psychiatry 151: 1006-1009
91. Cummings JL (1988) Depression in vascular dementia. Hillside J Clin Psychiatry 10: 209-231
92. Erkinjuntti T (1987) Types of multi-infarct dementia. Acta Neurol Scand 75: 391-399
93. Huber SJ, Shuttleworth EC, Paulson GW (1986) Dementia in Parkinson's disease. Arch Neurol 43: 987-990
94. Lieberman A, Dziatolowski M, Kupersmith M, Serby M, Goodgold A, Korein J, Goldstein M (1979) Dementia in Parkinson disease. Ann Neurol 6: 355-359
95. Menza MA, Cocchiola J, Golbe LI (1995) Psychiatric symptoms in progressive supranuclear palsy. Psychosomatics 36: 550-554
96. Pujol J, Leal S, Fluvia X, Conde C (1989) Psychiatric aspects of normal-pressure hydrocephalus: a report of five cases. Br J Psychiatry 154[Suppl 4]: 77-80
97. Brown RG, MacCarthy B (1990) Psychiatric morbidity in patients with Parkinson's disease. Psychol Med 20: 77-87
98. Brown RG, MacCarthy B, Gotham AM, Der GJ, Marsden CD (1988) Depression and disability in Parkinson's disease: a follow-up study of 132 cases. Psychol Med 18: 49-55
99. Gotham AM, Brown RG, Marsden CD (1986) Depression in Parkinson's disease: a quantitative and qualitative analysis. J Neurol Neurosurg Psychiatry 49: 381-389
100. Huber SJ, Paulson GW, Shuttleworth EC (1988) Depression in Parkinson's disease. Neuropsychiatry Neuropsychol Behav Neurol 1: 47-51
101. Jankovic J, McDermott M, Carter J, Gauthier S, Goetz C, Golbe L, Huber, S, Koller W, Olanow C, Shoulson I, Stern M, Tanner C, Weiner W (1990) Variable expression of Parkinson's disease: a base-line analysis of the DATATOP cohort. The Parkinson Study Group. Neurology 40: 1529-1534
102. Fischer P, Simanyi M, Danielczyk W (1990) Depression in dementia of the Alzheimer type and in multi-infarct dementia. Am J Psychiatry 147: 1484-1487
103. Pearson JL, Teri L, Reifler BV, Raskind MA (1989) Functional status and cognitive impairment in Alzheimer's patients with and without depression. J Am Geriatr Soc 37: 1117-1121
104. Breen AR, Larson EB, Reifler BV, Vitaliano PP, Lawrence GL (1984) Cognitive performance and functional competence in coexisting dementia and depression. J Am Geriatr Soc 32: 132-137
105. Rovner BW, Broadhead J, Spencer M, Carson K, Folstein MF (1989) Depression and Alzheimer's disease. Am J Psychiatry 146: 350-353

106. Fitz AG, Teri L (1994) Depression, cognition, and functional ability in patients with Alzheimer's disease. J Am Geriatr Soc 42: 186-191
107. Starkstein SE, Preziosi TJ, Bolduc PL, Robinson RG (1990) Depression in Parkinson's disease. J Nerv Ment Disord 178: 27-31
108. Chan-Palay V, Asan E (1989) Alterations in catecholamine neurons of the locus ceruleus in senile dementia of the Alzheimer type and in Parkinson's disease with and without dementia and depression. J Comp Neurol 287: 373-392
109. Forstl H, Levy R, Burns A, Luthert P, Cairns N (1994) Disproportionate loss of nora-drenergic and cholinergic neurons as cause of depression in Alzheimer's disease – a hypothesis. Pharmacopsychiatry 27: 11-15
110. Mayeux R, Stern Y, Sano M, Williams JB, Cote LJ (1988) The relationship of serotonin to depression in Parkinson's disease. Mov Disord 3: 237-244
111. Zubenko GS, Moossy J (1988) Major depression in primary dementia: clinical and neuropathologic correlates. Arch Neurol 45: 1182-1186
112. Davis KL, Hollander E, Davidson M, Davis M, Mohs RC, Horvath TB (1987) Induction of depression with oxotremorine in patients with Alzheimer's disease. Am J Psychiatry 144: 468-471
113. Mayberg HS, Starkstein SE, Sadzot B, Preziosi T, Andrezejewski PL, Dannals RF, Wagner HN Jr, Robinson RG (1990) Selective hypometabolism in the inferior frontal lobe in depressed patients with Parkinson's disease. Ann Neurol 28: 57-64
114. Mayberg HS, Starkstein SE, Peyser CE, Brandt J, Dannals RF, Folstein SE (1992) Paralimbic frontal lobe hypometabolism in depression associated with Huntington's disease. Neurology 42: 1791-1797
115. Sultzer DL, Mahler ME, Cummings JL, Van Gorp WG, Hinkin CH, Brown C (1995) Cortical abnormalities associated with subcortical lesions in vascular dementia. Clinical and positron emission tomographic findings. Arch Neurol 52: 773-780
116. Reifler BV, Teri L, Raskind M, Veith R, Barnes R, White E, McLean P (1989) Double-blind trial of imipramine in Alzheimer's disease patients with and without depression. Am J Psychiatry 146: 45-49
117. Fuchs A, Hehnke U, Erhart C, Schell C, Pramshohler B, Danninger B, Schautzer F (1993) Video rating analysis of effect of maprotiline in patients with dementia and depression. Pharmacopsychiatry 26: 37-41
118. Nyth AL, Gottfries CG (1990) The clinical efficacy of citalopram in treatment of emotional disturbances in dementia disorders. A Nordic multicentre study. Br J Psychiatry 157: 894-901
119. Olafsson K, Jorgensen S, Jensen HV, Bille A, Arup P, Andersen J (1992) Fluvoxamine in the treatment of demented elderly patients: a double-blind, placebo-controlled study. Acta Psychiatr Scand 85: 453-456
120. Dewan VK, Burke WJ, Roccaforte WH (1995) Selective serotonin reuptake inhibitors for treatment of depression and psychosis in dementia. In: New research and program abstracts: 1995 annual meeting. American Psychiatric Association, Washington DC
121. Volicer L, Rheaume Y, Cyr D (1994) Treatment of depression in advanced Alzheimer's disease using sertraline. J Geriatr Psychiatry Neurol 7: 227-229
122. Postma JU, Vranesic D (1985) Moclobemide in the treatment of depression in demented geriatric patients. Acta Ther 11: 1-4
123. Tariot PN, Cohen RM, Sunderland T, Newhouse PA, Yount D, Mellow AM, Weingartner H, Mueller EA, Murphy DL (1987) l-Deprenyl in Alzheimer's disease. Preliminary evidence for behavioral change with monoamine oxidase B inhibition. Arch Gen Psychiatry 44: 427-433

124. Flint AJ (1994) Epidemiology and comorbidity of anxiety disorders in the elderly. Am J Psychiatry 151: 640-649

125. Cassem EH (1990) Depression and anxiety secondary to medical illness. Psychiatr Clin North Am 13: 597-612

126. Alexopoulos GS (1990) Anxiety-depression syndromes in old age. Int J Geriatr Psychiatry 5: 351-53

127. Lindesay J, Briggs K, Murphy E (1989) The Guy's/Age Concern Survey: prevalence rates of cognitive impairment, depression and anxiety in an urban elderly community. Br J Psychiatry 155: 317-329

128. Regier DA, Boyd JH, Burke JD, Jr, Rae DS, Myers JK, Kramer M, Robins LN, George LK, Karno M, Locke BZ (1988) One-month prevelance of mental disorders in the United States: based on five Epidemiologic Catchment Area sites. Arch Gen Psychiatry 45: 977-986

129. Wands K, Merskey H, Hachinski VC, Fisman M, Fox H, Boniferro M (1990) A questionnaire investigation of anxiety and depression in early dementia. J Am Geriatr Soc 38: 535-538

130. Shader RI, Greenblatt DJ (1982) Management of anxiety in the elderly: the balance between therapeutic and adverse effects. J Clin Psychiatry 43[Suppl]: 8-18

131. Cohen-Mansfield J, Billig N (1986) Agitated behaviors in the elderly. I. A conceptual review. J Am Geriatr Soc 34: 711-721

132. Cohen-Mansfield J, Marx MS, Rosenthal AS (1989) A description of agitation in a nursing home. J Gerontol 44: M77-M84

133. Yudofsky SC, Silver JM, Jackson W, Endicott J, Williams D (1986) The Overt Aggression Scale for the objective rating of verbal and physical aggression. Am J Psychiatry 143: 35-39

134. Ryden MB, Bossenmaier M, McLachlan C (1991) Aggressive behavior in cognitively impaired nursing home residents. Res Nursing Health 14: 87-95

135. Devanand DP, Brockington CD, Moody BJ, Brown RP, Mayeux R, Endicott J, Sackeim HA (1992) Behavioral syndromes in Alzheimer's disease. Int Psychogeriatr 4[Suppl 2]: 161-184

136. Ryden M (1988) Aggressive behavior in persons with dementia who live in the community. Alzheimer Dis Assoc Disord 2: 342-355

137. Lachs MS, Becker M, Siega AP, Miller RL, Tinetti ME (1992) Delusions and behavioral disturbances in cognitively impaired elderly persons. J Am Geriatr Soc 40: 768-773

138. Mann AH, Graham N, Ashby D (1984) Psychiatric illness in residential homes for the elderly: a survey in one London borough. Age Ageing 13: 257-265

139. Aronson MK, Post DC, Guastadisegni P (1993) Dementia, agitation, and care in the nursing home. J Am Geriatr Soc 41: 507-512

140. Chui HC, Lyness SA, Sobel E, Schneider LS (1994) Extrapyramidal signs and psychiatric symptoms predict faster cognitive decline in Alzheimer's disease. Arch Neurol 51: 676-681

141. Gold DP, Reis MF, Markiewicx D, Andres D (1995) When home caregiving ends: a longitudinal study of outcomes for caregivers of relatives with dementia. J Am Geriatr Soc 43: 10-16

142. Coyne AC, Reichmann WE, Berbig RMT (1993) The relationship between dementia and elder abuse. Am J Psychiatry 150: 643-646

143. Bridges-Parlet S, Knopman D, Thompson T (1994) A descriptive study of physically aggressive behavior in dementia by direct observation. J Am Geriatr Soc 42: 192-197

144. Nilsson K, Palmstierna T, Wistedt B (1988) Aggressive behavior in hospitalized psychogeriatric patients. Acta Psychiatr Scand 78: 172-175

145. Burnside IM (1980) Wandering behavior. In: Burnside IM (ed) Psychosocial nursing care of the aged, 2nd edn. McGraw-Hill, New York

146. Rabins PV, Mace NL, Lucas MJ (1982) The impact of dementia on the family. JAMA 148: 333-335

147. Ryan JP, McGowan J, McCaffrey N, Ryan GT, Zandi T, Brannigan GG (1995) Graphomotor perseveration and wandering in Alzheimer's disease. J Geriatr Psychiatry Neurol 8: 209-212

148. Okawa M, Mishima K, Hishikawa Y, Hozumi S, Hori H, Takahashi K (1991) Circadian rhythm disorders in sleep-waking and body temperature in elderly patients with dementia and their treatment. Sleep 14: 478-485

149. Snyder LH, Rupprecht P, Pyrek J (1979) Wandering. Gerontologist 18: 272

150. Buchner DM, Larson EB (1987) Falls and fractures in patients with Alzheimer-type dementia. JAMA 257: 1492-1495

151. Hewawasam L (1996) Floor patterns limit wandering of people with Alzheimer's. Nursing Times 92: 41-44

152. Hussian RA, Brown DC (1987) Use of two-dimensional grid patterns to limit hazardous ambulation in demented patients. J Gerontol 42: 558-560

153. Arno S, Frank DI (1994) A group for "wandering" institutionalized clients with primary degenerative dementia. Perspect Psychiatr Care 30: 13-16

154. Michel K, Kolakowska T (1981) A survey of prescribing psychotropic drugs in two psychiatric hospital. Br J Psychiatry 138: 217-221

155. Prien Y, Haber PA, Caffey EMJ (1975) The use of psychoactive drugs in elderly patients with psychiatric disorders: survey conducted in twelve Veterans Administration hospitals. J Am Geriatr Soc 23: 104-112

156. Rada RT, Kellner R (1976) Thiothixene in the treatment of geriatric patients with chronic organic brain syndrome. J Am Geriatr Soc 24: 105-107

157. Petrie WM, Ban TA, Berney S, Fujimori M, Guy W, Ragheb M, Wilson WH, Schaffer JD (1982) Loxapine in psychogeriatrics: a placebo- and standard-controlled clinical investigation. J Clin Psychopharmacol 2: 122-126

158. Chesrow EJ, Kaplitz SE, Vetra H, Breme JT, Marquardt GH (1965) Blind study of oxazepam in the management of geriatric patients with behavioral problems. Clin Med 72: 1001-1005

159. DeLemos GP, Clement WR, Nickels E (1965) Effect of diazepam suspension in geriatric patients hospitalized for psychiatric illnesses. J Am Geriatr Soc 13: 355-359

160. Beber CR (1965) Management of behavior in the institutionalized aged. Dis Nerv Syst 26: 591-595

161. Coccaro EF, Kramer E, Zemishlany Z, Thorne A, Rice CM, Giordani B, Duvvi K, Patel BM, Torres J, Nora R, Neufeld R, Mohs RC, Davis KL (1990) Pharmacologic treatment of noncognitive behavioral disturbances in elderly demented patients. Am J Psychiatry 147: 1640-1645

162. Tariot PN, Erb R, Leibovici A, Podgorski CA, Cox C, Asnis J, Kolassa J, Irvine C (1994) Carbamazepine treatment of agitation in nursing home patients with dementia: a preliminary study. J Am Geriatr Soc 42: 1160-1166

163. Chambers CA, Bain J, Rosbottom R, Ballinger BR, McLaren S (1982) Carbamazepine in senile dementia and overactivity-a placebo controlled double blind trial. IRCS Med Sci 19: 505-506

164. Gleason R, Schneider LS (1990) Carbamazepine treatment of agitation in Alzheimer's outpatients refractory to neuroleptics. J Clin Psychiatry 51: 115-118

165. Patterson JF (1987) Carbamazepine for assaultive patients with organic brain disease. Psychosomatics 28: 579-581

166. Patterson JF (1988) A preliminary study of carbamazepine in the treatment of assaultive patients with dementia. J Geriatry Psychiatry Neurol 1: 21-23

167. Lott AD, McElroy SL, Keys MA (1995) Valproate in the treatment of behavioral agitation in elderly patients with dementia. J Neuropsychiatry Clin Neurosci 7: 314-319

168. Roy A, Linnoila M (1988) Suicidal behavior, impulsiveness and serotonin. Acta Psychiatr Scand 78: 529-535

169. van Praag HM (1991) Serotonergic dysfunction and aggression control. Psychol Med 21: 15-19

170. Lawlor BA, Sunderland T, Mellow AM, Hill JL, Molchan SE, Murphy DL (1989) Hyperresponsitvity to the serotonin agonist m-chlorphenylpiperazine in Alzheimer's disease. Arch Gen Psychiatry 46: 542-549

171. Houlihan BJ, Mulsant DJ, Sweet RA, Rifai AH, Pasternak R, Rosen J, Zubenko GS (1994) A naturalistic study of trazodone in the treatment of behavioral complications of dementia. Am J Geriatr Psychiatry 2: 78-85

172. Lebert F, Pasquier F, Petit H (1994) Behavioral effects of trazodone in Alzheimer's disease. J Clin Psychiatry 55: 536-538

173. Sultzer DL, Gray KF, Gunay I, Berisford MA, Mahler ME (1997) A double-blind comparison of trazodone and haloperidol for treatment of agitation in patients with dementia. Am J Geriatr Psychiatry 5: 60-69

174. Dehlin O, Hedenrud B, Jansson P, Norgard J (1985) A double-blind comparison of alaproclate and placebo in the treatment of patients with senile dementia. Acta Psychiatr Scand 71: 190-196

175. Geldmacher DS, Waldman AJ, Doty L, Heilman KM (1994) Fluoxetine in dementia of the Alzheimer's type: prominent adverse effects and failure to improve cognition. J Clin Psychiatry 55: 161

176. Goad DL, Davis CM, Liem P, Fuselier CC, McCormack JR, Olsen KM (1991) The use of selegiline in Alzheimer's patients with behavior problems. J Clin Psychiatry 52: 342-345

177. Schneider LS, Pollock VE, Zemansky MF, Gleason RP, Palmer R, Sloane RB (1991) A pilot study of low dose l-deprenyl in Alzheimer's disease. J Geriatr Psychiatry Neurol 4: 143-148

178. Burke WJ, Roccaforte WH, Wengel SP, Bayer BL, Ranno AE, Willcockson NK (1993) l-Deprenyl in the treatment of mild dementia of the Alzheimer type: results of a 15-month trial. J Am Geriatr Soc 41: 1219-1225

179. Hermann N, Eryavec G (1993) Buspirone in the management of agitation and aggression associated with dementia. Am J Geriatr Psychiatr 1: 249-253

180. Sakauye KM, Camp CJ, Ford PA (1993) Effects of buspirone on agitation associated with dementia. Am J Geriatr Psychiatry 1: 82-84

181. Williams KH, Goldstein C (1979) Cognitive and affective response to lithium in patients with organic brain syndrome. Am J Psychiatry 136: 800-803

182. Greendyke RM, Kanter DR (1986) Therapeutic effects of pindolol on behavioral disturbances associated with organic brain disease: a double blind study. J Clin Psychiatry 47: 423-426

183. Greendyke RM, Kanter DR, Schuster DB, Verstreate S, Wootton J (1986) Propranolol treatment of assaultive patients with organic brain disease. A double-blind crossover, placebo-controlled study. J Nerv Ment Dis 174: 290-294

184. Kaufer DI, Cummings JL, Christine D (1996) Effect of tacrine on behavioral symptoms in Alzheimer's disease: an open-label study. J Geriatry Psychiatry Neurol 9: 1-6

185. Cummings JL, Petry S, Dian L, Shapira J, Hill MA (1990) Organic personality disorder in dementia syndromes: an inventory approach. J Neuropsychiatr Clin Neurosci 2: 261-267

186. Burns A, Folstein S, Brandt J, Folstein M (1990) Clinical assessment of irritability, aggression, and apathy in Huntington and Alzheimer disease. J Nerv Mental Dis 178: 20-26

187. Starkstein SE, Mayberg HS, Preziosi TJ, Andrezejewski P, Leiguarda R, Robinson RG (1992) Reliability, validity, and clinical correlates of apathy in Parkinson's disease. J Neuropsychiatry Clin Neurosci 4: 134-139

188. Bozzola FG, Gorelick PB, Freels S (1992) Personality changes in Alzheimer's disease. Arch Neurol 49: 297-300

189. Reichman WE, Coyne AC, Amirneni S, Molino B Jr, Egan S (1996) Negative symptoms in Alzheimer's disease. Am J Psychiatry 153: 424-426

190. Rubin EH, Morris JC, Berg L (1987) The progression of personality changes in senile dementia of the Alzheimer's type. J Am Geriatr Soc 35: 721-725

191. Marin RS (1991) Apathy: a neuropsychiatric syndrome. J Neuropsychiatry Clin Neurosci 3: 243-254

192. Marin RS, Firinciogullari S, Biedrzycki RC (1993) The sources of convergence between measures of apathy and depression. J Affect Disord 28: 117-124

193. Cummings JL (1995) Anatomic and behavioral aspects of frontal-subcortical circuits. Ann NY Acad Sci 769: 1-13

194. Cummings JL (1994) Vascular subcortical dementias: clinical aspects. Dementia 5: 177-180

195. Lilly R, Cummings JL, Benson DF, Frankel M (1983) The human Kluver-Bucy syndrome. Neurology 33: 1141-1145

196. Cummings JL, Benson DF (1987) The role of the nucleus basalis of Meynert in dementia: review and reconsideration. Alzheimer Dis Assoc Disord 1: 128-145

197. Whitehouse PJ, Price DL, Struble RG, Clark AW, Coyle JT, Delon MR (1982) Alzheimer's disease and senile dementia: loss of neurons in the basal forebrain. Science 215: 1237-1239

198. Kluver H, Bucy P (1937) "Psychic blindness" and other symptoms following bilateral temporal lobectomy in rhesus monkeys. Am J Physiol 119: 352

199. Terzain H, Ore G (1955) Syndrome of Kluver and Bucy. Neurology 5: 373-380

200. Sourander P, Sjogren H (1970) The concept of Alzheimer's disease and its clinical implications. In: Wolstenholme G, O'Connor M (eds) Alzheimer's disease and related conditions. Churchill Livingstone, London, pp 11-31

201. Cummings JL, Duchen L (1981) Kluver-Bucy syndrome in Pick disease: clinical and pathologic conditions. Neurology 31: 1415-1422

202. Forstl H, Burns A, Levy R, Cairns N, Luthert P, Lanto P (1993) Neuropathological correlates of behavioural disturbance in confirmed Alzheimer's disease. Br J Psychiatry 163: 364-368

203. Janati A (1985) Kluver-Bucy syndrome in Huntington's chorea. J Nerv Ment Dis 173: 632-635

204. Marin RS, Fogel BS, Hawkins J, Duffy J, Krupp B (1995) Apathy: a treatable syndrome. J Neuropsychiatry Clin Neurosci 7: 23-30

205. Leo RJ, Kim KY (1995) Clomipramine treatment of paraphilias in elderly demented patients. J Geriatr Psychiatry Neurol 8: 123-124

206. Tiller JWG, Dakis JA, Shaw JM (1988) Short-term buspirone treatment in disinhibition with dementia. Lancet 2(8609): 510 (letter)

Non-Cognitive Symptoms and Alterations in Demented Patients

W.C. McCormick

Promising new medical therapies for Alzheimer's disease are being tested [1] that may be useful in preventing or delaying the cognitive impairment characteristic of Alzheimer's disease. So far these agents have been used in persons already at a fairly advanced stage of illness. To effectively prevent symptoms from progressing to advanced stages, therapies will probably need to be applied as early as possible, when symptoms are barely detectable [2]. Yet, Alzheimer's disease can be difficult to diagnose with certainty ante mortem, especially early in the course of the illness [3], and it is sometimes not recognized in elderly individuals even when advanced [4]. The disease usually manifests initially with very subtle symptoms that increase insidiously over several years [5]. Diagnostic tests (chemical markers [6] or brain imaging [7]) may some day allow definitive diagnosis, perhaps even prior to the onset of symptoms. Currently, however, codified clinical diagnostic criteria such as those developed by the NINCDS-ADRDA [8] or those in the DSM-IIIR [9] are the research standards for diagnosing Alzheimer's disease ante mortem. These criteria are based on typical symptoms of the illness and have good reliability [10] and validity [11] when judged against post mortem pathological diagnoses.

Hence we need to learn as much as possible about symptomatology in the earliest stages of Alzheimer's disease (both cognitive symptoms and non-cognitive symptoms) to be as accurate as we can in diagnosis and treatment. In order to do so, cohorts of subjects should be studied prospectively [12, 13] in order to understand the content and patterns of complaints and symptoms among patients with probable Alzheimer's disease and other forms of dementia compared to other groups of patients who are not to be demented and remain cognitively intact in advanced age. Morris [3] has suggested that elderly persons with memory complaints who do not meet rigorous criteria for dementia may have very early, subclinical dementing illness, so it is important to understand how symptoms among demented patients may somehow be different from those among non-demented subjects with memory complaints and non-demented subjects without memory complaints. In doing such studies, it is important to control for comorbidity by comparing the aggregate number and severity of common diseases between demented and non-demented subjects. These types of investigations allow us to learn whether common symptoms of elderly persons (especially symptoms not clearly related to dementia) are reported early in the course of Alzheimer's disease prior to diagnosis. Do patients with very early, unrecognized Alzheimer's disease under-report symptoms (due to memory impairment) or over-report symptoms (perhaps due to perseveration, often found in patients with Alzheimer's disease)? In the latter case, might over-reported symptoms not clearly related to memory problems (e.g., vague abdominal discomfort) be "proxies" for symptoms of underlying cognitive impairment? Certainly, complaints of age-related memory problems are common among elderly persons; the key is to learn how these complaints are reflected in the course of illness, and if differences in complaint patterns hold clues useful in early evaluation, diagnosis, and management of persons with Alzheimer's disease. This is of importance for case ascertainment and will become increasingly important as therapeutic agents are developed.

At the beginning of this decade, colleagues at the University of Washington (UW) in Seattle, Washington, USA, sought to begin to answer some of these questions through a longitudinal study of subjects enrolled in the Alzheimer's Disease Patient Registry (ADPR) between April 1987 and July 1991 [14]. The population base for epidemiological study under the auspices of the ADPR and another companion study (Genetic Differences in Alzheimer's Cases and Controls) was all persons age 60 or above who are members of a large health maintenance organization (HMO; Group Health Cooperative of Puget Sound) [13]. This was a stable, geographically defined population of approximately 23,000 persons; attrition not due to mortality was very low (< 1%). Sex, age and race distributions were similar to those of the general population in the Seattle area. The ADPR was formed in 1987 to establish an epidemiological and clinical database allowing quantification of the incidence of dementing illness (particularly Alzheimer's disease) and was modeled after cancer registries in form and purpose [12]. From the HMO base population, potential cases of dementia were identified, examined, diagnosed, and followed. Subjects were diagnosed as probable or possible Alzheimer's disease, as other forms of dementia (e.g., multi-infarct dementia), or as "not demented."

For comparison, a group of cognitively intact normal controls was obtained from the same HMO. Potential subjects for the ADPR were obtained from several sources, including referrals from physicians, emergency room logs, radiology records (head CT scans), hospital discharge data, and physician visit forms. Methods of recruitment and further details regarding these studies can be found elsewhere [13]. All ADPR enrollees were evaluated by a study physician using a standardized medical history, physical, and neurological examination, and received neuropsychological testing and laboratory testing (including head CT scans). Diagnosis was based on consensus (using standardized NINCDS-ADRDA [8] and DSM-IIIR [9] diagnostic criteria) among members of an interdisciplinary team, including internists, geriatricians, neuropsychologists, psychiatrists, and neurologists, as well as an epidemiologist, study nurse, and psychometrist. All subjects received annual follow-up with repeated cognitive testing, behavioral assessment, and interval medical history. The study was approved by the University of Washington Human Subjects Committee.

Symptomatology, visit frequency, and comorbidity were assessed for three groups of subjects. Cases were 154 ADPR subjects with probable Alzheimer's disease based on NINCDS-ADRDA and DSM-IIIR criteria. Two other groups were made up of cognitively intact subjects: the "not demented" group consisted of 92 subjects initially identified by ADPR surveillance with symptoms potentially indicating dementia, but after complete evaluation (identical to cases) were found not to be demented, and controls were 129 cognitively intact controls selected randomly from the same defined population as the ADPR by frequency-matching for sex and age (within 2 years). The symptoms and diagnoses recorded for each physician visit were abstracted from medical records for the period beginning 2 years before study enrollment to at least 6 months after enrollment. Computerized utilization records of visits to physicians allowed quantitation of visits (in the 2 years prior to enrollment) to each subject's primary physician as well as visits to consultant physicians. Three medical records technicians (blinded to the purpose of the study) abstracted the data from charts; 94% of visits to physicians logged in computerized utilization records were found in the written medical records. Inter-rater reliability for technician abstracting was assessed on a 10% sample of charts; agreement between technicians was 96%. Comorbidity of each study subject was scored using methods developed by Charlson [15]. The Charlson Comorbidity Index is a commonly used, validated index that allows quantification of comorbidity based on the number and severity of common diseases of adults. To use the Index, comorbid conditions are ascertained for each subject, and each condition is assigned a weight; the comorbidity index score equals the sum of the weights for any individual subject.

The frequencies of the first recorded symptom in the chart suggestive of cognitive impairment were reported, as were the times before enrollment these symptoms appeared in the chart, whether patients were identified as demented in the chart, and if so, whether identification of dementia occurred before or after enrollment. Common symptoms not suggestive of cognitive impairment and the relative frequencies (of the symptom having ever occurred during the abstraction period) were then reported for the three groups. The 25 commonest (most frequently occurring) symptoms were examined: depression, weight loss, hypersomnia, gastrointestinal symptoms, joint pain, genitourinary symptoms, cough, vision problems, rash, hearing impairment, fever, chills, stress, dizziness, weakness, falls, chest pain, constipation, insomnia, headache, incontinence, shortness of breath, numbness, and low back pain. Statistical comparisons were by Chi squared analysis of 2×2 tables for categorical variables (e.g., symptoms) and t-tests for continuous variables (e.g., age and comorbidity). Sample size was sufficient to detect 20% differences in individual symptom frequency with power of 80% at alpha $= 0.05$. Adjustments for 25 multiple comparisons per experiment were done using the Bonferroni technique; to hold up to the Bonferonni correction for multiple comparisons [16], the individual P values should be less than 0.002 to remain significant using a per experiment alpha level of 0.05.

The comparison groups were similar in age, sex, and race (Table 1), although the not-demented group was significantly younger than the other two groups.

Table 1. Comparisons of demographic data

	Cases	Not demented	Controls
n	154	92	129
Age (years, ± SD)	78 ± 6	70 ± 11*	77 ± 6
Sex (% male)	32%	43%	37%
Race (% Caucasian)	89%	92%	96%
Marital status (% married)	49%	57%	64%*
Education (% above high school)	38%	57%*	59%*
Mini-Mental State Examination	20 ± 5	27 ± 2*	28 ± 2*
Visits to any MD (over 2 years, ± SD)	17 ± 13	19 ± 12	14 ± 10*
Visits to primary MD	12 ± 9	12 ± 7	8 ± 6*

* $P < 0.05$ vs cases

Cases were less educated and less often married than the other two groups. The Mini-Mental State Examination score obtained at the time of enrollment was assessed in the three groups, demonstrating the differences in cognitive impairment as assessed by this screening examination [17]. During the 2 years prior to enrollment, cases and subjects in the not-demented group visited both primary care physicians and consultant physicians more often than controls. The comparison groups had similar comorbidity scores using the Charlson Comorbidity Index [15] if the diagnosis of dementia was not included in calculation of the comorbidity index; if dementia was included in calculation of the index, cases had significantly higher comorbidity (Table 2). Cases, the not-demented group, and controls were similar even when the individual disease elements used in the Charlson Index were compared between groups after adjustment for multiple comparisons (Table 2).

For cases (subjects with probable Alzheimer's disease, $n = 154$), the following is a list of the first symptom suggestive of cognitive impairment found in the chart (followed by frequency in the group): "memory loss" (57), "memory problems" (16), "forgetfulness" (30), "confusion" (21), "forgetting things" (3), "word finding difficulty" (3), "disoriented" (2), "agitation" (2), and "worried about Alzheimer's," "got lost," "helplessness," "suspicion," "senility," "memory poor," "memory changes," "thinking problems," "decreased chores," "difficulty writing," "behavior change," and "poor judgment" (1 each). In sum, symptoms were found in 95% of these charts prior to ADPR enrollment. Symptoms were recorded an average of 7.7 months (median = 6 months, SD = 6.2) prior to enrollment. The majority of cases (106 of 154, or 69%) were identified as having dementia by their doctors in the medical record. Of the 106 so identified, 76 were identified as demented before ADPR enrollment and 30 after enrollment in the ADPR.

Among the ADPR enrollees with no dementia (the not-demented group, $n = 92$), symptoms of cognitive decline as they first appeared in the chart were as follows: "memory loss" (34), "memory problems" (9), "forgetfulness" (11), "confu-

Table 2. Comparisons of comorbidity

Charlson Comorbidity Index	Cases		Not demented	Controls		Weight[a]
Without dementia	0.7		0.7	0.5		
With dementia	1.7*		0.7	0.5		
Diseases used in the Charlson Comorbidity Index (%)						
Ischemic heart disease	12	(8)	13 (14)	27 (21)**		1
Congestive heart failure	12	(8)	3 (3)	6 (5)		1
Peripheral vascular disease	6	(4)	4 (4)	7 (5)		1
Cerebrovascular disease	4	(3)	10 (11)**	5 (4)		1
Dementia	154	(100)*	0 (0)	0 (0)		1
Chronic pulmonary disease	7	(5)	5 (5)	9 (7)		1
Connective tissue disease	1	(1)	1 (1)	4 (3)		1
Ulcer disease	1	(1)	2 (2)	5 (4)		1
Diabetes mellitus	9	(6)	5 (5)	5 (4)		1
Chronic renal failure	2	(1)	0 (0)	0 (0)		2
Diabetes (end organ damage)	3	(2)	4 (4)	0 (0)		2
Any tumor	16	(10)	11 (12)	8 (6)		2
Cirrhosis	0	(0)	1 (1)	0 (0)		3

* $P < 0.001$ vs other groups
**$P < 0.05$ vs cases per comparison, but not significant after multiple comparisons correction
[a] The assigned weights for each condition are summed for each individual subject; the total equals the subject's Charlson Comorbidity Index Score [15]

sion" (7), "memory changes" (3), "thinking problems" (3), and "calculation difficulties," "drop in mental acuity," "speech incoordination," and "agitation" (1 each). In sum, these symptoms were found in 77% of the charts prior to ADPR enrollment. Symptoms were recorded an average of 7.8 months (median 6 months, SD = 6.9) prior to enrollment. Of the 92, 16 (17%) were identified as demented in the chart, 12 before and 4 after enrollment, even though the ADPR evaluation showed that they did not meet diagnostic criteria for dementia.

Among controls ($n = 129$), symptoms of cognitive decline were rarely recorded in the medical record: "memory problems" (1), "forgetfulness" (2), "confusion" (1), "thinking problems" (2), and "agitation" (2). In sum, symptoms were found in 6% of the charts prior to enrollment. None of the subjects in this group was identified as demented in the chart. Symptoms were recorded an average of 14 months (median = 14, SD = 7.4) prior to enrollment.

Table 3 displays the three symptoms not obviously suggestive of cognitive decline that occurred more commonly in cases than in controls ($P < 0.05$ per comparison); these differences did not remain statistically significant after adjusting for multiple comparisons by the Bonferroni technique [16]. Seven symptoms in Table 3 occurred more commonly in controls than in cases with probable Alzheimer's disease. Five of these symptoms were significantly more common in controls than cases after correction for multiple comparisons. Cases differed from the ADPR subjects who were not demented only in the frequency of

Table 3. Comparison of the distribution of non cognitive symptoms in control and demented patients

	Cases ($n = 154$)	Controls ($n = 129$)
Symptoms more common in cases than in controls		
Depression	38 (25%)	14 (11%)
Weight loss	18 (12%)	3 (2%)
Hypersomnia	7 (5%)	0 (0%)
Symptoms more common in controls than in cases		
Gastrointestinal symptoms[a]	34 (22%)	57 (44%)
Joint pain[a]	49 (32%)	73 (57%)
Genitourinary symptoms	11 (7%)	24 (19%)
Cough[a]	19 (12%)	36 (28%)
Vision problems[a]	15 (10%)	31 (24%)
Rash[a]	10 (6%)	31 (24%)
Hard of hearing	9 (6%)	20 (16%)
	Not demented ($n = 92$)	**Controls ($n = 129$)**
Comparison of symptoms between not demented and controls		
Gastrointestinal symptoms	25 (27%)	57 (44%)
Stress	22 (24%)	13 (10%)
Depression[a]	32 (35%)	14 (11%)
Joint pain[a]	26 (28%)	73 (57%)
Genitourinary symptoms	5 (5%)	24 (19%)
Insomnia[a]	18 (20%)	7 (5%)
Vision problems[a]	5 (5%)	31 (24%)
Rash	7 (8%)	31 (24%)
Cough	11 (12%)	36 (28%)

[a] Statistically significant after correction for multiple comparisons

insomnia, which was more common in the not-demented group than in cases (20% vs 5%); this difference remained significant after correction for multiple comparisons. Table 3 also shows the symptom frequency comparing the not-demented group with controls. Four of the symptoms examined were more common in one group than the other even after adjustment for multiple comparisons; however, two were more common in the not-demented group (depression and insomnia), and two were more common in controls (joint pain and vision problems).

In summary, symptoms clearly related to failing cognition were evident in almost all of the medical records of patients with probable Alzheimer's disease.

Patients with probable Alzheimer's disease resembled non-demented patients in frequency of memory complaints in that the majority of patients in these two groups had such symptoms. Some have conjectured that perhaps common symptoms not obviously suggestive of cognitive decline may actually be "proxies" for symptomatology related to cognitive impairment, and that persons with early dementia might over-report such symptoms. While this may be a common belief, no objective evidence has been found for this phenomenon on a population basis. In fact, subjects with Alzheimer's disease in the UW study complained of common symptoms not having to do with cognitive impairment considerably less frequently than other non-demented elderly persons, even though individuals with Alzheimer's disease appeared to have similar major comorbidity compared to the two groups without dementing illnesses (the latter two groups were similar in regard to non-cognitive complaints). Comparison of individual symptoms (which could conceivably result from a variety of diagnoses) with individual diagnoses (which could cause a variety of symptoms) was specifically avoided in the UW study; such analyses would probably result in complex, multiple comparisons that would be difficult to interpret. Rather, a general, validated index of total comorbidity was used; the general pattern of symptom frequency is reported to highlight the disparity between higher comorbidity in the demented group, yet lower general symptom reporting. This finding supports the premise that patients with dementia of the Alzheimer's type may under-report symptoms and is consistent with earlier studies of elderly outpatients evaluated for dementia [18–20] in which previously unrecognized illnesses were found in nearly half of subjects. Hence elderly outpatients with dementia may be at special risk for occult medical illnesses, perhaps due to difficulty in expressing an accurate medical history during physician visits. In the population studied at UW, the discrepancy in symptom frequency could not be explained by difficulty obtaining medical care in the first place, given the visit-frequency patterns observed; patients with probable Alzheimer's disease visited their physicians as often as patients in the not-demented group, and more often than controls. Significant and comparable comorbidity was found in the three groups of subjects. This is in contrast to previous studies, which have suggested that persons with Alzheimer's disease are, in general, healthier than other non-demented elderly persons [21]. As noted by others [22–24], studies finding lower comorbidity in demented patients have suffered from selection and other biases and have not used validated measures of comorbidity to make comparisons. Given the population-based nature of the UW cohort of subjects, selection bias was minimized. Further, the relatively large size of the comparison groups decreased the probability of a type II error (not finding a difference in comorbidity when in fact there is one) [25].

At this point, it is reasonable to conclude that persons with early Alzheimer's disease do not have lower comorbidity (i.e., are not "healthier") than non-demented elderly persons, and more worrisome, that persons with Alzheimer's disease may under-report common symptoms. This finding is in keeping with, and may explain in part, findings in previous studies [18–20] showing under-recognition of common medical illnesses among elderly demented outpatients.

The challenge to clinicians represented by these findings is to be alert for the possibility of unreported illness and to attempt to obtain historical information from a knowledgeable informant whenever possible, since demented patients seem to bear the same burden of illness as non-demented persons, yet may under-report common symptoms. It will be important for clinicians caring for large numbers of elderly outpatients to develop effective surveillance strategies to enhance ascertainment of symptoms and signs of illness. If surveillance strategies are effective and do minimize excess disability in persons with Alzheimer's disease, these patients may be able to remain in the community longer, thereby avoiding some of the burden of institutionalization [26]. Registries such as the ADPR may be a suitable source of subjects with very early dementing illness for development of these surveillance strategies and for clinical trials to investigate the efficacy of drugs and other strategies to minimize the rate of decline or prevent the mental deterioration inherent in Alzheimer's disease.

Note. This article is adapted from McCormick et al. [14].

References

1. Davis KL, Thal LJ, Gamzu ER, et al (1992) A double-blind, placebo-controlled multicenter study of tacrine for Alzheimer's disease. Tacrine Collaborative Study Group. N Engl J Med 327: 1253-1259
2. Rubin EH, Morris JC, Grant EA, Vendegna T (1989) Very mild senile dementia of the Alzheimers type. Arch Neurol 46: 379-382
3. Morris JC, Fulling K (1988) Early Alzheimer's disease: diagnostic considerations. Arch Neurol 45: 345-349
4. Roca RP, Klein LE, Kirby SM, et al (1984) Recognition of dementia among medical inpatients. Arch Int Med 144: 73-75
5. Teri L, Hughes JP, Larson EB (1990) Cognitive deterioration in Alzheimer's disease: behavioral and health factors. J Gerontol 45: P58-P63
6. Farlow M, Ghetti B, Benson MD, Farrow JS, van Nostrand WE, Wagner SL (1992) Low cerebrospinal fluid concentrations of soluble amyloid beta-protein precursor in hereditary Alzheimer's disease. Lancet 340: 453-454
7. DeCarli C, Kaye J, Horwitz B, Rapoport S (1990) Critical analysis of the use of computer-assisted transverse, axial tomography to study human brain in aging and dementia of the Alzheimer's type. Neurology 409: 872-883
8. McKhann G, Drachman D, Folstein M, et al (1984) Clinical diagnosis of Alzheimer's disease: report of the NINCDS-ADRDA work group. Neurology 34: 939-944
9. American Psychiatric Association (APA) (1980) Diagnostic and statistical manual of mental disorders, 3rd edn, revised. APA, Washington DC
10. Kukull WA, Larson EB, Reifler BV, Lampe TH, Yerby M, Hughes JP (1990) Inter-rater reliability of Alzheimer's disease diagnosis. Neurology 40: 257-260
11. Kukull WA, Larson EB, Reifler BV, Lampe TH, Yerby M, Hughes JP (1990) The validity of 3 clinical diagnostic criteria for Alzheimer's disease. Neurology 40: 1364-1369
12. Hughes JP, van Belle G, Kukull WA, Larson EB, Teri L (1989) On the uses of registries for Alzheimer's disease. Alzheimer Dis Assoc Disord 3: 205-217

13. Larson EB, Kukull WA, Teri L, McCormick WC, Pfanschmidt M, van Belle G, Sumi M (1990) University of Washington Alzheimer's Disease Patient Registry (ADPR): 1987-1988. Aging 2: 404-408

14. McCormick WC, Kukull WA, van Belle G, Bowen JD, Teri L, Larson EB (1994) Symptom patterns and comorbidity in the early stages of Alzhiemer's disease. J Am Geriatr Society 42: 517-521

15. Charlson ME, Pompei P, Ales KL, MacKenzie CR (1987) A new method of classifying prognostic comorbidity in longitudinal studies: development and validation. J Chron Dis 40: 373-383

16. O'Brien PC, Shampo MA (1988) Statistical considerations for performing multiple tests in a single experiment. Mayo Clin Proc 63: 813-820

17. Folstein MF, Folstein SE, McHugh PR (1975) Mini-Mental State – a practical method for grading the cognitive state of patients for the clinician. J Psychiatr Res 12: 189-198

18. Larson EB, Reifler BV, Featherstone HJ, English DR (1984) Dementia in elderly outpatients: a prospective study. Ann Int Med 100: 417-423

19. Larson EB, Reifler BV, Sumi M, Canfield CG, Chinn NM (1986) Diagnostic tests in the evaluation of dementia: a prospective study of 200 elderly outpatients. Arch Int Med 146: 1917-1922

20. Teri L, Wagner AW (1991) Assessment of depression in patients with Alzheimer's disease: concordance among informants. Psychol Aging 6: 280-285

21. Wolf-Klein GP, Silverstone FA, Brod MS, Levy A, Foley CJ, Termotto V, Breuer J (1988) Are Alzheimer patients healthier? J Am Geriatr Soc 36: 219-224

22. Rosenfeld SA (1989) Are Alzheimer patients healthier than others? J Am Geriatr Soc 37: 486 (letter)

23. Pawlson LG (1989) Are Alzheimer patients healthier than others? J Am Geriatr Soc 37: 486 (letter)

24. Piccinin AM (1991) Uncertain diagnosis or uncertain entity? J Am Geriatr Soc 39: 1141-1142 (letter)

25. Detsky AS, Sackett DL (1985) When was a negative clinical trial big enough? Arch Intern Med 145: 709-712

26. Welch HG, Walsh JR, Larson EB (1992) The cost of institutional care in Alzheimer's disease: nursing home and hospital use in a prospective cohort. J Am Geriatr Soc 40: 221-224

Pharmacotherapeutic Approach to the Treatment of Alzheimer Disease

E. Giacobini

Drug Therapy for Alzheimer Disease

A systematic attempt to develop drugs to treat Alzheimer disease (AD) was initiated on a large scale 10 years ago following publication in the *New England Journal of Medicine* of the first successful results obtained with the cholinesterase inhibitor (ChEI) tacrine (THA, tetrahydroaminoacridine) by Summers et al. [1]. Tacrine is not the first ChEI to be tested clinically for treating AD. Numerous studies [2] had been performed previously, particularly in the USA, in small groups of patients, with physostigmine (physo) alone or in combination with lecithin. Physostigmine, like tacrine, showed definite but only short-lasting improvements in cognitive symptoms (attention, concentration, memory), which were accompanied with severe peripheral and central cholinergic side effects. These consisted mainly of gastrointestinal symptoms and drowsiness, but in the case of tacrine also of liver toxicity. It was soon realized that in spite of this first encouraging result, both physo and tacrine were far from an ideal drug for AD treatment. However, physo and tacrine represent important milestones in AD therapy, as they supported for the first time in the patient the pharmacological hypothesis formulated in the experimental animal [3–6] that a treatment improving function of the central cholinergic system obtained through an increase in brain acetylcholine (ACh) would also improve cognition in AD patients. Targeting the cholinergic system for AD therapy does not necessarily limit itself to use of a ChEI. Two other classes of cholinergic drugs might represent valid alternatives, such as nicotinic and muscarinic agonists alone or in combination with ChEI (Fig. 1). A second, non-cholinergic approach, is based on the classic pathological landmarks of the disease with the aim of decreasing beta-amyloid (beta-A4) deposition and amyloidogenic amyloid precursor protein (APP) release in brain (Fig. 1). A third kind of AD therapy has the aim of correcting events that are probably secondary to the disease process by means of estrogens, anti-oxidants, free-radical scavengers and anti-inflammatory agents (Fig. 1). Based on our present knowledge of beta-A4 synthesis, processing, accumulation and deposition in cortex and related genetic factors [7–10] (Fig. 2), we may think of at least five pharmacological interventions to decrease selectively extracellular concentration and deposition of

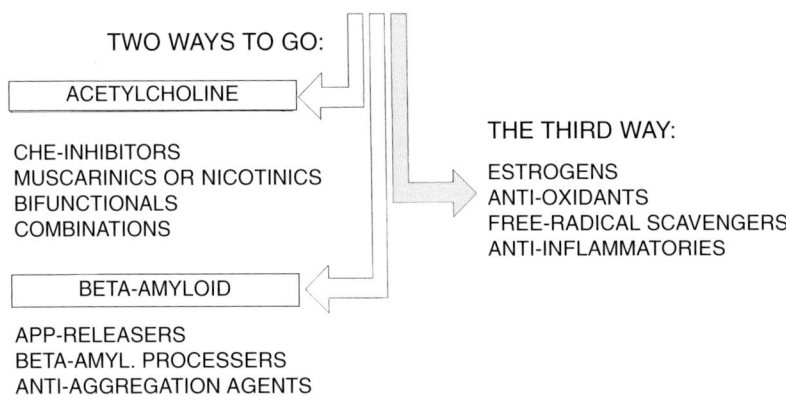

Fig. 1. Treatment of Alzheimer disease. Which way to go?

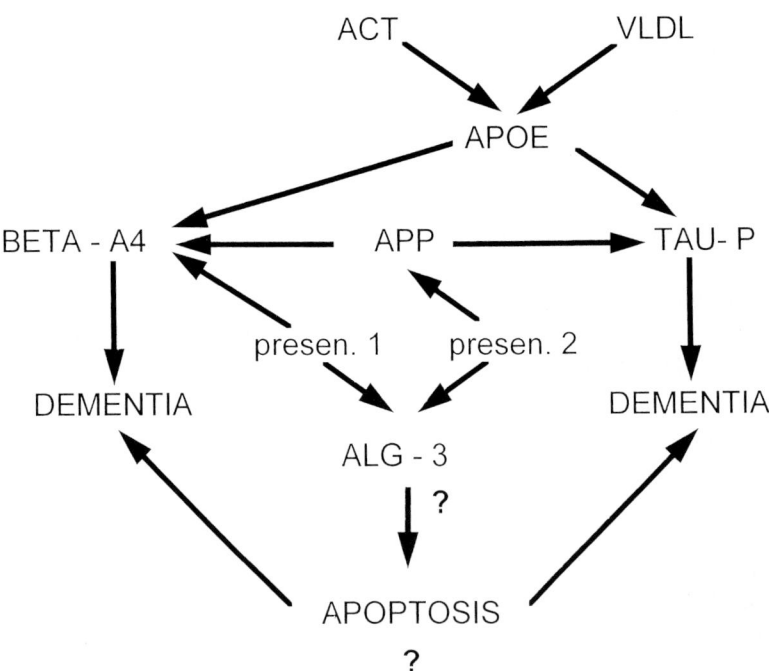

Fig. 2. Processing of the beta-amyloid (beta-A4) precursor protein (APP), its regulation in Alzheimer disease and its relationship to dementia and apoptosis. *ACT,* Alpha-antichymotrypsin; *ALG-3,* apoptosis-linked gene; *APOE,* apolipoprotein-E; *TAU-P,* phosphorylated tau protein; *VLDL,* very low density lipoprotein receptor

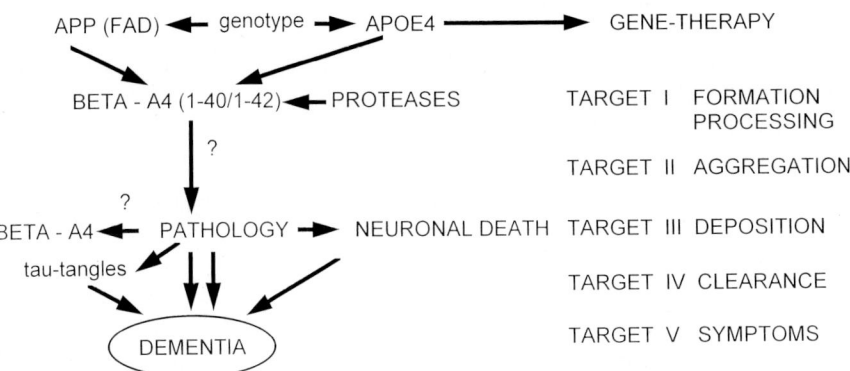

Fig. 3. Five pharmacological interventions to selectively decrease extracellular concentration and deposition of beta-A4 in brain and slow down dementia

beta-A4 in brain and slow down the disease process (Fig. 3). In theory, the same result could be achieved by blocking beta-A4 processing or slowing down secretion or deposition of amyloidogenic forms of APP. Processing and secretion of APP could be modified by inhibiting specifically one or several involved proteases (alpha, beta or gamma secretases) [7]. One might also think of reducing aggregation by enhancing removal of deposited beta-A4 (Fig. 4). There is evidence for considering apolipoproteins such as APOE-3 and APOE-2 as factors reducing beta-A4 aggregation and APOE-4 as promoting it (Fig. 4) [11, 12]. Presenilin 1, on the other hand, promotes both clearance and degradation of beta-A4 (Fig. 4) [8, 13]. The final goal is to block the progressive and irreversible degeneration of synapses and neurons, which represents the main feature of the disease. A better understanding of the mechanism of action of presenilins and APOE could lead to new therapeutical strategies of developing drugs to interact with these processes [11] or preventive and protective measures based on early identification of genet-

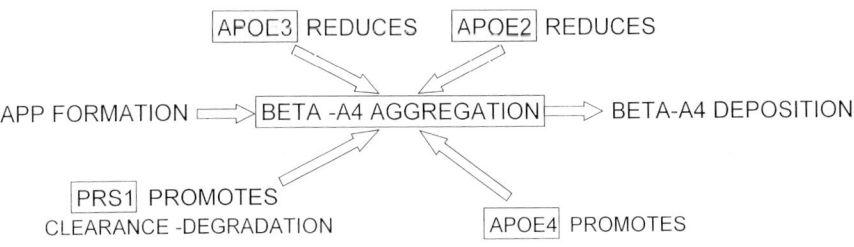

Fig. 4. Effect of APOEs and presenilin 1 on beta-4 aggregation

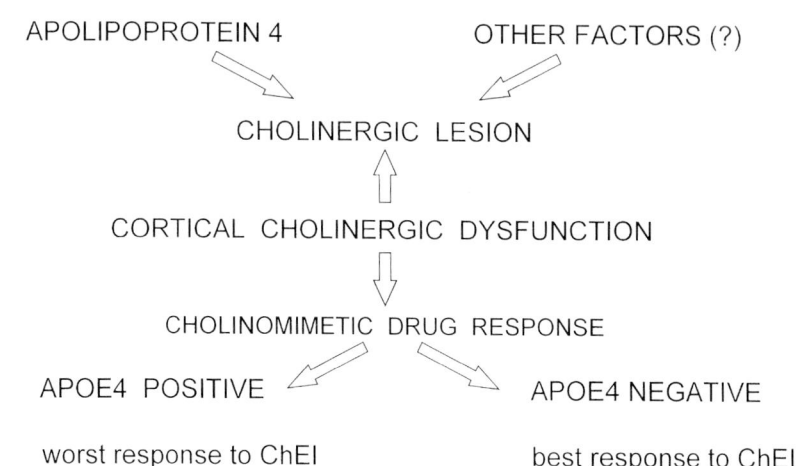

Fig. 5. Relationship between apolipoprotein 4, cholinergic lesions and cholinomimetic drug response to ChEI

ic risk factors (see low-fat diet and control of hypertension for cardiovascular diseases). Recent experimental and clinical data suggest an involvement of the cholinergic system in beta-A4 processing [14] (Fig. 5). Chemical cholinergic lesions in the nucleus basalis of the rat forebrain decrease ACh cortical levels and release and increase accumulation of APP in cerebrospinal fluid (CSF) and cortex (Table 1) [15, 16]. The effect of the lesion is mimicked by muscarinic antagonists and reversed by treatment with ChEI. The data of Soininen et al. [17] and Poirier et al. [18] show that the APOE-4 genotype exerts a deteriorating effect on the cholinergic system and might facilitate selective neuronal damage. This cholinergic hypofunction might condition the response to cholinergic drugs in AD patients [18] (Fig. 5). Other data suggesting an interaction between cholinesterase inhibition and brain metabolism as clinical evidence of neuroprotection are summarized in Table 2. These include the effect of ChEI (tacrine) on cortical glucose metabolism and cerebral blood flow monitored in AD patients with imaging tech-

Table 1. Cholinergic lesions and APP processing

A chemical lesion in the nucleus basalis of the rat [52] produces:

    A. A decrease in ACh cortical level and release
    B. A rapid and persistent accumulation in cortical and CSF APP

This response is:

    A. Age dependent
    B. Mimicked by muscarinic antagonists
    C. Reversed by treatment with ChEI [15, 16]

Table 2. Cholinesterase inhibitors: clinical evidence of neuroprotection

1. Increased cortical glucose metabolism and rCBF
   (18 months of THA, PET DFG [19, 53])

2. Increased cerebral nicotinic binding
   (12 months of THA, PET-^{11}C nicotine [53])

3. Stable regional cerebral rCBF
   (14 months of THA, SPECT 133 Xe [54])

4. Delay in time to nursing home placement
   (500 days of THA [21])

niques (PET and SPECT) utilizing deoxy-fluoroglucose (DFG) and ^{11}C-nicotine as biological markers of metabolic and cholinergic activity, respectively [19, 20]. A delay in time to nursing home placement is chosen as a clinical marker of delayed deterioration of the patient [21]. Although multiple non-cholinomimetic interventions are feasible (Fig. 1), this review will focus mainly on ChEI compounds for which we have the most detailed basic and clinical information (Tables 3, 4).

Table 3. Drugs for AD treatment in the clinical phase in Japan (1996)

Cholinesterase inhibitors

NIK247	Nikken
Rivastigmine	Sandoz
TAK147	Takeda
Donepezil	Eisai

Acetylcholine releasers

T-588	Toyama Kagaku

M1 agonists

YM976	Yamanouchi
AF102B	Snowbrand

Neuropeptides, peptidase inhibitors

TRH	Japan Tobacco Inc.
JTP-4819	Japan Tobacco Inc.
Epibizatide	Hoechst

Other: DBZ inv. agonists, nootropics, Ca^{++}-uptake inhibitors

TRH	Daiichi
S5810	Shiaroci
Nimodipine	Bayer

Due to the rapid evolution of the field, some of the data reported in the table, referred to 1996, may be outdated. Nonetheless, the table does illustrate the past and current efforts to develop new drugs in the field

Table 4. Cholinesterase inhibitors: AD clinical trials in the USA and Europe (1997)

Compound	Country	Company	Clinical phase[c]	Side effects
Physostigmine slow release	USA	Forest	III	N.A.
Rivastigmine	USA/Europe	Novartis	III Regist.	Low
Eptastigmine[a]	USA/Italy	Mediolanum	III	Low
Donepezil	USA/Japan	Eisai, Pfizer, Bracco	III Regist.	Low
MDL 73,745	USA/Europe	Marion Merrell Dow	II	Low
Metrifonate	USA/Germany	Bayer/Miles	III	Low
Tacrine (THA)[b]	USA/Europe	Warner-Lambert	IV Regist.	Hepatotoxicity
Velnacrine (HP029)[a]	USA/Europe	Hoechst-Roussel	II	Hematology[d]
Suronacrine (HP128)[a]				Hepatotoxicity
Galanthamine	Germany	Shire Pharm.	II	Low
	USA	Ciba-Geigy		
Huperzine A	China	Chinese Acad. Sci.	III	N.A.
NX-066	England/USA	Astra Arcus	II	N.A.
CP-118,954	USA	Pfizer	II	N.A.
KA-672	Germany	Schwabe	I	N.A.
NIK247	Japan	Nikken	III	Low
TAK147	Japan	Takeda	III	Low

Due to the rapid evolution of the field, some of the data reported in the table, referred to 1997, may be outdated. Nonetheless, the table does illustrate the past and current efforts to develop new drugs in the field
N.A., data not available; *Regist.*, registered in USA and/or in some European countries
[a] Withdrawn
[b] Other indications: HIV, tardive dyskinesia
[c] Clinical phase in USA
[d] Neutropenia or agranulocytosis

Cholinergic Therapy: Which Way to Go? Cholinesterase Inhibitors or Muscarinic Agonists? Future Development

The notion of the molecular structure of the catalytic site and two anionic sites of AChE would allow the synthesis of inhibitors highly selective for either AChE or BuChE. This property is particularly interesting since it is known that many cholinergic side effects are not associated with inhibition of BuChE. Moreover, AChE is strongly reduced in the central nervous systems (CNS) of AD patients, while BuChE is intact or even increased (Table 5). Therefore, an ideal inhibitor for AD treatment should be selective not only for membrane-anchored AChE, but should also inhibit BuChE, which hydrolyzes the ACh extrasynaptic pool. In addition, the hydrophobic tetrameric-tailed G-4 form of AChE anchored to the plasma membrane is dominant in brain synapses (Table 5). Based on our pharmacological knowledge, a ChEI selective for this molecular form could also be designed [22]. Compounds having such characteristics may constitute future alternatives of new ChEI molecules.

Table 5. Percent variation in enzyme activity in human cortex of AD patients compared to normal controls

Enzyme	Activity (% of normal)	Molecular form
AChE	10–15	Mainly G4
BuChE	120	Mainly G1
ChAT	10–15	

References [22, 55–58]

Table 6. Differences in clinical potential and side effects between cholinesterase inhibitors and muscarinic agonists suggest a differential action on subtypes of receptors and specific pre- and post-synaptic interactions

Class	Clinical effects	Side effects
Cholinesterase inhibitors	Predominantly cognitive	Low incidence with second-generation drugs
Muscarinic agonists	Predominantly behavioral	Significant cardiovascular and gastrointestinal side effects

A parallel line of research has been to develop drugs such as muscarinic agonists to stimulate selectively postsynaptic M1 receptors or antagonists to inhibit the effect of M2 presynaptic receptors to improve ACh synaptic release (Fig. 1, Table 3). Neither approach has so far produced highly selective drugs; therefore, compounds have seldom reached clinical trials or are still at early clinical phases. One major obstacle is represented by the severity of cholinergic toxicity, particularly gastrointestinal and cardiac side effects. Because of these limitations, future muscarinic agonists need to demonstrate higher brain-receptor selectivity. The number of muscarinic agonists presently in clinical trial is lower than that of ChEI, but some differences are emerging with regard to the clinical potential of these two classes of cholinergic drugs (Table 6). Cholinesterase inhibitors seem to exert predominanly cognitive effects (improve attention, memory, and concentration)[1], while muscarinic agonists also seem to act on behavioral symptoms. A combination of these properties may prove to be of benefit. In spite of differences in the severity of the toxicity in the two groups of drugs, for muscarinic agonists the main obstacle is to overcome autonomic side effects. For AChEI, both the severity and frequency of side effects seem to be less significant, particularly for the second generation of ChEI (Table 4).

---

[1] Some recent clinical trials with cholinesterase inhibitors show that certain drugs of this class may also exert significant effects on behavioral symptoms.

Cholinesterase Inhibitors: Do They Work in Alzheimer Disease?
How Do They Work? Is There a Difference?

A 1997 list of ChEI presently in clinical trials in Japan, the USA and Europe includes more than a dozen drugs, most of which have already advanced to clinical phase III (Tables 3, 4). The next 2-year period (1997–1999) will be crucial in selecting efficacious drugs. The second generation of ChEI to replace tacrine will have to fulfil specific requirements [12]. When analyzing the results from numerous trials throughout the world, the first question to be answered is obviously: do ChEI work in the patient and, if so, how do they work? A second question is: are there major differences between various compounds? These two fundamental questions can be examined and partially answered for the first time by comparing recent clinical data. Table 7 shows the effect of five ChEI on ADAS-cog test using ITT (intention to treat) criteria. The duration of the trials varied between 12 and 30 weeks. Based on these data, the answer to question one is affirmative. All five ChEI produce statistically significant improvements evaluated with specific scales of standardized and internationally validated measures of both cognitive and non-cognitive function. The cognitive items have been most widely used in these investigations. One first observation is the similarity in effect for all five drugs when expressed as a difference between drug- and placebo-treated patients. Differences consist generally of 2–4 points (ADAS-cog), depending on the type of analysis performed and on less than 0.5 on CIBIC (Clinician Interview-Based Impression of Change). The difference varies from a maximal gain of 5.3 points (tacrine high dose) to a minimum of 3.9 points (ADAS-cog, donepezil high dose). Does this trend suggest a ceiling effect of 4–5 ADAS-cog points on average for all ChEI? Or does it indicate the fact that drugs have not been tested at their maximal potential capacity? The high percentage of dropouts and side

Table 7. The effect of five cholinesterase inhibitors on ADAS-cog test (ITT)

Drug (%)	Dose (mg/day)	Duration (weeks)	Treatment difference from placebo	Improved patients (%)	Dropout (%)	Side effects (%)
Tacrine [24, 26]	80–160	30	4.0–5.3	30–50	50–73	40–58
Eptastigmine [59, 60]	45	25	4.7	30	12	35
Donepezil [26]	5	12	3.9	25	12	13
Rivastigmine [23]	12	26	4.9	18[a]	22	28
Metrifonate [27]	30	12	4.6	35	2	2–12

ADAS-cog, AD Assessment Scale-cognitive subscale; *ITT,* intention to treat
[a] A 4-point improvement

effects for tacrine seems to indicate a limit in drug effect. For other drugs (e.g. metrifonate), using a relatively high dosage which produces high ChE inhibition (up to 80%), the severity of side effects does not seem to be a limiting factor. In general, the percentage of improved patients varies from 17% (rivastigmine, low dose) to 50% (tacrine, high dose) with an average of 30% for most drugs. This indicates that approximately one third of treated patients demonstrate a positive response to these ChEI. This not very impressive figure becomes more relevant if we consider that the number of improved patients could be increased to 50% by using ChEI with fewer severe side effects. Looking at the 6-month data, one could say that patients treated with the active compound do not change significantly from the baseline determined at the beginning of treatment (6 months earlier). As an example, in a US study with rivastigmine, patients given placebo for 26 weeks would deteriorate approximately 4 points on the ADAS-cog compared to only 0.3 in patients given 6–12 mg/day of the drug [23]. The difference seen after 6 months between placebo-treated and drug-treated groups seems to depend more on the deterioration of the placebo group (3–4 points) than on a true additional improvement. This interpretation implies an anti-deterioration and slowing effect rather than a purely symptomatic effect of the drug. This presumed "anti-deterioration effect" could be related to an improvement in cholinergic function, as reflected by the cognitive improvement measured by ADAS-cog. Tacrine, velnacrine and donepezil, on the other hand, seem to show a small, but real initial improvement (2–3 points), as compared to the placebo group lasting 4–24 weeks, depending on the dose [24–26]. Whether or not this represents a real difference among drugs remains to be demonstrated. As indicated by studies of longer duration (up to 18 months) [27], it is possible that the difference between placebo- and drug-treated subjects could be maintained beyond the 6-month limit to at least 1 year. This would represent a significant gain for both patient and caregiver. Drugs could differ also in this respect. In comparing the results of clinical trials and the effect of different drugs, one should take into consideration the fact that studies may differ from one another as a result of differences in selection criteria, age of subjects, severity of disease, concomitant illnesses, variable instruments of assessment and side-effect evaluation, although using "completers" instead of ITT analysis could produce somewhat higher effects. Thus, studies are not totally comparable, and conclusions at this stage can only be indicative. Given these limitations, the immediate next goal to achieve for a ChEI would be an improvement of 5–6 ADAS-cog points and a 0.6–0.7 point on CIBIC during a 6-month treatment period. Most of this effect should still be present at 12 months and the dropout should be no more than 20%. Is this goal achievable?

What Makes the Difference Between Various Cholinesterase Inhibitors?

The relationship between the percent of peripheral ChE inhibition and cognitive (ADAS-cog) or global impression of change rated by the clinician (CGIC) effect is a relevant factor that is reported in Table 8.

Table 8. Relationship between percent ChE inhibition and effect on ADAS-cog or CGIC

Drug	Dose (mg/day)	Steady state (% inhibition)	Optimal (% inhibition)	Correlation ChEI/ ADAS-cog or CGIC
Physostigmine [61]	3–16	40–60 (BuChE)	30–40	U-shaped
Eptastigmine [59, 62]	30–60	13–54 (AChE)	30–35	U-shaped
Metrifonate [63]	30	35–75 (AChE)	65–80	U-shaped
Donepezil [26]	5	64 (ACHE)	60	Linear
Tacrine [24, 25, 64]	160	40 (BuChE) 60 (AChE)	30	Linear

ADAS-cog, AD Assessment Scale-cognitive subscale; *CGIC*, Clinician Global Impresssion of Change

The data presented in Table 8 support the concept of optimal brain ChE inhibition and functional ACh levels that might vary for each drug and relate to an optimal gain in cognitive and therapeutic effect [12, 28, 29]. This hypothesis is in accordance with pharmacological data in animals [3, 30] and in humans [28]. The level of peripheral enzyme inhibition that has been measured in the patient (AChE activity in erythrocytes or plasma BuChE activity) producing maximal effect on cognitive testing varies between 30 and 60%, depending on the kinetic and pharmacological characteristics of the compound (Table 8). However, for some drugs, (see donepezil and metrifonate) this can be as high as 80%. As predicted by pharmacological and behavioral data, there is a clear correlation between ChE inhibition (or drug plasma concentration) and cognitive effect [30, 31]. Drugs producing only mild side effects at high dosage and causing high brain ChE inhibition have the advantage that they can be tested clinically within their full range of therapeutic potential. For some drugs (physo, eptastigmine and metrifonate) the relationship between ChE inhibition and cognitive effect is inverse-U shaped, while for others (such as tacrine and donepezil) this relation seems to be linear. The U-shaped form can be explained by the fact that by increasing the dose of the inhibitor, one sees increasing efficacy as long as adverse effects do not become a limiting factor. Other reasons for the U-shaped curve depend on the specific inhibition kinetics of the inhibitor- and substrate-induced saturation effects of ChEs. The level of ACh brain elevation varies according to brain ChE inhibition [12, 32, 33]. With increased brain concentrations of ACh, substrate inhibition of enzyme activity is a phenomenon of particular importance that is observed in vitro and is probably present also in vivo [32]. Plotting the velocity of enzymatic reaction against substrate (ACh) concentration, a bell-shaped curve with a peak in the case of AChE and a sigmoid curve in the case of BuChE is observed. Thus, AChE is inhibited by a large excess of ACh, such as can be produced by a sudden high inhibition of brain AChE. This substrate elevation has the effect of decreasing the catalytic potency of the enzyme and subsequently its pharmacological (and perhaps therapeutical) effect. From this relationship it can be predicted that a high ChE inhibition reached rapidly during treatment (rapid passage of the drug into the brain and its accumulation) will not increase but reduce efficacy and increase

CNS-dependent side effects (drowsiness, nausea, vomiting, etc). It should be an advantage to use a slow-release type of ChEI to inhibit the brain enzymes at a slow pace and slowly reach steady-state levels of brain ACh. Kinetic mechanisms may explain the inverse U-shaped relationship seen for some drugs (eptastigmine, tacrine and metrifonate). There is excellent agreement between clinical and animal data for both physo and tacrine with regard to dose/behavioral effect relationships. Rupniak et al. [34], using two primate models (rhesus monkeys), found that both tacrine and physo improved visual recognition memory significantly. Both drugs showed a clear inverse U-shaped relationship with a maximal effect at around 0.0010–0.02 mg/kg i.m. for physo and 0.8–1 mg/kg for tacrine. Lower or higher doses did not improve performance. The central cholinergic side effects that may develop early in the treatment are not related directly to brain AChE inhibition, but mainly to elevation of ACh levels [32, 33]. In addition, peripheral side effects may occurr depending on rapid distribution of the drug between extra-CNS (peripheral organs) compartments and the CNS. A combination of pharmacokinetic and pharmacodynamic effects of the drug (including possible downregulation of muscarinic and nicotinic receptors) may be responsible for the tolerance to therapeutic effects developed by the patient.

Last but not least, the problem of ChEI therapy is the early identification of the patients most likely to benefit from therapy. The correlation of therapy with risk factors (APOE allele) represents an attempt in this direction (Fig. 5). Choosing the proper and possibly early stage of disease at which to start medication may be crucial for the success of the therapy.

Effects of Cholinesterase Inhibitors on Cortical Neurotransmitters

Clinical as well as experimental evidence indicates involvement and interaction between the cholinergic system and the biogenic amine system in the cognitive

Table 9. ChEI effects on ACh, NE, DA levels and ChE activity in rat brain cortex after s.c. administration

Compound[a]	Dose (mg/kg)	ChE (max. % inhibition)	ACh	NE (% increase)	DA
Physostigmine [65]	0.3	60	4000	75	120
Heptyl-physostigmine [65]	2	75	2500	25	75
Donepezil [65]	2	35	2100	100	80
MF-268 [66]	2	40	2500	100	60
MDL 73,745[67]	2	65	1020	120	370
Metrifonate [68]	80	70	1700	60	75

[a] *Donepezil*, (R,S)-1benzyl-4-(5,6 dimethoxy-1-idanon)-2-yl-methylpiperidine; *MF-268*, 2, 6-dimethylmorfolin-octyl-carbamoyl eseroline; *Metrifonate*, O,O-dimethyl-(1-hydroxy-2,2,2 trichloroethyl-phosphate); *MDL 73,745*, 2,2,2-trifluoro-1-(3-trimethylsilylphenyl) ethanone

impairment observed in AD. We have devoted particular attention to this problem [31].

Table 9 compares the effects on ACh, norepinephrine (NE) and dopamine (DA) levels as well as AChE inhibition after systemic administration of six clinically tested ChEI studied in our laboratory. Our results show a significant increase in the cortex for all three neurotransmitters and for all six ChEI investigated.

The results reported in Table 9 suggest that extracellular ACh levels in cortex are not only related to ChE inhibition, supporting the results of previous microdialysis studies showing comparable elevations of ACh levels in spite of different magnitudes of ChE inhibition [35]. This consideration may be of importance in predicting clinical effects and side effects (see dopaminergic effects) and setting dosages of various ChEI. The development of combinations of ChEI and monoamine-receptor agonists and antagonists is suggested by these studies [12, 31].

Cholinesterase Inhibitors as a Possible Mechanism to Slow Down Deterioration

The β-amyloid peptide (βA4), one of the major constituent proteins of neuritic plaques in the brain of AD patients, originates from a larger polypeptide-denominated APP [36]. APP is widely distributed throughout the mammalian brain, including rat brain with a prevalent neuronal localization [37]. APP can be processed by several alternative pathways. A secretory pathway is believed to generate non-amyloidogenic soluble derivatives (APPs) following cleavage within the βA4 segment [38]. Cholinergic agonists regulating processing and secretion of APPs by increasing, as demonstrated in vitro, protein kinase C (PKC) activity of target cells [39–42] could decrease potentially amyloidogenic derivatives. We suggested that long-term inhibition of ChE increasing levels of synaptic ACh may result in activation of normal APP processing in AD brain [32]. This effect could slow down the formation of amyloidogenic APP fragments. Lahiri et al. [43], using

Table 10. Effect of ChEI on β-amyloid processing in vitro and in vivo

Drug	Ref.	Concentration	APP Accumulation	APP Secretion	Preparation
Tacrine	[44]	500 mM	-	+	APP 770
Tacrine	[44]	<500 mM	+	-	APP 770
Tacrine	[43]	100 mg/ml	+	-	Cell lines
Phenserine	[16]	2.5 mg/kg	+	-	Rat CSF
Tacrine	[45, 67]	0.5 mM	+	-	Rat cortical slices
Tacrine	[45, 67]	0.1 mM	-	+	Rat cortical slices
Physostigmine	[45, 67]	0.1 mM	-	+	Rat cortical slices
Eptastigmine	[45, 67]	0.1 mM	-	+	Rat cortical slices
DDVP (metrifonate)	[45, 67]	0.02 mM	-	+	Rat cortical slices

Table 11. Drug-stimulated changes of basal APPs release and APP-KPI mRNA from rat brain [45]

Drug	Concentration (µM)	APPs release increase (% of basal)	ChE activity (% inhibition)	APP-KPI mRNA (% of basal)
Bethanechol	1	48	0	–
	100	53	0	–
Physostigmine	0.1	48	25	–
Heptyl-physostygmine	0.1	41	61	-35[a]
Dichlorvos	0.02	33	95	–
Phorbol myristate	0.1	–	–	+50

[a] From [29] (5 mg/kg s.c, 48 h)

nerve cell cultures, found that the level of secretion of APP derivatives into conditioned media was inhibited when treated with 100 µg/ml tacrine (Table 10). Chong and Suh [44] found a dose-dependent effect of tacrine on APP processing (Table 10). Tacrine at low concentrations (0.02–0.5 mM) enhanced, whereas concentrations above 0.5 mM blocked APP 770 processing in vitro (Table 10). Haroutunian et al. [16] reported that 1-week treatment with the ChEI phenserine decreased the levels of secreted beta-APP in the CSF of forebrain cholinergic lesioned rats, suggesting that secretion of beta-APP into the CSF and neurons can be modulated by ChEI (Table 1). To determine whether ChEI could alter the release of APP in brain we used superfused cortical slices of the rat [45] following the method described by Nitsch et al. [46]. Both short- and long-acting ChEI were tested for their ability to enhance the release of non-amyloidogenic soluble derivatives of APP [45]. These included: physo, heptastigmine and dichlorvos (DDVP), a metabolite of metrifonate, at concentrations producing ChE brain inhibitions ranging from 5 to 95%. All three ChEI elevated APPs release significantly above control levels (Table 11).

Using two different doses of tacrine (0.5 µM and 0.1 µM) we found that only the lower dose elevated the release of APPs in the cortical slices, which supports the data of Chong and Suh [44] of a dose-dependent modulation of APP secretion by AChE inhibition (Table 10). In our study, electrical field stimulation significantly increased the release of APPs within 50 min. A similar increase was observed after muscarinic receptor stimulation with bethanechol (BETHA), supporting results from in vitro experiments [41] (Table 11). Tetrodotoxin (TTX) completely blocked the effect of electrical stimulation [45].

The level of total APP RNAs in rat cortical slices did not change after incubation with BETHA, DDVP and physo, but activation of PKC with phorbol 12-myristate-13-acetate (100 nM) increased the level of total APP mRNA by 50% (Table 11) [29]. Physo and DDVP administration (0.3 mg/kg and 80 mg/kg s.c., respectively) for 3–48 h did not significantly change the levels of APP 695 and APP-KPI (Kunitz-type) protease inhibitor mRNAs (Table 11). Heptastigmine

Table 12. Effect of human AChE on APP processing.

1. AChE (but not BuChE) promotes β-A4 formation [48]
2. This effect is blocked by ChEI binding to the peripheral anionic binding site of AChE [48]
3. AChE accumulates within senile plaques and has non-intrinsic proteolytic activity on APP-770 in vitro [50]
4. AChE and its associated protease promotes amyloidogenic processing of APP-770 [44]

administration (5 mg/kg s.c., 3–48 h) decreased by 35% the level of APP-KPI mRNA in rat cerebral cortex [29] (Table 11). AD pathology has been associated with an increase in the KPI-containing forms of APP and the propensity across species to develop neuritic plaques in cortical regions [47]. Our findings suggest that administration of ChEI to AD patients by increasing secretion of APP and inhibiting formation of specific APP mRNAs may exert a neuroprotective effect, activating normal APP processing through a muscarinic mechanism and decreasing amyloid deposition in brain cells.

This effect of ChEI on APP processing could be reflected clinically by slowing down cognitive deterioration, as suggested by clinical data of short- and long-term treatments discussed in previous sections of this paper. In order to understand the mechanism of action of ChEI on APP processing, in addition to the muscarinic mechanism mentioned above, it is important to consider some new data related to specific properties of human AChE, as summarized in Table 12.

Inestrosa et al. [48] demonstrated that recombinant human AChE accelerates beta-A4 formation. This effect is not shared by BuChE. Thus, in vitro data suggest that AChE could play a role in amyloid deposition. It is also interesting to note that this effect is blocked by ChEIs binding to the peripheral anionic binding site of AChE. This could represent an interesting property for future developmental peripheral-site blocker ChEI.

It has been suggested that the proteolytic activity associated with AChE could play a role in APP processing, acting as an alpha-secretase and participating in the non-amyloidogenic cleavage of APP [49]. According to recent work of Funes [50], only the AChE derived from the brain of Alzheimer patients has such a proteolytic activity. Inhibition of the esteratic activity, however, does affect the proteolytic activity of AChE as well as amyloidogenic cleavage of APP [51]. These results do not suggest a protective effect of ChEI per se in AD related to inhibition of AChE-associated proteases.

Conclusions: the Future of Alzheimer Disease Therapy

Cholinesterase inhibitors are presently the drugs of choice for AD. In less than 10 years, starting from non-specific first-generation drugs, they have been developed into a second generation of more selective molecules. The latest data from clinical trials suggest that optimization and maintenance of clinical effects for 1 year or more is possible. This progress will depend on our knowledge of pharmacodynamic effects of long-term treatment. Cholinesterase inhibitors, particularly second generation (post-physo and post-tacrine compounds), affect cortical as well as subcortical neurotransmitters other than ACh. Their effects on NE and DA are of particular clinical interest. A newly demonstrated feature of ChEI is their ability to enhance the release of non-amyloidogenic soluble derivatives of APPs in vitro and in vivo and possibly to slow down formation of amyloidogenic compounds in brain. This process might also slow down cognitive deterioration of the patient as indicated by the analysis of recent clinical data. A critical question is how long the effect of ChEI will be clinically significant. Clinical trials extending beyond 12 months' duration should be able to demonstrate whether or not the pharmacological effect on APP metabolism is of significance. Cholinomimetic alternatives other than ChE inhibition are being explored pharmacologically and clinically. The drugs most investigated are those showing direct stimulation of postsynaptic M1 and M3 muscarinic receptors. This line of therapy has not produced convincing results so far, but may be more promising as a combination therapy.

Depending on the success of ChEI, one can see potential indications for applications to different stages of AD such as: (a) preclinical presymptomatic stages in individuals at risk with MCI (minimal cognitive impairment), (b) early AD patients with manifest symptoms (CDR 0.5–1), presently the most treated group, (c) late AD (CDR 2) patients with behavioral symptoms. A combination of ChEI with muscarinic or nicotinic agonists or with beta-A4 processers or APP-releasers and estrogens (Figs. 1-3) represents an interesting alternative if tolerance to ChEI monotreatment occurs.

Acknowledgements. The author wishes to thank the Japan Foundation for Aging and Health, which provided financial support for this study during his stay in Japan in July–August 1996. Thanks are also due to Professor Yasuyuki Nomura, Faculty of Pharmaceutical Sciences, Department of Pharmacology, University of Hokkaido, Sapporo, for making this study possible while the author was a visiting professor in his department. The author thanks Christine Mesmer for typing and editing the manuscript.

References

1. Summers WK, Majovski LV, Marsh GM, Tachiki K, Kling A (1986) Oral tetrahydroaminoacridine in long term treatment of senil dementia, Alzheimer type. N Engl J Med 315: 1241-1245
2. Becker R, Moriearty P, Unni L (1991) The second generation of cholinesterase inhibitors: clinical and pharmacological effects. In: Becker R, Giacobini E (eds) Cholinergic basis for Alzheimer therapy. Birkhäuser, Boston, pp 263-296
3. Mattio T, McIlhany M, Giacobini E, Hallak M (1986) The effects of physostigmine on acetylcholinesterase activity of CSF, plasma and brain. A comparison of intravenous and intraventricular administration in beagle dogs. Neuropharmacology 25: 1167-1177

4. Giacobini E (1987) Models and strategies of cholinomimetic therapy of Alzheimer disease. In: Dowdall MJ, Hawthorne JN (eds) Cellular and molecular basis of cholinergic function. Horwood, New York, pp 882-901

5. Hallak M, Giacobini E (1986) Relation of brain regional physostigmine concentrations to cholinesterase activity, and acetylcholine and choline levels in rat. Neurochem Res 11: 1037-1048

6. Giacobini E, Mussini I, Mattio T (1986) Aging of cholinergic synapses: fiction or reality. In: Hanin I (ed) Dynamics of cholinergic function, vol 30. Plenum, New York, pp 177-190

7. Checler F (1995) Processing of the b-amyloid precursor protein and its regulation in Alzheimer's disease. J Neurochem 65: 1431-1444

8. Van Broeckhoven C (1995) Presenilins and Alzheimer disease. Nat Genet 11: 230-232

9. Sandbrink R, Hartmann T, Masters CL, Beyreuther K (1996) Genes contribution to Alzheimer's disease. Mol Psychiatry 1: 27-40

10. Vito P, Lacaná E, D'Adamio L (1996) Interfering with apoptosis: Ca^{2+}-binding protein ALG-2 and Alzheimer's disease gene ALG-3. Science 271: 521-525

11. Roses A (1995) Perspective on the metabolism of apolipoprotein E and the Alzheimer disease. Exp Neurol 132: 149-156

12. Giacobini E (1996) Cholinesterase inhibitors do more than inhibit cholinesterase. In: Becker R, Giacobini E (eds) Alzheimer disease: from molecular biology to therapy. Birkhäuser, Boston, pp 187-204

13. Yankner B (1996) New clues to Alzheimer's disease: unraveling the roles of amyloid and tau. Nat Med 2: 850-852

14. Soininen H, Riekkinen J Sr (1996) Apolipoprotein E, memory and Alzheimer's disease. Trends Neursosci 19: 224-228

15. Haroutunian V, Greig NH, Gluck R, Fiber E, Davis KL, Wallace WC (1995) Selective attenuation of lesion-induced increases in secreted b-APP by acetylcholinesterase inhibitors. Soc Neurosci 21 208.1: 502

16. Haroutunian V, Greig N, Pei XF, Utsuki L, Acevedo LD, Gluck R, Davis KL, Wallace D (1996) Pharmacological modulation of Alzheimer's b-amyloid precursor protein levels in the CSF of rats with forebrain cholinergic system lesions. Soc Neurosci 22 461.7: 1169

17. Soininen H, Kosunen O, Helisalmi S, Mannermaa A, Paljärvi L, Talasniemi S, Ryynänen M, Riekkinen P (1995) A severe loss of choline acetyltransferase in the frontal cortex of Alzheimer patients carrying apolipoprotein e4 allele. Neurosci Lett 187: 79-82

18. Poirier J, Delisle MC, Quirion R, Aubert I, Farlow M, Lahiri D, Hui S, Bertrand P, Nalbantoglu J, Gilfix BM, Gauthier S (1995) Apolipoprotein E4 allele as a predictor of cholinergic deficits and treatment outcome in Alzheimer disease. Proc Natl Acad Sci USA 92: 12260-12264

19. Nordberg A (1993) Clinical studies in Alzheimer patients with positron emission tomography. Behav Brain Res 57: 215-224

20. Nordberg A (1995) Long term treatment effects on progression of Alzheimer's disease as determined by functional brain studies. In: Ikbal K, Mortimer JA, Winblad B, Wiesniewski HM (eds) Research advances in alzheimer's disease and related disorders. Wiley, Chichester, pp 293-298

21. Gracon S, Smith F, Hoover T (1996) Long-term tacrine treatment: effect on nursing home placement and mortality. In: Becker R, Giacobini E (eds) Alzheimer disease: from molecular biology to therapy. Birkhäuser, Boston, pp 205-209

22. Ogane N, Giacobini E, Struble R (1992) Differential inhibition of acetylcholinesterase molecular forms in normal and Alzheimer disease brain. Brain Res 589: 307-312

23. Anand R, Hartman RD, Hayes PE (1996) An overview of the development of SDZ ENA 713, a brain selective cholinesterase inhibitor. In: Becker R, Giacobini E (eds) Alzheimer disease: from molecular biology to therapy. Birkhäuser, Boston, pp 239-243

24. Farlow M, Gracon SI, Hershey LA, Lewis KW, Sadowski CH, Dolan-Ureno JA (1992) Controlled trial of tacrine in Alzheimer's disease. J Am Med Assoc 268: 2523-2529

25. Knapp ZJ, Knopman DS, Solomon PR (1994) A 30 week randomized controlled trial of high-dose tacrine in patients with Alzheimer's disease. JAMA 271: 985-991

26. Roger SL, Friedhoff T (1996) The efficacy and safety of donepezil in patients with Alzheimer's disease: results of a US multicentre, randomized, double-blind, placebo-controlled trial. Dementia 7: 293-303

27. Becker R, Moriearty P, Unni L, Vicari S (1996) Cholinesterase inhibitors as therapy in Alzheimer's disease: benefit to risk considerations in clinical application. In: Becker R, Giacobini E (eds) Alzheimer disease: from molecular biology to therapy. Birkhäuser, Boston, pp 257-266

28. Giacobini E, Becker R, McIlhany M, Kumar V (1988) Interacerebroventricular administration of cholinergic drugs: preclinical trials and clinical experience in Alzheimer patients. In: Giacobini E, Becker R (eds) Current research in Alzheimer therapy. Taylor and Francis, New York, pp 113-122

29. Giacobini E, Mori F, Buznikov A, Becker R (1995) Cholinesterase inhibitors alter APP secretion and APP mRNA in rat cerebral cortex. Soc Neurosci Abstr 21: 988

30. Giacobini E, DeSarno P, Clark B, McIlhany M (1989) The cholinergic receptor system of the human brain – neurochemical and pharmacological aspects in aging and Alzheimer. In: Nordberg A, Fuxe K, Holmstedt B (eds) Progress in brain research. Elsevier, Amsterdam, pp 335-343

31. Giacobini E, Cuadra G (1994) Second and third generation cholinesterase inhibitors: from preclinical studies to clinical efficacy. In: Giacobini E, Becker R (eds) Alzheimer disease: therapeutic strategies. Birkhäuser, Boston, pp 155-171

32. Giacobini E (1994) Cholinomimetic therapy of Alzheimer disease: does it slow down deterioration? In: Racagni G, Brunello N, Langer SZ (eds) Recent advances in the treatment of neurodegenerative disorders and cognitive dysfunction, vol 7(23). International Academy of Biomedical Drug Research. Karger, New York, pp 51-57

33. Giacobini E (1995) Cholinesterase inhibitors: from preclinical studies to clinical efficacy in Alzheimer disease. In: Quinn D, Balasubramaniam AS, Doctor BP, Taylor P (eds) Enzymes of the cholinesterase family. Plenum, New York, pp 463-469

34. Rupniak NMJ, Field MJ, Samson NA, Steventon MJ, Iversen SD (1990) Direct comparison of cognitive facilitation by physostigmine and tetrahydroaminoacridine in two primate models. Neurobiol Aging 11: 609-613

35. Messamore E, Warpman U, Williams E, Giacobini E (1993) Muscarinic receptors mediate attenuation of extracellular acetylcholine levels in rat cerebral cortex after cholinesterase inhibition. Neurosci Lett 158: 205 208

36. Kang J, Lemaire HG, Unterbeck A, Salbaum JM, Master CL, Grzeschil KH, Multaup G, Beyreuther K, Muller-Hill B (1987) The precursor of Alzheimer disease amyloid A4 protein resembles a cell-surface receptor. Nature 325: 733-736

37. Beeson JG, Shelton ER, Chan HW, Gage FH (1994) Differential distribution of amyloid protein precursor immunoreactivity in the rat brain studied by using five different antibodies. J Comp Neurol 342: 78-96

38. Sisodia SS, Koo EH, Beyreuther K, Unterbeck A, Price DL (1990) Evidence that beta amyloid protein in Alzheimer disease is not derived by normal processing. Science 248: 492-495

39. Nitsch RM, Slack BE, Wurtman RJ, Growdon JH (1992) Release of Alzheimer precursor derivatives stimulated by activation of muscarinic acetylcholine receptors. Science 258: 304-307

40. Buxbaum JD, Oishi M, Chen HI, Pinkas-Kramarski R, Jaffe EA, Gandy SE, Greengard P (1992) Cholinergic agonists and interleukin 1 regulate processing and secretion of the Alzheimer βA4 amyloid protein precursor. Proc Natl Acad Sci USA 89: 10075-10078

41. Buxbaum JD (1995) Post-translational control of the amyloid b-protein precursor processing. In: Nitsch RM, Growdon JM (eds) Pathobiology of Alzheimer's disease. Academic, New York, pp 98-114

42. Nitsch RM, Growdon JH (1994) Role of neurotransmission in the regulation of amyloid beta-protein precursor processing. Biochem Pharmacol 47: 1275-1284

43. Lahiri DK, Lewis S, Farlow MR (1994) Tacrine alters the secretion of the beta-amyloid precursor protein in cell lines. J Neurosci Res 37: 777-787

44. Chong YH, Suh YH (1996) Amyloidogenic processing of Alzheimer's amyloid precursor protein in vitro and its modulation by metal ions and tacrine. Life Sci 59: 545-557

45. Mori F, Lai CC, Fusi F, Giacobini E (1995) Cholinesterase inhibitors increase secretion of APPs in rat brain cortex. Neuroreport 6: 633-636

46. Nitsch RM, Farber SA, Growdon JH, Wurtman RJ (1993) Release of amyloid beta-protein precursor derivatives by electrical depolarization of rat hippocampal slices. Proc Natl Acad Sci USA 90: 191-193

47. Anderson JP, Refolo LM, Wallace W, Mehta P, Krishnamurthi M, Gotlib J, Bierer L, Haroutunian V, Perl D, Robakis NK (1989) Differential brain expression of the Alzheimer's amyloid precursor protein. EMBO J 8: 3627-3632

48. Inestrosa NC, Alvarez A, Perez C A, Moreno RD, Vicente M, Linker C, Casanueva O, Soto C, Garrido J (1996) AChE accelerates assembly of amyloid-b peptides into Alzheimer's fibrils: possible role of the peripheral site of the enzyme. Neuron 16: 881-891

49. Small DH, Moir RD, Fuller SJ, Michaelson S, Bush A, Qiao-Xin L, Milward E, Hilbich C, Weidemann A, Beyreuther K, Colin Masters L (1991) A protease activity associated with acetylcholinesterase releases the membrane-bound form of the amyloid protein precursor of Alzheimer's disease. Biochemistry 30: 10795-10799

50. Funes S (1996) Acétylcholinestérase et maladie d'Alzheimer. Thèse présentée pour obtenir le grade de docteur de l'Université de Louis Pasteur, Strasbourg

51. Chong YH, Jung JM, Choi W, Park CW, Sou Choi K, Suh YH (1994) Bacterial expression, purification of full length and carboxyl terminal fragment of Alzheimer amyloid precursor protein and their proteolytic processing by thrombin. Life Sci 54: 1259-1268

52. Wallace WC, Bragin V, Robakis NK, Sambamurti K, VanderPutten D, Merril CR, Davis KL, Santucci AC, Haroutunian V (1991) Increased biosynthesis of Alzheimer amyloid precursor protein in the cerebral cortex of rats with lesions of the nucleus basalis of Meynert. Mol Brain Res 10: 173-178

53. Nordberg A (1996) Pet imaging of nicotinic receptors in Alzheimer's disease – implication with diagnosis and drug treatment. In: Becker R, Giacobini E (eds) Alzheimer disease: from molecular biology to therapy. Birkhäuser, Boston, pp 439-444

54. Minthon L, Nilsson K, Edvinsson L, Wendt PE, Gustafson L (1995) Long-term effects of tacrine on regional cerebral blood flow changes in Alzheimer's disease. Dementia 6: 245-251

55. Perry EK, Perry RH, Blessed G, Tomlinson BE (1978) Changes in brain cholinesterases in senile dementia of Alzheimer type. Neuropathol Appl Neurobiol 4: 273-277

56. Davies P (1979) Neurotransmitter-related enzymes in senile dementia of the Alzheimer Type. Brain Res 171: 319-327

57. Atack JR, Perry EK, Bonham JR, Candy JM, Perry RH (1987) Molecular forms of

butyrylcholinesterase in the human neocortex during development and degeneration of the cortical cholinergic system. J Neurochem 48: 1687-1692

58. Arendt T, Brückner M, Lange M, Bigl V (1992) Changes in acetylcholinesterase and butyrylcholinesterase in Alzheimer's disease resemble embryonic development – a study of molecular forms. J Neurochem 21: 381-396

59. Canal I, Imbimbo BP (1996) Clinical trials and therapeutics: relationship between pharmacodynamic activity and cognitive effects of eptastigmine in patiens with Alzheimer's disease. Clin Pharmacol Ther 60: 1-11

60. Imbimbo BP (1996) Eptastigmine: a cholinergic approach to the treatment of Alzheimer's disease. In: Becker R, Giacobini E (eds) Alzheimer disease: from molecular biology to therapy. Birkhäuser, Boston, pp 223-230

61. Thal L, Fuld PA, Masur DM, Sharpless NS (1983) Oral physostigmine and lecithin improve memory in Alzheimer disease. Ann Neurol 13: 491-496

62. Imbimbo BP, Lucchelli PE (1994) A pharmacodynamic strategy to optimize the clinical response to eptastigmine (MF-201). In: Giacobini E, Becker R (eds) Alzheimer disease: therapeutic strategies. Birkhäuser, Boston pp 103-107

63. Becker R, Colliver J, Elble R (1990) Effects of metrifonate, a long-acting cholinesterase inhibitor, Alzheimer disease: report of an open trial. Drug Dev Res 19: 425-434

64. Knapp S, Wardlow ML, Albert W, Waters D, Thal J (1991) Correlation between plasma physostigmine concentrations and percentage of acetylcholinesterase inhibition over time after controlled release of physostigmine in volunteer subjects. Drug Metab Dispos 19: 400-404

65. Giacobini E, Zhu XD, Williams E, Sherman KA (1996) The effect of the selective reversible acethylcholinesterase inhibitor E2020 on extracellular acetylcholine and biogenic amines levels in rat cortex. Neuropharmacology 35: 205-211

66. Zuh XD, Cuadra G, Brufani M, Maggi T, Pagella PG, Williams E, Giacobini E (1996) Effects of MF268, a new cholinesterase inhibitor, on acetylcholine and biogenic amines in rat cortex. J Neurosci Res 43: 120-126

67. Mori F, Cuadra G, Giacobini E (1995) Metrifonate effects on acetylcholine and biogenic amines in rat cortex. Neurochem Res 20: 1081-1088

68. Zhu XD, Giacobini E, Hornsperger JM (1995) Effect of MDL 73,745 on acetylcholine and biogenicamine levels in rat cortex. Eur J Pharmacol 276: 93-99

Non-Pharmacological Treatment of Alzheimer's Disease and Related Disorders

O. Zanetti and M. Trabucchi

Introduction

Notwithstanding the fact that multidisciplinary effort has quickly transformed the area of Alzheimer's disease (AD) and related disorders, leading to better understanding of the clinical presentation, risk factors, and treatment, definitive knowledge of the pathogenetic mechanisms is still lacking, thus heavily influencing pharmacological and non-pharmacological interventions. The new drugs recently placed on the market, as well as those in the advanced stage of pharmacological trials, constitute significant progress, but mainly in the area of the old cholinergic hypothesis, where only partial efficacy can be expected. Non-pharmacological approaches could therefore be of paramount importance. Even if effective pharmacological treatment could be readily available in the short term, non-pharmacological interventions [1] are necessary in order to face the various clinical aspects – cognitive, functional, behavioral and affective – of AD and related disorders. The present review will deal mainly with AD, but also includes other types of dementia.

Until recently there has been widespread nihilism as far as the idea of non-pharmacological treatment for the demented patient is concerned. Most physicians believed that there was little reason to spend money or effort for diseases such as dementia with no known risk factors, preventive measures, or treatment and whose evolution is characterized by a progressive course. But it should be kept in mind that this course may last 8–9 years! A second element that has negatively influenced the idea of non-pharmacological approaches for dementias is constituted by the primary impairment of intellectual functions in this disease. Patients affected by AD show impairment of memory, concentration, perceptual skills, and apraxia that can prevent full cooperation with rehabilitation techniques, many of which are complex tasks [2]. Once more, however, it should be underlined that the impairment of cognitive functions in AD is not an "all-or-nothing" phenomenon and, above all during the first stages of the disease, relatively preserved cognitive functions, i.e., procedural memory, could be the target of stimulation and rehabilitative interventions [3–5].

The negative attitude toward a rehabilitative approach for the dementias has led to the use of prosthetic approaches or non-specific global stimulation, nourished also by the failure of various memory and cognitive interventions whose effectiveness and long-lasting impact in normal elderly is well demonstrated [4, 6]. It should be underlined that until 10 years ago the rehabilitative approaches for dementias were based upon poor theoretical and methodological bases that can be grouped under the slogan "use it or lose it."

Recently, this pessimistic attitude toward rehabilitative approaches in AD has been partially smashed by the demonstration that AD patients, although in the milder stages of the disease, can perform motor, perceptual, and cognitive tasks, thanks to the preservation of procedural memory [7–12]. The great majority of the studies support the view that significant results can be achieved if rehabilitative interventions are targeted at spared or relatively spared cognitive functions, at least in the mild or mild-moderate stages of AD or related disorders.

Along with "non-specific cognitive stimulation" that can be useful in the moderate stage, it is now possible to adopt rehabilitative interventions more tailored to the rehabilitation of memory functions in AD. The rehabilitative interventions supported by clinical trial or rigorous testing are the following: reality orientation therapy (ROT) and procedural memory stimulation. Moreover, external memory aids also seem promising in the milder stages of AD [10].

The key general principles of rehabilitation apply also to dementia. Rehabilitation is concerned with lessening the impact of disabling conditions; rehabilitation is a complex set of processes usually involving several professional disciplines and aimed at improving the quality of life of older people facing daily living difficulties caused by chronic disease. The key purposes of rehabilitation can be summarized succinctly: (a) realization potential; (b) re-ablement; (c) resettlement; (d) role fulfillment; (e) readjustment [2]. Usually the rehabilitation process focuses predominantly on physical functioning. In demented patients successful rehabilitation requires a broader perspective – one that allows social and psychological problems to be identified and addressed. Other success factors for rehabilitation have been long established: a positive attitude and approach, individual assessment of patient and caregiver, involvement of patient and caregiver, and team partecipation.

The promotion of a better quality of life for patients and caregivers can be achieved by special and general therapeutic techniques, and optimizing the environment. It should be underlined that treatment of patients affected by dementia is difficult and often involves maintenance rather than rehabilitation.

Caregivers play a key role in the management of demented patients (nearly 80% of patients are cared for at home!); therefore, the contribution of caregivers needs to be positively acknowledged by purposefully allocating time to understand their perspective and needs (the caregiver may be depressed or anxious and need treatment, and routine discussion of respite care is recommended). Rehabilitative interventions, dealing with a progressive disorder such as dementia, is essentially management of problems and impairments produced by the illness with the primary objectives of slowing down the course of the disease and reducing excess disability.

The main outcome for the demented patient consists in modifying the natural history of the disease. We, therefore, need faithful assessment techniques that allow the detection of subtle clinical changes, although moderate, that can improve the patient's and caregiver's quality of life. The demonstration of observable changes represents an imperative in intervention trials for dementia; in fact, in today's economy empirical evidence is required to support the use of treatment modalities.

Patients need thorough assessment in order to understand the amount of dysfunction, promote better functional status, and evaluate the impact of the training. While all problems identified may not be controllable, having a good overview of how well the patient functions may help to explain puzzling and seemingly unrelated behaviors and may lead to better management. The treatment approaches need to meet the changing needs and responses of the patients. The functional areas affected and major treatment approaches used in dementing illnesses are listed in Table 1. Among these, however, cognitive impairment – and above all memory – have received relatively more concern by research, clinicians and rehabilitation therapists. Tables 2 and 3 list the main strategies and methodological approaches for cognitive impairment.

Because the patient's needs will change over time it is necessary to modify the approaches during progression of the illness. The disease stages – mild-moderate-severe – are rarely predictable; patients will progress at variable speeds. Stages may be passed through quickly or slowly or in some cases may be skipped; skill retention or loss may not be orderly or predictable. The observation and assessment skills of the clinician or the therapist will be important in ensuring relevant treatment.

Many non-pharmacological/rehabilitative approaches are available for the demented patient; however, for many empirical evidence to support their use is lacking. The great majority of these approaches, i.e., behavior-oriented, emotion-oriented, and cognition-oriented approaches, have not been subjected to randomized clinical trials, but are supported by case-control studies or by single-case studies and are in widespread clinical use. Other non-pharmacological approaches, such as stimulation-oriented treatment, have only modest support from clinical trials, but common sense supports their use as part of the humane care of patients with dementia [14].

While these behavior-oriented, emotion-oriented, cognition-oriented, and stimulation-oriented treatments differ in philosophy, focus, and methods, they have the broadly overlapping goals of improving quality of life and maximizing function in the context of existing deficits. Many have as an additional goal the improvement of cognitive skills, mood, or behavior. Therefore, many of these treatments share overlapping goals and apparent non-specificity of action.

The selection of appropriate treatment for patients with progressive dementing illness should be based on the patient's clinical characteristics – mainly cognitive and functional impairment; the second step is deciding how the deficits can be treated and at what point in the disease the patient will need these treatments.

Table 1. Treatment approaches for patients with dementing illness. (Adapted from [13])

Functional areas	Treatment approaches
Cognitive function	Counseling (early stages) Cognitive facilitation Environmental structuring Sensorimotor therapy Relaxation Structured communication Structured activities Self-care activities
Sensory-perceptual function	Physical aids (glasses, hearing aids) Environmental structuring Adaptative equipment and techniques Sensorimotor therapy Structured activities
Affective functioning	Counseling (early stages) Relaxation Structured activities Structured socialization Spiritual expression Exercise and movement Milieu therapy
Social functioning	Structured activities Structured socialization Milieu therapy
Language and communication	Structured verbal and non-verbal communication techniques Cognitive facilitation Environmental structuring Behavior-modification program
Self-care activities	Adaptative intervention and techniques Exercise and sensorimotor therapy Environmental structuring Milieu therapy Behavior-modification program Memory training
Physical health and sleep	Relaxation Exercise and sensorimotor therapy Structured activities Environmental structuring Medication monitoring
Nutrition and elimination	Adaptative equipment and techniques Environmental structuring Behavior-modification program Exercise and movement
Motor function	Environmental structuring Adaptative equipment and techniques Exercise and sensorimotor therapy

Table 2. Rehabilitative strategies for cognitive impairment

1. Environmental adaptation
2. ROT
3. External aids
4. Internal aids (mnemotechnics)
5. Procedural memory training

Table 3. Rehabilitative approaches for cognitive impairment

- "Global" approaches
    - ROT
    - 3R therapy (ROT, reminiscence, and remotivation)
    - Reminiscence therapy
    - Validation therapy

- Selective approaches to memory functions
    - Spaced-retrieval
    - Prospective memory training
    - Procedural memory (cognitive, motor, and perceptual) training

- Mnemotechnics
    - "Loci" method
    - Noun-face association
    - Pegwords
    - Signposts
    - Electronic devices

As for cognitive stimulation intervention, the use of internal or external memory aids (mnemotechnics), and the stimulation of procedural memory (motor, perceptual, and cognitive) can be adopted in the milder stages of AD; ROT in the mild-to-moderate stages; reminiscence and remotivation therapy during the mild and moderate stages of the disease; lastly, validation therapy in the moderate-to-advanced stages of the disease. During the course of AD the possibility to use the patient's cognitive functions decreases and, in proportion, the role of environmental and behavioral interventions grows. Rehabilitative interventions, either targeted at memory impairment or at environmental manipulation, should be therefore be selected on the basis of the remaining cognitive/functional resources.

Because the patient's and family's needs change over time, the ideal situation is a continuum of care [13] (i.e., as wide a variety of services and providers as possible for a variety of needs levels). It is this continuum that allows maximization of functioning at any given point in the progression of the illness. However, optimal care depends on many factors: family social support system, availability of health and social services, family finances, the patient's rate of decline and clinical characteristics, and cultural factors.

Along with cognitive rehabilitative interventions devoted at improving or slowing down the progression of the cognitive domain of the disease, of paramount importance is the management of non-cognitive symptoms because they can negatively influence the cognitive training and related outcomes. The management of depressive symptoms, above all, is important for their influence on cognitive performance and on social relationships. Cognitive-behavioural interventions have been suggested for the treatment of depression in AD [15]. Supportive psychotherapy is used by some practitioners to address issues of loss in the early stages of dementia, reminiscence therapy has some modest research support for improvement of mood and behavior, and validation therapy and sensory integration have less research support. None of these modalities has been subjected to rigorous testing [14].

A final general issue must be addressed: who can provide cognitive rehabilitation? Unfortunately, there are no specific therapist operators. There is an urgent need for therapist education and training in order to face this growing need.

Cognition-Oriented Treatments

Cognition-oriented treatments cover a number of techniques (Table 3). Some of these techniques – reality orientation, memory training, and skills training (focused on specific cognitive deficits) – deal mainly with memory function and some, such as reminiscence and 3R (ROT, reminiscence, and remotivation), deal with affective problems as well, because a patient's emotional state has a strong influence on memory and cognitive functioning.

For higher-level patients, a cognitive stimulation group or one-on-one session may help. Tasks must be adapted to the functional level of the patient and may include material that taps long-term memory, such as well-known life facts or associative material. The tasks may involve verbal and/or written responses and may be multisensory, stimulating visual, olfactory, gustatory, auditory, and tactile memory. Other cognitive skills, such as judgment, sequencing, language and spelling, math, and abstract thinking, may also be included when appropriate.

Reality Orientation Therapy

The reality orientation therapy (ROT) programme was initiated in 1958 at the Veterans Administration in Topeka, Kansas, and was later refined by Taulbee and Folsom [16–18]. ROT is a psychosocial intervention used in the rehabilitation of persons having memory loss, episodes of confusion, and time-place-person disorientation. ROT is a technique for multimodal (verbal, visual, written, pictorial, musical…) presentation and reinforcement of basic orientation information. This can be refined for higher-level patients or made quite simple for lower-level patients. ROT methods appear to have their roots based in cognitive theories. Cognitive theories are insight-oriented approaches that emphazise recognition and changing negative thoughts and maladaptive beliefs [19]. ROT seems to be based on cognitive theory, as its goal is to halt confusion in the elderly patient by re-orienting him to reality. The main objective of ROT intervention is to re-orient the patient by means of continuous stimulation with repetitive orientation to environment. This is done in two ways that are meant to be complementary. The first, informal ROT, is meant to go on all the time and requires care staff to convey to the patient again and again basic information as to who and where he is and what time it is. As an adjunct or alternative to this 24-h method, ROT has also developed a more formal didactic group therapy (class ROT or formal ROT), in which patients meet daily in a special classroom for sessions of about 30–45 min. Patients should be divided into three separate groups, basic, standard, or advanced, according to their level of disorientation. All three levels are designed to help the client succeed, be oriented, and communicate. For more practical details on running ROT groups see [20] and [21]. Presently, ROT is one of the most widely applied psychosocial approaches for the demented elderly [22–24]. Despite the widespread use of ROT in long-term care facilities and in day centers, controversy surrounds its actual clinical value [25–27].

Specific ROT procedures have been found to improve demented patients' orientation and memory for personal facts [28] as well as memory for appointments [29]. Nowadays, ROT represents one of the few examples of cognitive training yielding positive results in AD [4]. The enthusiasm with which ROT has been accepted may be due to several reasons: ROT is a simple, unexpensive technique and can be performed by relatively unspecialized staff members. Moreover, ROT gives the staff (and the family caregivers) a sense of "doing something" with patients who have bleak futures. In the past 20 years, numerous researchers have analyzed the efficacy of ROT. A review of the literature concluded that ROT produced some improvement in verbal orientation but changes were minimal in other areas of functioning [30, 31]. Hanley et al. [25] and Citrin and Dixon [32] have found that while class ROT does improve cognitive function to a certain extent, it has no effect on other behaviors, such as activities of daily living skills. It has been speculated that the success of ROT in any particular treatment setting may depend upon the staff's belief and enthusiasm for the technique and that the benefit of ROT relies on the change in the interactional process between staff and patients [30]. Few studies have analyzed the permanence of the effect and attempted some form of follow-up [33]; moreover, there have been few attempts to outline the kind of patient most likely to benefit from ROT [31]. Brook et al. [34] and Zepelin et al. [35] found that the least deteriorated subjects did best, but Hanley et al. [25] were unable to demonstrate any influence attributable to severity of dementia. ROT seems more effective when formal or "in class" therapy is combined with informal treatment. This ranges from modifications of the environment in which the elderly individual spends his time during the day (with the introduction of precise temporospatial points of reference) to the training and involvement of institution or nursing home staff or family members [36–40]. Therefore, it is best to use reality orientation as part of a total program of cognitive facilitation and build information aids into the environment. Other authors have reported achieving some improvement in the mental status of demented elderly individuals by comparing [41] or combining ROT [42] with other methods of mental stimulation, such as reminiscence (past events and related objects are used to stimulate memory through recollection), remotivation (testing and stimulation of the individual's intellectual and cognitive characteristics through discussion, thought, and thought deduction [43]) and reactivating occupational therapy [44]. However, ROT, formal and/or informal, combined with other environmental or staffs, is presently the most well-known and widely used therapeutical intervention for the confused demented elderly.

Recently Zanetti et al. [45], conducted a longitudinal, controlled study with the aim of evaluating the effects of a long-term program of formal didactic group therapy (class reality-orientation therapy) in AD. The study was conducted in the Day Hospital of an Alzheimer's Dementia Research and Care Unit (Brescia, Italy), a multidisciplinary care center providing diagnostic evaluation and treatment for elderly patients with cognitive impairment. The criteria for the inclusion of patients in the study were: mild-moderate cognitive impairment (Mini-Mental State Examination - MMSE - between 11 and 24/30), and absence of major aphasia, blindness, and overt behavioral disturbances such as wandering or agitation. Sixteen patients constituted the experimental group and 12 the control group. The last cognitive, functional and affective evaluation in the experimental and control group was performed 8.2 months and 8.5 months after baseline assessment, respectively. The experimental group had repeated cycles of 1-month ROT classes, and 5–7 weeks were allowed between each cycle. The differential change for the MMSE score between the two groups was significant. In the experimental group there was mild improvement in MMSE score (0.68 points) at the last assessment, whereas the control group declined (-2.58 points). This treatment effect on MMSE score (3.27 points) was controlled for potential confounders in a multiple regression analysis. Adjusted treatment effect, including age, education, baseline MMSE, disease duration, disease severity, number of diseases other than Alzheimer's, and time elapsing from baseline to last assessment, was very slightly lower: 3.12. In the experimental group, treatment effect was evaluated by comparing ROT cycle changes and resting period changes. A clearly significant treatment effect was found for MMSE and Verbal Fluency. The results of this study, in accordance with previous research, show that formal ROT could positively influence cognitive performance in Alzheimer's disease elderly patients. The effect of ROT training on cognitive performance seems to counteract the expected decline observed in the control group. In fact, it is well known that, notwithstanding the recognized heterogeneity in Alzheimer's patients, the expected yearly decrease in MMSE score ranges between 1.8 and 4.2 points [46]. These results can be interpreted as indicative of the effectiveness of ROT programs in slowing, at least temporarily, the disease process. Given that the normal course of Alzheimer's disease is one of progressive decline (as confirmed by our control group), even a lack of decline – though not an actual improvement – can be viewed as evidence of the success of training.

While little reliable information exists about the characteristics of subgroups of patients who might benefit from ROT, it can be suggested that patients with mild-to-moderate cognitive impairment and without overt behavioral disturbances, blindness or major aphasia might be good candidates for ROT, underlining the importance of the patient's active participation in the rehabilitative intervention [31].

As reported by Dietch et al. [27], ROT and other attempts to orientate patients to reality can also have adverse effects. Patients may become emotionally upset during the ROT classes when attempts are made to bring them into the here and now. Therefore, in some patients, approaches other than ROT are needed. However, the low risk of adverse reactions and the cognitive improvement and global beneficial impact on patients and caregivers seem to justify the adoption of ROT in the treatment of demented patients.

The intrinsic limitation of many study designs and of ROT as rehabilitative approach needs to be considered in order to evaluate the results critically. A learning effect or a contamination of the results by therapy content should not be excluded. Also, taking care of a patient and his caregivers can favorably affect their quality of life. Moreover, it can be hypothesized that caregivers change their behavior towards the patients. In light of these observations the specific cognitive effects of ROT cannot be clearly differentiated from those of the sociorelationship environment and from the psychosocial effect due to "care" by the ward staff.

Memory Training

AD patients fail to benefit from a variety of different forms of environmental and cognitive rehabilitative interventions conducted in order to improve cognitive functions [4]. Some interesting exceptions to this negative pattern have been demonstrated, mainly within the realm of reality-orientation therapy and reactivating occupational therapy [42, 44, 45]. Even though these results are interesting, the methodology and ecological impact have been greatly criticized. The first problem is the contamination of the mental tests with therapy "content": patients learn the same overpresented material without acquiring a real compensatory strategy. A second problem is the lack of functional correlates of psychometric improvement. A third problem is related to the methods of stimulation that require a considerable amount of cognitive effort and associative skills that are severely impaired in AD [47].

Recently, more specific and tailored mental-stimulation strategies for demented patients have been proposed that rely on the stimulation of procedural learning [3, 8, 9, 12, 48], which is part of implicit memory and has been shown to be relatively spared in mild AD [7, 12, 49–53].

Largely amnesic patients may show a normal performance on skills learning [12], while patients with basal ganglia or cerebellar diseases may display a selective impairment of skill learning without memory deficits [54]. These observations support the view that memory is engaged in at least two functionally and anatomically different systems, i.e., declarative and procedural memory [55]. Procedural memory is expressed implicitly by changes in performances as a result of prior experience. Three subclasses, i.e., motor, perceptual, and cognitive procedural memory, have been distinguished [56]. Evidence has accumulated that motor skill learning (pursuit motor task) and perceptual skill learning (mirror reading), visual serial reaction, and reading skill are preserved in AD patients.

Hirono et al. [12] compared motor, perceptual, and cognitive skill learning abilities of mild AD with matched controls. In those who completed the task, skill learning was as good as in normal controls. The results support the view that patients with mild AD can acquire motor, perceptual, and cognitive skills and that the neural system subserving procedural skill is not related to the neural systems for declarative memory.

The medial temporal region, diencephalon, and basal forebrain are critical sites for episodic memory, and the role of the basal ganglia and cerebellum on procedural learning has also been emphazised. A functional network between the frontal association cortex, basal ganglia, and cerebellum has been proposed as a neural substrate for procedural learning. Perani et al. [57], in a study of 18 AD patients, demonstrated that glucose metabolism in the left and right basal ganglia, frontal association cortex, and cerebellum were related to skill learning. Neuropathological changes in AD, including neurofibrillary tangles, senile plaques, and neuronal and synaptic loss, affect predominantly the association areas, hippocampus, and amygdala, resulting in severe cognitive and episodic memory impairment. The structures subserving procedural memory, i.e., the basal ganglia and cerebellum, were less involved in early AD.

Since cognitive functions in patients with AD are generally defective, the capacity to solve the tasks is apparently critical for acquiring the skills. If the task requires more than the patient' s residual cognitive function, i.e., if it is too difficult, and the patient's cognitive impairment is too severe, skill learning is unlikely. Relatively preserved cognitive function is probably more crucial in cognitive or perceptual skill learning than in motor skill learning. Knopman and Nissen [58], in a study using serial visual reaction time tasks, demonstrated that patients with probable AD showed a preserved perceptual procedural learning ability.

Eslinger and Damasio [7] found preserved motor learning capacity in patients with AD and no correlation between procedural learning and global intellectual level.

Zanetti et al. [11] evaluated the efficacy of a procedural memory stimulation program in mild and mild-moderate AD. Twenty basic and instrumental activities of daily living were selected and divided into two groups comparable for difficulty. Ten normal elderly subjects (age: 68.0 + 4.8 years; MMSE score: 28.7 + 0.9; education: 7.6 + 3.5 years) were asked to perform in the two groups of daily activities, and the time required to perform the tasks in each group was recorded and used as a reference. Ten mild and mild-moderate AD patients (age: 77.2 + 5.3 years; MMSE score: 19.8 + 3.5; education: 7.3 + 4.7 years) without major behavioral disturbances constituted the experimental group. Patients were evaluated in all 20 daily activities and the time employed was recorded at baseline and after a 3-week training (1 hour/day, 5 days/week). Five patients were trained during the 3 weeks on half of the 20 daily activities and the other five patients were trained on the remainder. This procedure was adopted in order to detect separately the improvement in "trained" and "not trained" activities, allowing better control of the effects of the intervention. Assessment of the functional impact of the training was directly measured through the variation of time employed to perform tasks before and after training. After 3 weeks of training a significant improvement was observed for the trained activities, from 3.6 to 1.9 SD below the performance of the normal elderly controls ($P < 0.05$). AD patients improved also in not-trained activities from 3.5 to 2.7 SD below the control performance ($P < 0.05$). These findings indicate that rehabilitation of activities of daily living through developing procedural strategies may be effective in mild and mild-moderate AD patients. It is noteworthy that the improvement was present also in 'not trained' activities, suggesting that functional achievements may be independent of the learning context. The reduction in the time required to perform daily basic and instrumental tasks could have a beneficial impact on patients' everyday life. These findings confirm the improvement, induced by training, of functions that rely on procedural memory strategies, reflecting the possibility of new learning and possibly changes in the underlying cognitive structures.

Ermini-Funfschilling and Meier [48] at the Memory Clinic in Basel provide long-term memory training as part of their "milieu therapy." Mildly demented outpatients attended weekly group sessions of 60 min for a special semistructured memory training. The goal is to relearn and practice strategies that help to preserve or extend the phase of autonomy and thus quality of life and self-esteem. Efficiency of the training on mood, global mental status, short-term memory, cognitive flexibility, and quality of life was tested, comparing the performance of groups of patients attending the memory training to non-treated matched controls groups after 1 year. Authors found that depression scores improved in the treated groups and worsened in the control groups, independently of diagnosis. Global mental status, as measured by MMSE, and performance on a figural and verbal fluency task, remained stable for the memorytraining groups and declined in the control groups. Short-term memory declined in both groups, however, significantly only in the non-treated groups. The individual quality-of-life schedule showed stable or improved scores for the treated group and lower scores for the control groups. These results suggest that memory training is effective in mildly demented outpatients in delaying the progression of cognitive deficits.

In two placebo-controlled studies, Israel et al. [1] examined the effects of either 160 mg/day of ginkgo biloba extractum (GBE) or piracetam 2.4 or 4.8 g/day, combined with a memory-training program in non-demented patients complaining of memory problems. The results of both studies suggest that nootropic drug treatment and memory training have an effect on different cognitive functions and, hence, are complementary. Some functions, such as attention/perception in the GBE study and learning in the piracetam study, seem to benefit from both treatments, suggesting a mutually potentiating effect of drug treatment and training [59].

Performance of procedural motor learning tasks like grooming, bathing, dressing and other basic activities of daily living provides sensory stimulation, gross and fine motor coordination, reinforcement of older learning, reality orientation, and reinforcement of the self-image.

Studies on cognitive skills in AD patients are scarce, and results of those studies are inconsistent. For example, learning ability for jigsaw puzzle construction was reportedly preserved [57] or impaired [60]. The inconsistencies in these studies are likely to be attributable to cognitive heterogeneity in subjects and inappropriate tasks applied to cognitively impaired patients. This suggests that both selection of task and selection of patients are necessary for the study of skills learning in AD.

The relative sparing of implicit learning in AD constitutes good theoretical support for the stimulation of procedural skills. The theoretical background of study aimed at procedural memory stimulation settles into the lane of other widely used rehabilitation strategies aimed at improving the deficits associated with specific pathologies such as strokes, head injuries and, most extensively, aphasia [61–63].

Spaced-Retrieval Techniques

The spaced-retrieval technique is characterized by the retrieval of the same target information, such as face-name association, at successively longer retention intervals. If a retrieval failure occurs, the subject receives feedback and the retention interval is reduced to a previous interval at which retrieval was successful. Training with this method has been found to improve performance in AD patients across a variety of tasks [10]: object naming, face-name and object-location associations, and prospective memory for future actions. In general, the size of the gains has been quite impressive. It has been speculated that implicit memory processes (memory without awareness) may be involved in spaced-retrieval learning.

Use of External Memory Aids

The other category of cognitive intervention yielding positive results in AD concerns training by utilizing various external memory aids, such as diaries, signposts, alarm clocks, or other concrete reminders. Training with this method has been found to improve performance in demented patients in memory for personal facts, and memory for appointments [10].

In a recent home-based study, it has been shown that the use of external memory aids resulted in greatly improved conversation behaviors in Alzheimer's patients [64].

Reminiscence Therapy

Reminiscence therapy is defined as the act of thinking about past experiences, especially those considered personally significant [65]. In 1961, Butler was the first to propose the therapeutic aspects of reminiscence. Prior to that, reminiscing was believed to be pathological. Reminiscing is now believed to be instrumental in resolving past conflicts and in maintaining social roles, self-esteem and goal-directed behavior in old age.

McMahon and Rhudick [66] described three types of reminiscence: story-type reminiscence involves remembering factual memories for their own pleasure. Life review is a naturally occurring process characterized by the return to consciousness of the past experiences in order to resolve past conflicts. Halo-effect reminiscence involves the recollection of a particular situation over and over and over due to overwhelming guilt and despair.

Reminiscence can occur either as an individual or in group sessions. In individual reminiscence, the therapist aids the patient through an already occurring self-analysis in order to make it more conscious, deliberate, and efficient. Group reminiscence allows individuals to re-experience their past through the lives of others. Although many authors make reference to the therapeutic use of group reminiscence, none indicates specific guidelines to follow when running such groups [67–69].

Like validation, reminiscence therapy is strongly based on psychodynamic theory [70]. It is based on Erikson's eight stages of human development, ego integrity versus despair [71], which occurs when individuals review their life during adulthood and evaluate it as being either primarily positive or negative [72]. The ability to find pleasure and meaning in life upon reminiscing will determine whether an individual attains ego integrity or despair.

Modified reminiscing groups present the patient with a theme for discussion or a picture or object that is pertinent to the patient's life history. This technique usually involves physical objects that patients can see, hear, handle, taste, and/or smell while listening or discussing. The patient is encouraged or helped to verbalize, share, or just enjoy silently any memories that are elicited. Such experiences can help a patient rediscover missing parts of his or her personal history and can be used for both cognitive and affective dysfunction.

Validation Therapy

For the severely disoriented patients who fail to respond to the attempts at reorientation, an alternative approach can be the validation therapy (VT) proposed by Feil in 1967 [73–76]: through empathetic listening, the therapist attempts to discover the patient's view of reality in order to make meaningful emotional contact; the aim is not the awareness of reality but the understanding of the personal meaning underlying an individual's behavior. However, there have not been any well-designed studies assessing the effectiveness of VT.

Feil introduced VT in 1966 after finding that her attempts with reality orientation therapy with the old-old were causing the clients to withdraw even further [76]. VT is a process of communicating with the disoriented elderly by validating and respecting their feelings in whatever time and place is real to them, even though it may not correspond with our reality. VT focuses on the emotional rather than the factual content of what is said. VT can be carried out individually or in regular group sessions.

Individual VT sessions occur daily for 5–20 min, depending on the individual's level of disorientation. The therapists sit close to the patient and speak in a clear, warm, loving voice. They use exploring words such as "who," "what," "where," and "how" to elicit information; moreover, they use genuine empathy to validate only those feelings which are expressed.

In group VT, 5 to 10 disoriented patients gather for discussions meant to elicit the universal feeling of anger, separation, or loss. These topics allow the individuals to explore unresolved feelings and conflicts. By verbalizing their memories and thoughts, and by having them validated, the patient gains a feeling of being accepted. The group should meet at least once a week at the same time and place for between 20 and 60 min. Each group member will have a specific role for every meeting. All groups include four parts: music, talk, movement, and food.

VT is strongly influenced by psychodynamic theory [77], although Feil herself claims to be eclectic, and there is some evidence to suggest that VT adopts a humanistic approach. Humanistic theories believe that an individual's subjective view of the world is more important than objective reality [19]. This belief fits well with VT, as it focuses on the emotional rather than the factual content of what is said. Feil added a ninth stage, resolution versus vegetation, to Erikson's developmental framework in order to incorporate the old-old confused patient [76]. Feil describes this stage as an opportunity for individuals to achieve integrity by justifying and resolving past conflicts before dying.

Remotivation Therapy

Remotivation is a technique used for withdrawn patients with both cognitive and affective disorders[43]. It is done in a very small group or, initially, one-on-one, it is highly structured and repetitive, and it initially places little demand on the patient for active participation and interaction. It is short, uses a well-defined beginning and end, and involves presentation and limited discussion of a theme that reflects an aspect of reality. The main goal of this approach is to bring a patient out of social isolation [40].

Environmental Structuring

The environment is a powerful participant in shaping behavior. The basic purpose of environmental structuring, i.e., building memory aids into the environment, is to compensate for functional deficits by rearranging or changing the patient's physical environment [78, 79]. Environmental structuring may include something as simple as hanging a wall calendar or as complex as architecturally redesigning a facility. Effective use of this approach depends on an understanding of the patient's deficits and an analysis of how the physical environment can enhance or replace lost skills. The use of environmental adaptation can be broadened to include not only physical functioning, but also sensation and perception, memory, communication, and social interaction. The home or institutional environment can partially compensate for sensory-perceptual loss by providing response facilitators such as large print, better lighting, high-contrast coloring, sound-proofing and visual blocks to eliminate distraction. Memory aids can be built into an environment, such as a wall calendar, daily schedules, seasonal decorations, name plates, pictures of significant others and significant events, familiar furniture, favorite memorabilia, pictures or facsimiles of favorite pets, and favorite music. Common areas can be structured for socialization so that patients mix and face each other. The home environment may need adaptation just as much as institutional environments.

Milieu Therapy

The social environment also needs to be structured to enhance skills or delay loss of skills, as functional skills that have no opportunity for practice will deteriorate, needlessly hastening the general decline of the patient [48]. Milieu therapy looks at the non-physical environment of the patient and seeks to restructure it in order to provide opportunities to support functional behaviors [80, 81]. Individual environments will present specific advantages and disadvantages, but each environment can usually provide some opportunity for skill practice. The first step is to identify the range of tasks in which patients can be involved. These tasks will address a number of skill areas, including physical, cognitive, sensory, perceptual, affective, and social functioning. The tasks are then matched with a patient's needs, skills, and interest as closely as possible. The patient is given responsibility for a chosen task and is given assistance as necessary for successful completion of the task. The patient-task match will need to be reviewed regularly and adjusted to correspond to changes in the patient's condition. A milieu therapy program can be limited or quite complex, depending on the patient's needs and the resources of the caregiver or the facility.

Movement, Exercise, and Sensorimotor Treatment

Except with very impaired or immobilized patients, movement in its various forms provides an activity patients can participate in [82]. Simple movement can help work off excess, undirected energy that shows up as agitation or pacing behavior [13]. Walking can be performed as an activity in itself. Other physical activity can be generated through ADLs, such as baking and cleaning. Exercise offers a more focused approach to movement and requires greater ability to comprehend verbal and visual directions, to sequence, and to motor plan. Sensorimotor treatment may include automatic and spontaneous movement, planned movement, and multisensory stimulation in one-on-one or group sessions [83]. When performed in a group format, opportunities for spontaneous and structured social interaction will also be present.

Relaxation

Relaxation may be used specifically when anxiety and agitation are present and may also be included as part of an overall program [13] to address the problems, such as memory loss. Standard verbal and imagery techniques may be beyond the patient's grasp. Techniques need to include more non-verbal material, such as light massage of neutral body areas, quiet music, tapes of soothing environmental sounds, rocking chairs, slow-paced but repetitive movements, and car or van rides. Activities will need to be adapted to fit the patient's response and attention span.

Stimulation-Oriented Treatments

These interventions cover a number of less often used therapeutic approaches [84, 85]. These approaches include music therapy, dance therapy, and pet therapy, along with other formal and informal means of maximizing pleasurable activities for patients. They have modest support from clinical trials for improving function and mood, and common sense supports their use as part of the humane care of patients with dementia.

Leisure, recreational, and craft activities for the patient with dementing illness can be used as treatment goals in themselves or can be used as a means to achieve other goals [13]. It is apparent that activities must be geared to the needs, skills, and interests of the patient. The range of activities can be as broad as the range for other populations, and the activities can be performed in a variety of settings.

Behavioral Therapy

The principle of behavioral therapy offers a powerful way of maintaining positive behaviors and decreasing negative behaviors [86]. The behavior-oriented approaches identify the antecedents and consequences of problem behaviors and institute changes in the environment that minimize precipitants and/or consequences. Even when rational thought and verbal communication are impaired, these principles may still be able to shape behavior. These approaches have not been subjected to randomized clinical trials [14], but are supported by single-case studies and are in widespread clinical use.

References

1. Israel L, Melac M, Milinkevitch D, Dubos G (1994) Drug therapy and memory training programs: a double-blind randomized trial of general practice patients with age-associated memory impairment. Int Psychogeriatr 6: 155-170
2. Young J (1997) Rehabilitation and older people. BMJ 313: 677-681

3. Camp CJ, McKitrck LA (1992) Memory interventions in Alzheimer's-type dementia populations: methodological and theoretical issues. In: West RL, Sinnot JD (eds) Everyday memory and aging: current research and methodology. Springer, Berlin Heidelberg New York, pp 155-172
4. Backman L (1992) Memory training and memory improvement in Alzheimer's disease: rules and exceptions. Acta Neurol Scand 39[Suppl 1]: 84-89
5. Camp CJ, Foss JW, Stevens AB, Reichard CC, McKitrick LA, O'Hanlon AM (1993) Memory training in normals and demented elderly populations: the E-I-E-I-O model. Exp Aging Res 19: 277-290
6. Floyd M, Scogin F (1997) Effects of memory training on the subjective memory functioning and mental health of older adults: a meta-analysis. Psychol Aging 12: 150-161
7. Eslinger PJ, Damasio AR (1986) Preserved motor learning in Alzheimer's disease: implications for anatomy and behavior. J Neurosci 6: 3006-3009
8. Josephsson S, Backman L, Borell L, Bernspang B, Nygard L, Ronnberg L (1993) Supporting everyday activities in dementia: an intervention study. Int J Geriatr Psychiatry 8: 395-400
9. Zanetti O, Magni E, Binetti G, Bianchetti A, Trabucchi M (1994) Is procedural memory stimulation effective in Alzheimer's Disease? Int J Geriatr Psychiatry 9: 1006-1007
10. Backman L (1996) Utilizing compensatory task conditions for episodic memory in Alzheimer' s disease. Acta Neurol Scand 165[Suppl]: 109-113
11. Zanetti O, Binetti G, Magni E, Rozzini L, Bianchetti A Trabucchi M (1997) Procedural memory stimulation in Alzheimer's disease: impact of a training programme. Acta Neurol Scand 95: 152-157
12. Hirono N, Mori E, Ikejiri Y, Imamura T, Shimomura T, Ikeda M, et al (1997) Procedural memory in patients with mild Alzheimer's disease. Dement Geriatr Cogn Disord 8: 210-216
13. Szekais B (1991) Treatment approaches for patients with dementing illness. In: Kiernat JM (ed) Occupational therapy and the older adult. A clinical manual. Aspen, Maryland, pp 192–219
14. American Psychiatric Association (APA) (1997) Practice guidelines. Work group on Alzheimer's disease and related disorders (Rabins P, Chair): practice guideline for the treatment of patients with Alzheimer's disease and other dementias of late life. Am J Psychiatry [Suppl] 154: 3
15. Teri L, Gallagher-Tompson D (1991) Cognitive-behavioral interventions for treatment of depression in Alzheimer' s disease. Gerontologist 31: 413-416
16. Taulbee LR, Folsom JC (1966) Reality orientation for geriatric patients. Hosp Community Psychiatry 17: 133-135
17. Folsom JC (1967) Intensive hospital therapy of geriatric patients. Curr Psychiatr Ther 7: 209-215
18. Folsom JC (1968) Reality orientation for elderly mental patient. J Geriatr Psychiatry 1: 291-307
19. Weiten W, Lloyd M, Lashley R (1990) Psychology applied to modern life: adjustment in the 90s, 3rd edn. Brooks/Cole, Pacific Grove
20. Holden UP, Woods RT (1988) Reality orientation, psychological approaches to the 'confused elderly', 2nd edn. Churchill Livingstone, Edinburgh
21. Woods RT, Holden U (1981) Reality orientation. In: Isaacs B (ed) Recent advances in geriatric medicine, vol 2. Churchill Livingstone, Edinburgh, pp 281-296
22. Taulbee LR (1984) Reality orientation and clinical practice. In: Burnside I (ed) Working with the elderly: group process and thechniques. Wardsworth Health Sciences Division, Monterey, pp 177-186

23. Edelson JS, Lyons WH (1985) Institutional care of the mentally impaired elderly. Van Nostrand Reinhold, New York, pp 50-76
24. Donahue EM (1984) Reality orientation: a review of the literature. In: Burnside I (ed) Working with the elderly: group process and techniques. Wardsworth Health Sciences Division, Monterey, pp 165-176
25. Hanley IG, McGuire RJ, Boyd WD (1981) Reality orientation and dementia: a controlled trial of two approaches. Br J Psychiatry 138: 10-14
26. Gubrium JF, Ksander M (1975) On multiple realities and reality orientation. Gerontologist 15: 142-145
27. Dietch JT, Hewett LJ, Jones S (1989) Adverse effects of reality orientation. J Am Geriatr Soc 37: 974-976
28. Hanley I (1981) The use of signposts and active training to modify ward disorientation in elderly patients. J Behav Ther Exp Psychiatry 12: 241-247
29. Hanley I, Lusty K (1984) Memory aids in reality orientation: a single-case study. Behav Res Ther 22: 709-712
30. Woods RT, Britton PG (1985) Clinical psychology with the elderly. Aspen Systems, Rockville, pp 215-249
31. Powell-Proctor L, Miller E (1982) Reality orientation: a critical appraisal. Br J Psychiatry 140: 457-463
32. Citrin CS, Dixon DN (1977) Reality orientation: a milieu therapy used in an institute for the aged. Gerontologist 17: 39-43
33. Baldelli MV, Pirani A, Motta M, Abati E, Mariani E, Manzi V (1993) Effects of reality orientation therapy on elderly patients in the community. Arch Gerontol Geriatr 17: 211-218
34. Brook P, Degun G, Mather M (1975) Reality orientation, a therapy for psychogeriatric patients: a controlled study. Br J Psychiatry 127: 42-45
35. Zepelin H, Wolffe CS, Kleinplatz F (1981) Evaluation of a year long reality orientation program. J Gerontol 36: 70-77
36. Salter CL, Salter CA (1975) Effects of an individualized activity program on elderly patients. Gerontologist 15: 404-406
37. Woods RT (1979) Reality orientation and staff attention: a controlled study. Br J Psychiatry 134: 502-507
38. Johnson CH, McLaren SN, McPherson FM (1981) The comparative effectiveness of three version of classroom reality orientation. Age Ageing 10: 33-35
39. Reewe W, Ivison D (1985) Use of environmental manipulation and classroom and modified informal reality orientation with institutionalized confused elderly patients. Age Ageing 14: 119-121
40. Williams R, Reeve W, Ivison D, Kavanagh D (1987) Use of environmental manipulation modified informal reality orientation with institutionalized confused elderly subjects: a replication. Age Ageing 16: 315-318
41. Baines S, Saxby P, Ehlert K (1987) Reality orientation and reminiscence therapy – a controlled cross-over study of elderly confused people. Br J Psychiatry 151: 222-231
42. Koh K, Ray R, Lee J, Nair A, Ho T, Ang PC (1994) Dementia in elderly patients: can the 3R mental stimulation programme improve mental status? Age Ageing 23: 195-199
43. Janssen JA, Giberson DC (1988) Remotivation therapy. J Gerontol Nurs 14: 31-34
44. Bach D, Bach M, Bohmer F, Fruhwald T, Grilc B (1995) Reactivating occupational therapy: a method to improve cognitive performance in geriatric patients. Age Ageing 24: 222-226
45. Zanetti O, Frisoni GB, De Leo D, Dello Buono M, Bianchetti A, Trabucchi M (1995) Reality orientation therapy in Alzheimer's disease: useful or not? A controlled study. Alzheimer Dis Assoc Disord 9: 132-138

46. Galasko D, Corey-Bloom J, Thal LJ (1991) Monitoring progression in Alzheimer's disease. J Am Geriatr Soc 39: 932-941
47. Breuil V, De Rotrou J, Forette F, Tortrat D, Ganansia-Ganem A, Frambourt A, Moulin F, Boller F (1994) Cognitive stimulation of patients with dementia: preliminary results. Int J Geriatr Psychiatry 9: 211-217
48. Ermini-Funfschilling D, Meier D (1995) Memory training: an important constituent of milieu therapy in senile dementia. Z Gerontol Geriatr 28: 190-194
49. Corkin S, Growdon JH, Nissen MJ, Huff FJ, Freed DM, Sagar HJ (1984) Recent advances in the neuropsychological study of Alzheimer's disease. In: Wurtman RJ, Corkin S, Growdon JH (eds) Alzheimer's disease: advances in basic research and therapy. Center for Brain Science and Metabolism Trust, Cambridge, MA
50. Schacter DL (1985) Multiple form of memory in humans and animals. In: Weinbergher N, Mc Gaugh JL, Lynch G (eds) Memory systems of the brain: animal and human cognitive processes. Guilford, New York
51. Heindel WC, Salmon DP, Shults CW, Walicke PA, Butters N (1989) Neuropsychological evidence for multiple implicit systems: a comparison of Alzheimer's, Huntington's and Parkinson's disease patients. J Neurol Sci 9: 582-587
52. Schacter DL (1990) Perceptual representation systems and implicit memory: toward a resolution of the multiple memory systems debate. In: Diamond A (ed) Development and neural bases of highest cognitive functions. Ann NY Acad Sci 608: 543-571
53. Keane MM, Gabriele DE, Growdoon JH, Corkin S (1994) Priming perceptual identification of pseudowords is normal in Alzheimer's disease. Neuropsychologia 32: 343-356
54. Pascual-Leone A, Grafman J, Clark K, Stewart M, Luo JS, Hallet M (1993) Procedural learning in Parkinson's disease and cerebelar degeneration. Ann Neurol 34: 594-602
55. Squire LR (1987) Memory and brain. Oxford University Press, New York
56. Squire LR, Frambach M (1990) Cognitive skills learning in amnesia. Psychobiology 18: 109-117
57. Perani D, Bressi S, Cappa SF, Vallar G, Alberoni M, Grassi F, Caltagirone C, Cipollotti L, Franceschi M, Lenzi GL, Fazio F (1993) Evidence of multiple memory systems in the human brain: a [18F] FDG PET metabolic study. Brain 116: 903-919
58. Knopman DS, Nissen MJ (1987) Implicit learning in patients with probable Alzheimer's disease. Neurology 37: 784-788
59. Debert W (1994) Interaction between psychological and pharmacological treatment in cognitive impairment. Life Sci 55: 2057-2066
60. Grafman J, Weingartner H, Newhouse PA, Thompson K, Lalonde F, Litvan I, et al (1990) Implicit learning in patients with Alzheimer's disease. Pharmacopsychiatry 23: 94-101
61. Cohadon F, Richer E (1987) Recovery of motor function after severe traumatic coma. Scand J Rehabil Med 17: 75-85
62. Basso A (1989) Therapy of aphasia. In: Boller F, Grafman J (eds) Handbook of neuropsychology, vol 2. Elsevier, Amsterdam, pp 67-82
63. Holland AL (1989) Recovery in aphasia. In: Boller F, Grafman J (eds) Handbook of Neuropsychology, vol 2. Elsevier, Amsterdam, pp 83-90
64. Burgeois MS (1990) Enhancing conversation skills in patients with Alzheimer's disease using a prosthetic memory aid. J Appl Behav Anal 23: 29-42
65. Gagnon DL (1996) A review of reality orientation, validation therapy, and reminiscence therapy with the Alzheimer's client. Phys Occup Ther Geriatr 14: 61-77
66. McMahon A, Rhudick P (1964) Reminiscing. Arch Gen Psychiatry 10: 292-298
67. Harwood K (1989) The effects of an occupational therapy reminiscence group: a single case study. Phys Occup Ther Geriatr 7: 43-57

68. Lewis M, Butler R (1974) Life-review therapy: putting memories to work in individual and group psychotherapy. Geriatrics 29: 165-173
69. Poulton J, Strassbeng D (1986) The therapeutic use of reminiscence. Int J Group Psychother 36: 381-397
70. Kovach C (1990) Promise and problems in reminiscence research. J Gerontol Nurs 16: 10-14
71. Boylin W, Gordon S, Nehrke M (1976) Reminiscing and ego integrity in institutionalized elderly. Gerontologist 16: 118-124
72. Santrock J (1992) Life-span development, 4th edn. Brown, Dubuque
73. Feil NW (1967) Group therapy in a home for the aged. Gerontologist 7: 192-195
74. Day CR (1997) Validation therapy. A review of the literature. J Gerontol Nurs 23: 29-34
75. Toseland RW, Diehl M, Freeman K, Manzanares T, Naleppa M, McCallion P (1997) The impact of validation group therapy on nursing home residents with dementia. J Appl Gerontol 16: 31-50
76. Feil N (1992) Validation. The Feil method. Feil, Cleveland
77. Goudie F, Stokes G (1989) Understanding confusion. Nursing Times 85: 35-37
78. Liebowitz B, Lawton MP, Waldman A (1979) Evaluation: designing for confused elderly people. Am Inst Arch J 68: 59-61
79. Skolaski-Pellitteri T (1984) Environmetal intervention for demented person. Phys Occup Ther Geriatr 3: 55-59
80. Coons DG (1978) Milieu therapy. In: Reichel W (ed) Clinical aspects of aging. Williams and Wilkins, Baltimore, pp 115-127
81. Szeikas B (1985) Using the milieu: treatment-environment consistency. Gerontologist 25: 15-18
82. Bower HM (1967) Sensory stimulation and treatement of senile dementia. Med J Austr 1: 1113-1119
83. Richman L (1969) Sensory training for geriatric patients. Am J Occup Ther 13: 254-257
84. Glickstein JK (1988) Therapeutic interventions in Alzheimer's disease. Aspen, Gaithesburg
85. Damon J, May R (1986) The effects of pet facilitative therapy on patients and staff in an adult day center. In: Foster P (ed) Therapeutic activities with the impaired elderly. Harworth, New York, pp 117-131
86. Eisdorfer C, Cohen D, Preston C (1981) Behavioral and psychological therapies for the older patient with cognitive impairment. In: Cohen G, Miller N (eds) Clinical aspects of Alzheimer's disease and senile dementia. Raven, New York, pp 209-226

GENTLECARE: the Prosthetic Life Care Approach to Providing Dementia Care at Home or in Institutions

M.J.D. Jones

Mother Theresa has been quoted in the London *Observer* (1971) as saying: "The biggest disease today is not leprosy or tuberculosis, *but rather the feeling of being unwanted.*"[1]

Elderly people living in our youth-oriented societies often feel ignored and unwanted. The specter of diminishing physical capabilities strikes fear in the hearts of young and old alike. Elderly people with dementing illnesses further disturb our perceptions of old age and how it should unfold. Cognitively impaired elders defy our expectations, disrupting and challenging our socially determined norms. They don't fit into the mold of "normal aging." More importantly, they don't fit into existing healthcare systems. For people with dementing illnesses there is no refuge ... not in their homes, in their communities, or in social or healthcare organizations. They become submerged in an overwhelming world of fear and anxiety, unable to affect the outcome of events, impotent to change their behavior, expelled from normal roles and relationships. They are condemned to a conspiracy of silence.

They do not suffer alone. Family members struggle for years with embarrassment and lack of information, helpless to assist their loved ones appropriately. Many have their lives destroyed by overwhelming guilt and exhaustion because they cannot alter the devastating outcomes of dementia.

Professional care providers do not escape the impact of dementing illnesses. They struggle with personal stress, physical injury, and job dissatisfaction. The escalating number of people requiring care, combined with decreasing levels of resources, requires that staff work more efficiently. To do more with less is the mantra of the 1990s and the emerging century. Overriding all of this effort is a miasma of nihilism – the troubling belief that nothing can be done to help the person with dementia. These issues are the challenges of dementing illnesses such as Alzheimer's disease.

The standard or traditional approach to the care of elderly people has evolved from a custodial model that focused on sustaining physical health for as long as possible, to a current-day biomedical model. The focus of the biomedical paradigm is centered on pathology and interventions that relieve symptoms and cure disease. This emphasis defines and shapes the systems we create and the attitudes and behaviors we exhibit as practitioners within this model. Reliance on the biomedical model, or its close neighbor, the psychiatric model of care, to explain and manage the experience of dementing illness has resulted in the *medicalization* of dementia.

Attempts are made to control the effects of dementia on human behavior with medication, when currently, despite exhaustive research, no pharmaceutical treatment has proven very helpful, and many exacerbate the primary behavioral problems. The biomedical approach overlooks the social construction of dementia in terms of the afflicted person's functional deterioration due to dementia, and the subsequent dynamics of the caregiving relationships, outside *and within* the health care system. Both the biomedical and psychiatric paradigms fail fully to take into account the impact of treatment contexts, environmental pressures, and demands on client performance.

Viewing dementia as a medical problem that must be "fixed" has led us to focus on what the person *can't do*, rather than placing emphasis on the individual and his/her residual abilities, and how these can be supported and re-enforced throughout the disease process. The medical perspective keeps us focused on the "problem behavior" and its management. Episodes of catastrophic behavior are viewed as an inevitable outcome of cognitive impairment rather than a controllable or preventable phenomenon. This narrow focus of the biomedical model causes many people experiencing dementia to fail to meet the performance requirements of most current healthcare systems and become the *unwanted* clients of those systems.

On the edges of this biomedical monolith, small, but hopeful trends in dementia care are emerging. Although many new and innovative care practices remain unevaluated, a substantial body of experiential and anecdotal evidence exists to support the belief that *individualized therapeutic care in specially designed environments* is a most effective way to help people live through the experience of dementia (Fig. 1). The **GENTLECARE Prosthetic Life Care System** is a powerful innovative approach that does exactly that. It provides a reasonable quality of life for the person afflicted with dementing illness, while at the same time it does much to protect the health of the family carer and the professional care provider. The system is based on the hypothesis that a "prosthesis of care" can be developed and implemented that compensates for deficits, while supporting and re-enforcing residual abilities of the afflicted person across the continuum of the illness.

Prosthesis... a device or artificial structure, either external or implanted, that substitutes or supplements a missing or deficit part.

Fundamental to the **GENTLECARE** system is a *prosthesis of care* consisting of three interacting elements – people, programs, and physical space. Operating in a synergetic fashion, these three elements provide maximum support of remaining skills and compensation, to the degree possible, for deficits in brain function. *For example*: The person experiencing dementia is often required to live in unfamiliar congregate living arrangements with strangers. Bedrooms, frequently with two or three occupants, are arranged in long corridors with each room identical to another. A common problem linked to dementing illness occurs when the

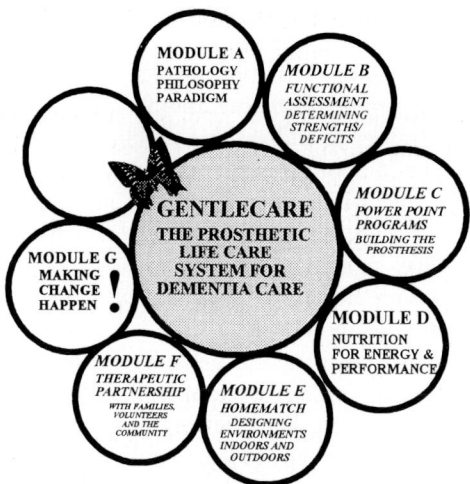

Fig. 1. Each circular module represents different aspects of the **GENTLECARE Prosthetic Life Care System**. This division helps facilitate learning by both lay and professional staff. (By courtesy of Moyra Jones Resources LTD.)

afflicted person cannot distinguish his/her room or bed from that of others. This causes untold difficulties with other clients, family members, and professional caregivers who exhaust time and resources redirecting people, de-escalating altercations, changing beds, and providing explanations to the public.

A **GENTLECARE** prosthesis to address this problem would consist of the following strategies. *All* personnel who might come in contact with the person experiencing dementia are expected to provide orientation and way-finding information. They also are responsible for ensuring transitional activities – that is, the person with dementia must be directed to and engaged in some activity, not merely given directions. The physical space of the bedroom would be clearly defined, either by color or familiar objects, or both.

Outside each door is a large (8" × 10") life panel consisting of a powerful picture of the person in a time frame appropriate to the stage of the person's disease, e.g., a wedding picture or family grouping, together with the person's name in large letters. The bedroom is developed as a bed-sitting room rather than a hospital room, with one, two, or three beds arranged against the walls, leaving common space available for use by all the occupants. Beds are also arranged in familiar positions so that the person's movement patterns from bed to bathroom to corridor are similar to patterns of performance before admission to a healthcare facility. Within the room, personal memorabilia are assembled to make the room as homelike as possible. These objects are not on display, but rather placed in use about the room similar to life-long habits in one's home. Objects include living plants, books, afghans, pictures, desks, chairs, footstools, favorite objects from home, tables, newspapers, bookshelves, etc. It is recommended that memorabilia be copied or duplicated, so that all occupants of the room may use the objects interchangeably. Precious pictures, jewelry or treasures should be replicated to prevent loss. This simple strategy of supportive people and a familiar environment combined with interesting activities can significantly alleviate the problem of cognitively impaired people mistaking another person's room for theirs. An additional successful strategy to address this problem is the use of the half-door or Dutch door. This type of door gives the occupant(s) of the rooms added security while allowing persons with dementia to "visit" or see the interior of a room without disturbing the occupants. Another strategy which discourages entry into the wrong room involves hanging vertical stripes of material in the entrance. These creative strategies on the part of family or staff help the impaired person to use their environment more appropriately.

GENTLECARE theorists postulate that behavior, the outcome of personal competence tested by disease and by environmental demands, can be shaped, changed, generalized, or extinguished. Viewing behavior of people experiencing dementia as *responses* to their environment enables care providers to relinquish performance expectations, and focus instead on their role as therapeutic agents. A GENTLECARE concept referred to as "mind bridging" is a strategy that encourages a carer, whether family or professional, to assist the person with their deficits in functioning, by substituting the *carer's own competence* for that of the afflicted person's, thus allowing the person to utilize fully his/her remaining function. *For example*: If the person with dementia is experiencing problems with memory or disorientation, all personnel, professional and lay, would be expected to provide information and orientation to compensate for the person's deficits. This prosthesis must be in place continuously. This requires a concerted and coordinated effort on the part of everyone. A "mind-bridge" needs to be available whenever the person is in distress or challenged by the environment.

This strategy is so simple and elegant that it frequently is considered too easy to be effective and is discarded or trivialized. When the concept is applied, however, observers often remark, "this is just common sense!" Indeed, the reality of the experience is that when people with dementia are appropriately assessed in terms of strengths and deficits, when they are surrounded by empathetic knowledgeable people, when provided with meaningful and interesting life activities, and housed in a residential environment that feels a little like their own home, magic happens! The prosthetic paradigm challenges not only the relentless loss of the person to dementia, but also the unnecessary dysfunction and expense of iatrogenic illness.

People

The GENTLECARE system begins with assembling and training *the people* who assume important roles in the afflicted person's life ... family members, professional care providers and community volunteers. Family members are not viewed as visitors or caregivers, but are instead encouraged to play out normal family roles.

"She is not my caregiver. She is my wife!" (A person with Alzheimer's disease).

They are engaged in a therapeutic partnership with staff and are viewed as integral to the therapeutic process. Volunteers and community organizations are encouraged to develop "mini-communities" within healthcare facilities with a view to maintaining the afflicted person's former life habits and activities.

Staff personnel, augmented by community volunteers, are considered principle therapeutic agents. Their skills, attitudes, language and creativity form the critical prosthetic component. The **GENTLECARE** treatment team includes all personnel, leisure skills staff, social workers, chaplains, rehabilitative personnel, housekeeping, nutritional services, administrative, and nursing care staff. In the **GENTLECARE** system *everyone participates* in direct dementia care. When the burden of care is shared equitably, significant positive outcomes can be achieved. Rather than relying on direct care staff to provide all aspects of care, non-nursing personnel engage in communication, orientation, walking programs, support of family and socialization with people experiencing dementia. They anchor or focus small groups of clients, freeing care staff to engage in individualized personal care.

The family carer, usually the spouse, is the most significant member of the support team. **GENTLECARE** focuses a great deal of energy on engaging their help and preserving their health throughout the process of the disease. Early referral to support networks and acquisition of knowledge about dementia at the beginning stages of illness are critical. Family members need to know as much about dementia as professional care providers, and they need to acquire the same skills. They are helped to design a modified prosthesis of care. In the home setting, *the people element* consists of the nuclear family, neighbors, community associates, home care personnel, and other health care professionals. **GENTLE-CARE** recommends the use of therapeutic companions to share the burden of caring for someone with dementia at home. All of these people must have special training and current knowledge about dementing illnesses. In terms of program, families are encouraged to keep their family member engaged in old familiar patterns of activity – housework, gardening, sports, and church activities. Friends and neighbors are encouraged to visit and engage the person in communication and exercise. Activities need to be modified according to the level of dysfunction. Every effort must be made to make the afflicted person feel useful, involved, and in control to the degree possible. Family caregivers are helped to avoid "smothering" or doing everything for the person, rather than helping them accomplish whatever aspects of the project that they can. The focus is on involvement rather than the end product. *For example*: In early stages of dementia the afflicted person may be able to continue with a beloved hobby like gardening. They may require help in initiating various steps in this complex activity, but often they can continue to plant and weed and mow lawns for several years. Eventually the person may not be able to use equipment or to prune, or plant the garden, but may be able to do more basic tasks of digging, watering, raking or mowing. As the disease progresses, the person may only be able to walk in the garden and converse about favorite plants or trees. As the disease nears final stages, the garden may need to be brought indoors in the form of plants, bouquets of flowers or fruit or vegetables. This simplification of any favorite activity over time can help the person with dementia to participate and enjoy life in spite of cognitive impairment.

The *physical space* of the person's home, of course, is the most therapeutic environment possible. Families should be encouraged to safe-proof the house by removing area rugs, placing gates in front of stairs, removing fuses from stoves, replacing precious objects with less fragile copies, where possible. Locks can be relocated higher on doors, and outdoor areas secured with attractive fences. This safe-proofing needs to be accomplished without removing familiar and comforting objects or condemning the person to hours of boredom. The exterior perimeter of the home should be secured to alleviate concerns about the person wandering into unsafe places. This allows the caregivers to relax. A person with dementia does not need access to all parts of the home. Areas like basements, guest bedrooms and dining rooms can be closed off. The basic areas of the house such as the bedroom, bathroom, kitchen, living room and outdoor garden should be accessible and made as interesting and comfortable as possible. Often the spouse needs a private space where he/she can regain energy and equilibrium. Not infrequently this is a separate bedroom.

With appropriate knowledge and assistance, families can often provide dementia care for many years in the home, sometimes even to the end of the disease, but they cannot do it alone. Every effort must be made to set up a system of home care at the time of diagnosis if the family caregiver is to survive the ordeal. The efforts of families and professional care providers alone simply are not enough to deliver effective dementia care. Individualized personal care requires trained and knowledgeable volunteers as well. Whether they be neighbors who volunteer to take the demented person for a walk, or someone who brings his/her pets into a facility, volunteers are indispensable! They can engage the persons in conversation, take them for walks, give massages, provide music, or demonstrate crafts, help with gardening, help create a homelike atmosphere, assist with cooking, take them for drives, read books, help them go to church, tell stories, assist with dances, hymn sings, send them mail, decorate for events, raise money, advocate for people who no longer can, assist the family caregiver, etc., etc., etc! Volunteers do not replace staff, but they do provide services that staff or the family carers do not have time to do. Volunteers help cognitively impaired people to retain old roles and familiar patterns of activity that help keep the person with dementia connected to his community.

Program

The **GENTLECARE** system focuses on every activity that engages the afflicted person's attention during his/her waking hours (Fig. 2). Based on Maslow's theory of motivation and personality [2], and taking into consideration Reisberg's Global Deterioration Scale [3], **GENTLECARE** utilizes a model of an ascending and re-enforcing spiral of activity, consisting of four levels of function.

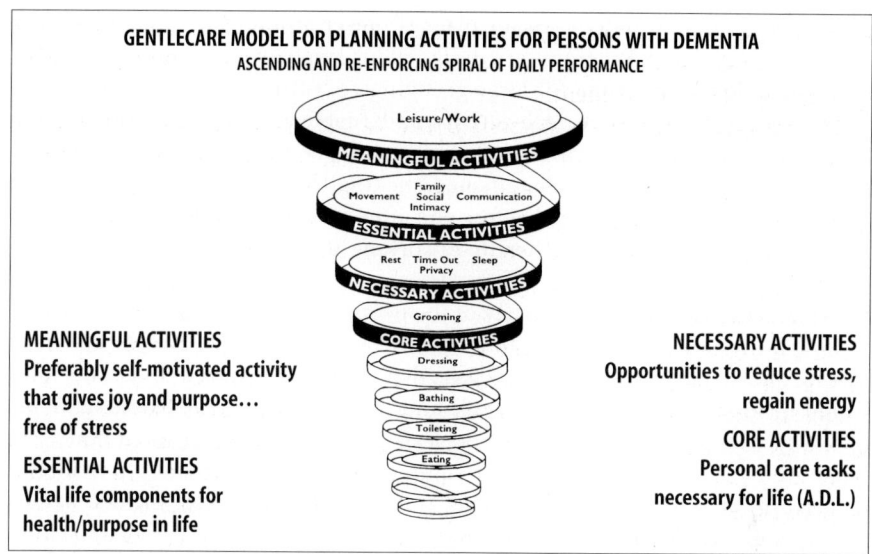

GENTLECARE MODEL FOR PLANNING ACTIVITIES FOR PERSONS WITH DEMENTIA
ASCENDING AND RE-ENFORCING SPIRAL OF DAILY PERFORMANCE

Leisure/Work
MEANINGFUL ACTIVITIES

Family
Movement Social Communication
Intimacy
ESSENTIAL ACTIVITIES

Rest Time Out Sleep
Privacy
NECESSARY ACTIVITIES

Grooming
CORE ACTIVITIES
Dressing
Bathing
Toileting
Eating

MEANINGFUL ACTIVITIES
Preferably self-motivated activity
that gives joy and purpose...
free of stress

ESSENTIAL ACTIVITIES
Vital life components for
health/purpose in life

NECESSARY ACTIVITIES
Opportunities to reduce stress,
regain energy

CORE ACTIVITIES
Personal care tasks
necessary for life (A.D.L.)

Fig. 2. The ascending and re-enforcing spiral divides activities into four levels of accomplishment, from the most basic activities of daily living (core activities) to the highest level of achievement (meaningful activities). Each subsequent level is supported by the preceding ones. (By courtesy of Moyra Jones Resources LTD.)

Within the most basic level of development are the five *core activities* of toileting, grooming, bathing, dressing and eating. The afflicted person may manage these activities, with or without help. The *necessary activities* of natural sleep, rest/relaxation or time out/privacy are interspersed between each of the *core activities*, reducing stress and allowing the afflicted individuals time to conserve energy and perform at their highest level of function. The *essential activities* represent the next higher level of challenge. They are communication, movement and social interaction including intimacy or touch. Essential activities give meaning to life. **GENTLECARE** practitioners overlay these undertakings over the basic activities of daily living depending on the client's ability to respond to challenges. The *meaningful activities* fall within the highest order of function. They include work, "doing things," play, and recreation. Afflicted people are helped to participate on whatever level they can manage that will bring them some joy and comfort. In advanced stages of dementia participation in meaningful activities is beyond the individual's ability.

The individual's daily care plan therefore emphasizes and supports self-care activities, communication, social relationships, legitimate intimacy, relaxation, and stress reduction. Activities are chosen that will support the person's remaining competence and former life roles. Diversionary, "keep busy" and token activities that are imposed on the lives of people with dementia are avoided. Such programs challenge the person inappropriately and diminish their spirit. **GENTLE-CARE** practitioners search for the essence of the individual, the elusive residual parts of functioning ability ... for the place of power or motivation that still exists within the individual. Programs are designed for small groups, similar to a family constellation, rather than for large groupings. Scrutiny is given to the level of energy required to perform each activity, and people are not challenged beyond their ability to perform. Activities such as eating, communication, bathing and walking are emphasized and lengthened, making them the central focus of the person's day. Rest is encouraged before and after each undertaking. Staff adjust to the individual's biorhythm rather than working to a pre-determined organizational schedule or program. The focus is on people rather than tasks. **GENTLE-CARE** practitioners join the journey of the people for whom they care.

Simple rather than complex activities are made available. Simple activities involve one-step, repetitive, physical tasks that emphasize learned skills. Examples would include cleaning, sweeping, polishing, raking, digging, and folding. Complex activities, conversely, require new learning, sequencing, abstract concepts, large-group social interaction, use of equipment and supplies, and the challenge to perform at levels that are no longer possible. Such activities include handicrafts, games, cooking, gardening, and sports. Rather than urge clients to participate by assembling them all in a room, activities are organized so that each individual can self-select what he/she would like to do. If staff or family members participate in an activity, they can gently urge the afflicted person to join in. People experiencing dementia appear to want to engage in activity, but lack the ability to initiate movement, communication, or ideas. Often encouraging them to assist or "help out" will provide the necessary impetus. They also enjoy simple watching activity without the pressure to participate or perform. The **GENTLECARE** system advocates a *5 + 10 Formula,* a listing of critical activities of daily living (ADL), which covers all four levels of activity, and are offered over a 24-h period. The five major ADL are eating, dressing, grooming, bathing, and toileting. The other ten activities include: walking, dancing, talking, singing, touching, "working," laughing, reminiscing, watching, and praying (if appropriate). These 15 activities, organized daily for each individual, form the care plan. Often individuals with similar interests and needs can be organized into small groups of three or four people to share the experience.

Sound nutrition is considered a powerful way to combat stress (Fig. 3). Once determined, each person's nutritional requirements are responded to with a *24-h nutritional clock.* The objective is to stabilize the person's metabolism, which reduces catastrophic behavior and improves performance. Eight small meals or "nutritional boosts" are offered over a 24-h period. Meals are served family style from heated containers. Each individual is offered a choice in food. Presentation of food on trays is discouraged as not re-enforcing familiar patterns of dining or providing necessary visual cuing. Clients are divided into similar functioning groups with an emphasis on harmonizing social combinations. Meals are served to small groups of people in several seatings rather than offering meals to a large group of people all at the same time. **GENTLECARE** emphasizes hydration programs offering 6–8 cups of water to individuals throughout the 24-h period. Ethnic and cultural practices are integrated into mealtimes.

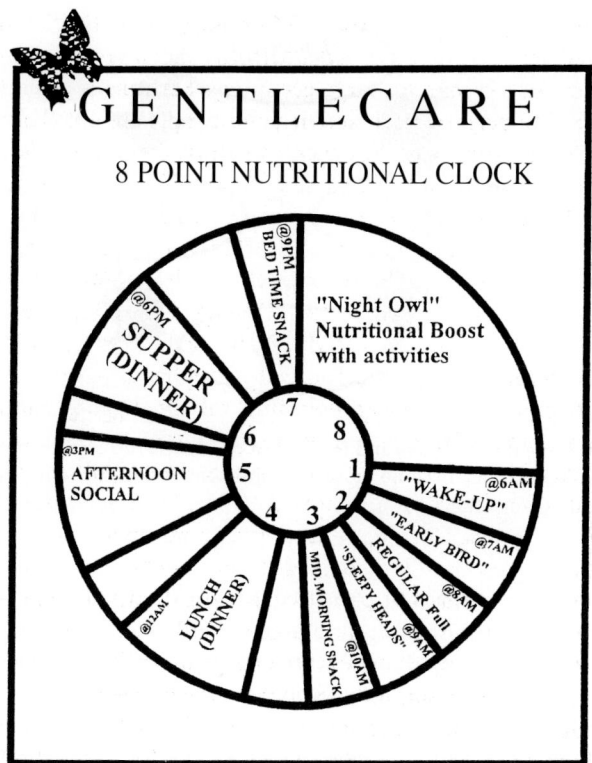

Fig. 3. Individual nutrition requirements are divided into 5, 6, or 7 small meals. The circle represents "nutritional boosts" aimed at maximizing nutrition, stabilizing metabolism and giving pleasure. (By courtesy of Moyra Jones Resources LTD.)

Since mealtimes are a major source of pleasure to elders, they are lengthened and enhanced for maximum therapeutic effect. Of the eight meals suggested by the **GENTLECARE** 24-h nutritional clock, three, the wake-up snack, breakfast, with its choices, and the nutritional snack for those who wake in the night, are extremely helpful to the client. Family participation and the social aspects of eating are incorporated into the dining experience. Eating becomes the principle therapy of the day.

Physical Space

Home is where the heart is. In each of our memories there is a special place that we regard as home. It may be the place where we grew up. It may be the first home we lived in after we were married, or the place we raised our children. It may even be a new location at the end of a busy lifetime. Home, in fact, may not be a build-

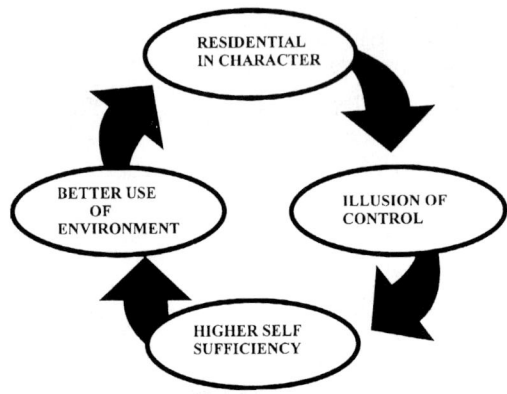

Fig. 4. The more the facility looks and feels like a residence, the greater the illusion of control by the residents. They become more self-sufficient and make better use of their environment

ing at all, but rather a space or treasured possession or a feeling of comfort and ownership.

The physical space of most healthcare organizations is designed as work space for staff. The challenge for elderly people *to live out their lives* in such an environment is daunting indeed. Excessive disability, dependency, and hopelessness result as vulnerable people struggle to combat the noise, rush, glare, confusion, and demands of an alien physical space.

Just as surely as architecture can dehumanize an environment, it can be a powerful healing force. The concept of creating *an enviro-match* with the individual is paramount in **GENTLECARE**.

The major assumption supporting **GENTLECARE** prosthetic environment is that clients would rather live in a place that reminds them of home than in a place that does not (Fig. 4). The more residential the ambiance, the greater will be the illusion of control the individual will feel in that environment. The more in control individuals feel, the higher their level of self-sufficiency will be. People in control function at higher levels than those who feel stressed and impotent. People who function at their highest possible level of competence tend to make better use of their environment. The greater the use of an environment, the more physical space will seem like a home. A re-enforcing cycle emerges that promotes health and well being.

On the other hand, the more institutional or technological in character an environment appears, the more an individual will feel a sense of diminished control (Fig. 5). This lack of control leads directly to a sense of reduced self-sufficiency. As individuals experience a loss of function, they tend to use the environment less effectively. As residents of institutions use space less frequently, that space begins to be used by staff members as work space. As more and more space is surrendered to staff to use, the building becomes more and more institutional in nature. The cycle of activity leads inevitably to problems of iatrogenic illness.

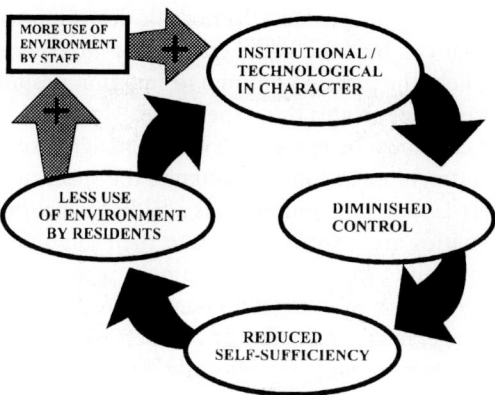

Fig. 5. If you design the staff in you will design the residents out. Space becomes work space rather than living space

To create an environmental prosthesis, a "zone of control" is developed for each person that resembles as closely as possible his/her former home. This control may extend over a bedroom, a piece of garden, a favorite chair, a special duty, or a familiar object. It must be readily available to comfort the person and be free of staff control. Facilities are designed or adapted to be residential in nature, and to resemble a place called **HOME.**

Clients are encouraged *to live in and use the space.* Staff, on the other hand, are required to work discreetly around the perimeter as if offering services in a person's home. The physical environment supports rather than challenges the dysfunctional person. The person's personal memorabilia serve as the basis of programming and daily activity. The environment as prosthesis encompasses secured parameters, easy access to an outdoor therapeutic garden, enhanced mobility and way-finding, and beautiful objects that are always available and free of staff control. Warm, mid-range colors and textures are recommended. Glare, shine, and patterns that cause distortions due to perceptual deficits are avoided. The environment encourages privacy, comfort, flexibility, choices, opportunity for change, and participation. There is lots of undocumented evidence that very little disruptive behavior exists in a prosthetic environment. In the **GENTLE-CARE** system, the environment is regarded as a "non-human caregiver," a form of care and interaction that does not require the intervention of staff and is always available.

On a personal level prosthetic environment consists of a carefully crafted *infrastructure* comprised of beautiful, familiar objects that we all enjoy in our daily lives. Books, magazines, plants, newspapers, antique objects, pictures, hats, jewelry, family photographs, cushions, hats, purses, clothing, quilts, tapestries, afghans, and baskets of interesting items are readily available throughout the environment to provide comfort to the person with dementia.

There is little that can be done to slow the horrific destruction of the Alzheimer victim's memory. What can be done is the creation of *an experience as powerful as memory* that can help the person continue to function despite his/her illness. The prosthetic environment creates this experience.

The prosthesis of care must be responsive to the changing support required by the client. The environment must contract, often to the space of a wheelchair or a bed. The focus of care is then directed towards avoiding sensory stimuli and providing comfort. Activities are broken down into their simplest components. As discussed, a life-long habit of gardening, for example, is continued by caring for a small plant by the bedside or enjoying a bouquet of fresh flowers. The person is supported with massage, music, pets, family contact, and spiritual comfort. The scale of the physical environment is significantly diminished to match the needs of the afflicted person.

Summary

The prosthetic model of life care for dementia offers some tantalizing outcomes that need to be researched. People offered this type of care appear to function at higher levels and for longer periods than others who do not receive this prosthetic care. Often they recover some of their ability to perform activities of daily living such as eating or dressing. They socialize more appropriately and experience fewer episodes of catastrophic behavior. Fewer altercations and incidents of noncompliance are observed. Family health improves, and satisfaction with care levels is reported. Families function more effectively as advocates and participate more freely in care. Professional care providers report feeling more competent, empowered, and satisfied with their jobs. Fewer staff injuries and less time away from work is reported. There are increased incidents of volunteer and community involvement. Cost savings are reported in terms of reduced staff requirement, decreased use of medications, reduced staff injury and time lost from the workplace. Despite these promising outcomes, there remains an urgent need for funding of projects to study innovative care strategies and their effectiveness.

The experience of living and working with people suffering from Alzheimer's disease or related dementia is challenging and often filled with despair, but there are moments of insight and clarity that open new pathways of direction or action. More often new thinking is inspired by the very people experiencing the disease rather than those of us who only observe their struggle. Diana Friel McGowin in her book *Living in the labyrinth* gives us such an insight on care requirements: "The Alzheimer patient asks nothing more than a hand to hold, a heart to care, and a mind to think for them when they cannot; someone to protect them as they travel through the dangerous twists and turns of the labyrinth" [4] (of dementia).

A great deal has changed due to the astonishing increase in knowledge about Alzheimer's disease and related dementias. And yet far too much remains the same. People with progressive dementias are discriminated against in healthcare systems that were designed for other purposes. Families are distanced and destroyed by lack of information and appropriate help. Professional personnel struggle to bring quality of life to elders in impossible organizational circumstances. Those of us who advocate for people experiencing dementia can only hope that the chaos swirling around dementia care will eventually evolve into humane and effective ways of responding to the needs of these vulnerable people. It is no longer a question of not knowing what to do, but rather asking ourselves whether the will exists to do things differently.

The Roman philosopher Seneca, dramatist and statesman suggested that things are difficult because we do not dare to change.

References

1. London Observer Newspaper, London, 3 October 1971
2. Maslow AH (1954) Motivation and personality. Harper & Roe, New York, p 80
3. Reisberg B (ed) (1983) Alzhemer's disease. The Free Press, New York, pp174-175
4. Friel McGowin D (1993) Living in the labyrinth. Delacorte, New York, p viii

Economics of Dementia

C. GEROLDI, A. BIANCHETTI, and M. TRABUCCHI

Introduction

In the last decade, direct and indirect costs of disease have been of great interest. Because of this we recognize that every choice in the health field is not an independent variable, but is strongly related to the economic resources available to a society and the historical context [1]. The great majority of studies, particularly those focused on chronic diseases, have suffered from important methodological problems. In studying the costs, one should not consider a disease to be an abstract entity, as if the clinical modifications were "in vitro" occurrences instead of a complex reality constituted by a puzzle of vital events. Therefore, a correct methodological basis is essential to compute the realistic costs of a chronic disease.

The costs of dementia have received growing attention in the last few years, because of the increasing number of persons afflicted. The prevalence of dementia in Western countries has not been precisely established. However, it is well known that it increases with increasing age, approximatively doubling every 5 years, ranging from 0.7% to 1% among people aged 60–64 years to about 40% among people 90 years of age and over.

In Italy, Alzheimer's disease (AD) is prevalent in 2.6% of the population over 60 years of age, and it occurs more frequently in women (3.4% vs 1.5% in men) [2]. A calculation based on the resident population in Italy in 1991 indicated that there are about 283,000 people who have the disease, a number that will increase in the future if the present demographic trends and the prevalence of AD do not change.

Dementia and, in particular, AD need specific care at the diagnostic, therapeutic and rehabilitative, and assisted-care level and the associated costs may be divided up accordingly. It should be stressed that although a diagnosis involves direct medical costs that are almost entirely paid by the National Health Care System, therapy and rehabilitation, defined as procedures aimed at optimizing the functional reserve of the patient at each time point of the illness, include direct and indirect costs that are sometimes difficult to quantify. The costs for therapy and rehabilitation have to be further divided among hospitals, rehabilitation facilities, nursing homes, and home care. Distribution of patients among these different categories varies with the severity of illness and may differ significantly from country to country and within a country, according to the services and resources locally available. It should be stressed that in Italy the general attitude is to keep the patient at home as long as possible with a share of costs between the family, social services, and National Health Care System.

Therefore, recognizing AD as a disease carries important consequences, as it affects the share of expenses between health and social services for both private and public health insurance programs [3]. Moreover, it affects the amount of resources allocated for research, the identification of risk factors, and the design of prevention programs.

At present, the resources are still somewhat scattered; there are no defined guidelines to help with the diagnosis or provide care for the affected person. Families and general practitioners are not aware of the resources and various structures available today to take care of AD patients. In Italy, a patient waits an average of 1.5 years after the first symptoms appear before seeing a physician. Then he is evaluated by various specialists before the diagnosis of AD, which in general is made by a neurologist or a geriatrician. This process is distressing and expensive for the families. Since resources are going to become ever more limited, models should be designed to make the diagnostic procedure easier and to organize the care system for the patient. An analysis of the benefit/cost ratio of each intervention will be mandatory, which also makes clear the importance of assessing the costs of AD.

Methodological Problems

Methodological issues have been defined in a recent review ([4]; see Table 1). In an economic analysis it is important to identify the perspective. The approach in a study on the costs of dementia may be selected from a society perspective or from a specific point of view (patient and his family, municipality or country council). Different perspectives may strongly vary the results of the analysis.

Furthermore, many factors are likely to add complexity to analysis of the costs of Alzheimer's dementia. Even if the criteria for the diagnostic recognition of AD are defined and the disease is clinically well characterized, AD patients present a great variety of symptoms and rates of progression [5]. The duration of the disease (from onset of the first symptoms to death), the appearance and the worsening of cognitive and functional deficits and behavioral disturbances are variable and only partially predictable in each patient. Therefore, events independent of the neurological disease can influence clinical progression. Concomitant somatic diseases, sociodemographic variables (age, gender, education), the personality of the patient and relatives, the caregiver's knowledge of the disease and the social and health care network have all been reported to have an impact on the onset of disability and behavioral disturbances and on the natural history of the disease [6].

The diagnosis of dementia is made roughly 1.5 years after onset of the illness, 4–5 days are needed for the initial diagnosis, and each year 40,000 new patients

Table 1. Methodological problems of analysis of the costs of dementia

1. Classification of the costs

a. Perspective
   – Society perspective
   – Committee perspective (patient, National Health Care System, social system)
   – Service provider perspective (health service company, home service enterprise)

b. Category
   – Direct costs
   – Indirect costs

2. Typology of the study

   – "Cost-of-illness" (COI) studies: studies focused on the total economical burden of
     dementia. These studies cannot analyze the value of different economical systems, but
     they are useful in analysis of health policy.
   – Studies of different systems of care: studies analyzing the costs of different systems
     of care and therapy.
   – Cost-effectiveness and cost-benefit studies: studies analyzing the costs related to clin-
     ical or economical outcomes.

3. Problems in comparing different studies

   – Diagnostic criteria for the disease
   – Features of the sample (number of subjects and inclusion criteria)
   – Study aspects (direct assessment versus estimation; duration of observation; catego-
     ry of costs)
   – Comparison of costs in different countries (change of money values; presence of ser-
     vices)

4. Methods for computation of informal costs

   – "Opportunity cost" method
   – "Replacement cost" method

are diagnosed as having AD. This amounts to an annual cost of US $ 3500 per
patient, an amount that can be significantly reduced in specialized centers allowing outpatient/day-hospital procedures for the diagnosis [7]. It should be stressed
that expenses are partially determined by the local availability of services. If no
specialized centers based on a day-hospital assessment program are available, the
patient will more likely be referred to a general hospital, increasing the costs.

Once the patient has been diagnosed as having AD, the costs will be greatly
determined by the severity of the illness, as more and more care is required at
later stages, finally leading to a totally dependent patient. The greater burden of
hospital care is due to concurrent diseases that increase dramatically the costs. In
fact, it should be emphasized that AD patients require more care than the average
age-matched non-demented patient. It has been calculated that the average AD
patient on each occasion of a concurrent disease will spend a period of 15–70 days

in the hospital. The expenses for monitoring patients who have already been diagnosed are calculated based on the assumption that 1 day every 6 months is needed to assess the progression of the disease. Behavioral disturbances also represent
a significant cost since they frequently require assessment and hospitalization in
a psychogeriatric or mental hospital. As dementia units in nursing homes are
being developed today, the housing and social services will require standards
higher (30%) than those for non-demented, non-self-sufficient patients, in particular for nursing staff [8]. The costs are roughly divided in half between medical and social services. The annual costs of home care for AD patients greatly
exceed the other associated costs. Costs can be divided into direct and indirect.
The direct costs consist mostly of social and nursing services. Indirect costs are
calculated as the equivalent number of hours spent for care by relatives at an
hourly cost of 6 US $. The loss of resources is calculated based on the fact that rel-
atives lose or quit their jobs, or their productivity at the working place decreases.
The time that is lost from daily activities, such as taking care of children, cleaning, cooking, etc. or volunteer services in the community, is also taken into consideration. Not included are costs that cannot be estimated, i.e., those linked to
pain, anxiety, suffering, or social distress. They are obviously of great importance
but cannot be evaluated in terms of money. Costs increase in the course of the illness, i.e., medical costs increase four times from phase I to phase II [8] (annual
total costs from US $ 156 million to > US $ 700 million), while social and indirect
costs increase up to over 40 times (annual total costs from US $ 130 million
to > US $ 5 billion). The expenses for drugs are relatively minor. For example, the
money spent for aids (such as diapers) equals that spent for drugs, representing
less than 1% of total expenses. Twenty percent of the cost allocated for drugs is
currently spent for sedatives and not for substances aimed at curing the illness. It
should be stressed that these data are for the pre-tacrin era.

New Models for Providing Care to AD Patients in Italy

It is predictable that providing help to families will allow a greater proportion of
long-term hospital-ward AD patients to be returned home instead of being sent
to nursing homes. What type of instruments is the Health Care System developing for the future to offer sufficient and qualified support for the medical needs
of demented people? In 1995, the local government of Lombardy (8 million inhabitants) launched a dementia care program characterized by 9 Alzheimer regional
centers, 60 dementia units of 20 beds in nursing homes, and 40 homecare organizations. The goal of the system is to provide a better and less costly diagnosis, to
help families to take care of the patient as long as they can, and to warrant a better quality of life after the institutionalization. The special care units (SCUs) are
specifically devoted to the long-term care of dementia patients affected by severe
behavioral problems, providing a physical environment adapted in a selected
area, an increase in staff, and a daily geriatric medical evaluation. A preliminary
study on the outcomes of SCUs showed that each SCU resident received 2.5 h/day

of nursing home care and 2.1 h/day of activity programs (occupational and physical therapy), whereas traditional nursing home residents received only 1.9 h/day of nursing home care, 1 h/day of activity programs, and a physician evaluation once every 2 days. Furthermore, after a 6-month stay in an SCU, 16 patients transferred from a traditional hospital were observed to have significantly fewer behavioral disturbances, and there was less need for physical restraint and psychotropic drugs. The mean daily cost for Italian Nursing Home patients is about 150,000 Italian lira (approximately US $ 100). For the SCU patients, a further 17,000 lira (approximately US $ 10), an increase of only 11%, has been paid by the National Health System, reaching significant clinical results [8].

Table 2 shows an analysis of the costs of the first Italian Alzheimer Regional Center, a multidisciplinary experimental care center that provides diagnostic evaluation and treatment mainly for elderly patients with recent onset of mental impairment or long-lasting dementia. The unit is designed to provide comprehensive assessment of medical, psychological, and social problems of the demented elderly and to provide therapy, rehabilitation, counseling, social, and legal and ethical support to the patient and family. The Regional Center is responsible for educational programs directed at staff nurses and caregivers, and for research programs. Involvement of the Regional Center in research should be stressed. It is our conviction that resources invested in research will bring a return in terms of awareness, new directions for diagnosis, detection of risk factors and prevention, and design of new drugs. The last 10 years of AD research have already shown progress. Finally, we believe that estimating the costs of AD care and the benefit/cost ratio of any intervention does not reflect a selfish attitude of a society toward a disabled patient, but is instead a desire to provide the best care in spite of limited resources and coping with medical ethics and economic constrictions.

Table 2. Analysis of annual costs (in US $) for a 40-bed regional center in Italy (inflated to 1993 value)

Staff (full time)	1,175,000
2 geriatricians; 2 neurologists; 1 psychiatrist;	
1 neuropsychologist; 2 biologists; 12 professional nurses;	
1 social worker; 2 rehabilitation therapists;	
6 occupational and cognitive therapists; 16 blue collar workers	
Diagnostic procedures	
Blood analysis	125,000
Neuroimaging	48,750
Other	15,600
Food and lodging	91,250
General functions	287,500
Total	1,743,100

The Results of CODEM (Cost of Dementia) Study

To provide detailed information on the problem of the costs of AD in Italy [7], a longitudinal study was conducted at the Alzheimer Center of Brescia, with the aim of measuring the direct and indirect costs in a sample of 100 community-dwelling AD patients followed for 1 year. The study was based on information from 103 AD patients (25 male and 78 female; mean age 77.7 ± 7.3) and their caregivers consecutively recruited from July 1994 to September 1995. The inclusion criteria were: (a) diagnosis of probable AD according to the NINCDS-ADRDA criteria [9], and (b) living at home at the time of inclusion in the study. The sample was balanced for Clinical Dementia Rating scale level to obtain a comparable proportion of patients in each of the three levels of the scale (0.5–1: questionable or mild dementia; 2: moderate dementia; 3: severe dementia) [10]. Patients were evaluated at baseline and after 6 and 12 months to assess cognitive and functional status, behavioral symptoms, and concurrent diseases. Cognitive function was assessed using the Mini-Mental State Examination (MMSE) [11]; functional status using the Activity of Daily Living (ADL) scale [12], and the Instrumental Activity of Daily Living (IADL) scale [13]. Depressive symptoms were evaluated using the Geriatric Depression Scale [14]. Behavioral symptoms were assessed using the Cohen-Mansfield Agitation Inventory [15]. Concurrent diseases were assessed using the Greenfield Index [16]. The patient's families were evaluated at baseline in order to obtain information on structure, socioeconomic status, employment, and house modifications. A social worker visited the caregivers every week for the first 3 months and every 2 weeks for the following 9 months, collecting data on all

Table 3. Mean weekly drug expenditure for AD patients at baseline (in US $)

	Daily mean	Annual cost	Proportion of the total expenditure
Nootropics	0.13	48.59	4.75%
Benzodiazepines	0.11	39.69	3.88%
Neuroleptics	0.11	40.15	3.93%
Antidepressant drugs	0.28	104.02	10.17%
Cardiovascular disorder drugs	0.32	118.17	11.56%
Pulmonary disorder drugs	0.05	17.57	1.72%
Gastrointestinal disorder drugs	0.25	92.62	9.06%
Vitamins	0.13	49.05	4.80%
Endocrine disorder drugs	0.03	11.86	1.16%
Non-steroidal anti-inflammatory drugs	0.06	21.22	2.07%
Genitourinary disorder drugs	0.01	2.28	0.22%
Topical drugs	0.08	30.11	2.95%
Antibiotics	0.88	320.52	31.35%
Other drugs	0.24	87.60	8.57%
Antitumoral	0.11	39.01	3.82%
All drugs	2.80	1022.46	100%

Table 4. Mean weekly and annual informal and formal caregiving hours and relative costs (in US $) based on 1-year longitudinal observations

Typology of care	Caring activities	Hours/week	Estimated weekly costs	Estimated annual costs
Informal care	Instrumental activities of daily living[a]	18.3	126.12	6,556.25
	Activities of daily living[b]	5.4	39.62	2,061.25
	Surveillance	81.0	690.00	35,906.25
	Nursing activities	–	–	–
	Total	104.7	855.62	44,453.12
Formal care	Instrumental activities of daily living[a]	2.4	16.81	875.00
	Activities of daily living[b]	0.5	9.08	471.87
	Surveillance	3.6	54.37	2,830.62
	Nursing activities	0.1	2.44	126.87
	Total	6.6	83.12	3,834.37
Total care	Instrumental activities of daily living[a]	20.7	144.19	7,500.00
	Activities of daily living[b]	5.9	48.75	2,533.12
	Surveillance	84.6	745.00	38,737.50
	Nursing activities	0.1	2.44	126.87
	Total	111.3	938.75	48,713.12

[a] Activities include: housekeeping, cooking, transportation
[b] Activities include: bathing, grooming, eating, mobility, drugs monitoring, medications

Table 5. Annual costs per person for AD in different countries (values in US $)

Authors	Ostbye [17][a]	Ernst [18]	Stommel [19]	Max [20]	Wimo [21]	Bianchetti [8]
Country	Canada	USA	USA	USA	Sweden	Italy
Year	1991	1991	1988	1988–1990	1991	1995–1996
No. of subjects	1,125	National estimation	182	93	National estimation	103
Duration of observation	One interview	Meta-analysis	3 months	12 months	Meta-analysis	12 months
Formal homecare	4,970	3,140	2,800	Na	27,500	3,608
Informal homecare	5,130	20,900	12,400	34,517	16,200	41,838
Daily centers	Na	Na	Na	Na	12,000	668
Nursing homes	19,100	7,570	Na	5,542	43,800	2,489
Acute ward	0	1,600	Na	Na	51,100	1,273
Rehabilitation ward	Na	Na	Na	Na	Na	1,951
Day hospital	30	Na	Na	Na	51,100	503
Drugs	240	Na	Na	Na	100	1,022 [b]
Parapharmaceuticals	Na	Na	Na	Na	Na	1,230

Na, data not available

[a] The cost has been computed as the difference between estimated costs in elderly people without dementia and elderly people with dementia

[b] The costs of the drugs prescribed for the dementia are US $ 232

direct and indirect costs (hours spent for caregiving, drugs and aids use, physician visits, examinations, hospitalization or day services use, other type of health or socialservices). All costs were calculated in Italian lira (for conversion to US $ we estimated 1 $ = 1,600 Italian lira). For determination of caregiving costs, we used information obtained from the national contract for domestic workers and the mean costs sustained by public health services in northern Italy (hourly wages for domestic workers: $ 6.25; for surveillance: $ 4.90; for patient care: $ 7.25; for nursing activities: $ 18.70). The paper is based on the information collected at baseline and during the first year of longitudinal observation.

The mean weekly drug expenditure at baseline was US $ 10.8 + 10.7; US $ 0.77 + 4.1 were spent for nootropics drugs, 1.48 + 1.9 for neuroleptics or benzodiazepines, and 2.2 + 4.3 for antidepressants (Table 3). The weekly total costs for patient care is 1,502,000 Italian lira (see Table 4). Data based on 1-year longitudinal observation show that 111.3 h/week were spent caring for the patients: 20.7 for housework, 5.9 for care of the person, 84.6 for patient supervision.

Table 5 shows the results of the CODEM study compared with other studies on AD costs carried on in different countries (Canada, USA, Sweden). The differences observed reflect the different periods of observation (in a span of time from 1988 to 1996), the different methodologies and, most of all, the differences in organization and administration of health and social services in different countries. The CODEM study provided the first analysis of the social costs of AD in Italy, with longitudinal direct observation of their determinants. The CODEM data showed that the mean annual costs for each AD patient in Italy were 93 million lira, and that informal care represented about 76% of these costs, whereas direct health costs were only about 11%. This clearly demonstrates the relevance of indirect costs (loss of personal resources, loss of employment, house modifications for informal care, etc.) in determining the costs of AD and shows that these represent a direct influence on the families.

Acknowledgements. The CODEM study has been accomplished thanks to a grant by Bayer SpA Italia.

References

1. Levorato A, Rozzini R, Trabucchi M (1994) I costi della vecchiaia. Mulino, Bologna
2. Rocca WA, Hofman A, Brayne C, et al (1993) Frequency and distribution of Alzheimer's disease in Europe: a collaborative study of 1980-1990 prevalence findings. The EURODEM-Prevalence Research Group. Ann Neurol 30: 381-390
3. National Institute on Aging (1993) Progress report on Alzheimer's disease. NIH publication no 93-3409
4. Wimo A, Ljunggren G, Winblad B (1997) Costs of dementia and dementia care. Int J Geriatr Psychiatry 12: 841-856
5. Geldmacher DS, Whitehouse PJ (1996) Evaluation of dementia. N Engl J Med 335: 330-336
6. Bianchetti A, Trabucchi M (1994) L'impatto economico per la demenza: una ipotesi per l'Italia. Giorn Geront 42: 687-692

7. Trabucchi M, Govoni S, Bianchetti A (1995) Socio-economic aspects of Alzheimer's disease treatment. In: Giacobini E, Becker R (eds) Alzheimer disease: therapeutic strategies. Birkhäuser, Boston, pp 459-463

8. Bianchetti A, Trabucchi M (1996) Dementia care. J Am Geriatr Soc 44: 1277-1278

9. McKhann G, Drachman D, Folstein M, Katzman R, Price D, Stadlan EM (1984) Clinical diagnosis of Alzheimer's disease. Neurology 34: 939-944

10. Hughes CP, Berg L, Danziger WL, et al (1982) A new clinical scale for the staging of dementia. Br J Psychiatry 140: 566-572

11. Folstein MF, Folstein SE, McHug PR (1975) Mini-Mental State: a practical method for grading the cognitive state of patients for the clinician. J Psychiatr Res 12: 189-198

12. Katz S, Ford AB, Moskowitz RW, Jackson BA, Jaffee MW (1963) The index of ADL: a standardized measure of biological and psychosocial function. JAMA 185: 914-919

13. Lawton MP, Brody E (1969) Assessment of older people: self maintaining and instrumental activities of daily living. Gerontologist 9: 179-186

14. Yesavage JA, Brink TL, Rose TL, Andrey M (1983) Development and validation of a geriatric depression screening scale. J Psychiatr Res 17: 37-49

15. Cohen-Mansfield J (1986) Agitated behavior in the elderly. II. Preliminary results in the cognitively deteriorated. J Am Geriatr Soc 34: 722-727

16. Greenfield S, Blano DM, Elashoff RM (1987) Development and testing of a new index of comorbidity. Clin Res 35: 346A

17. Ostbye T, Crosse E (1994) Net economic costs of dementia in Canada. Can Med Assoc J 151: 1457-1464

18. Ernst RL, Hay JW (1994) The US economic and social costs of Alzheimer's disease revisited. Am J Public Health 84: 1261-1264

19. Stommel M, Collins CE, Given BA (1994) The costs of family contributions to the care of persons with dementia. Gerontologist 34: 199-205

20. Max W (1993) The economic impact of Alzheimer's disease. Neurology 43[Suppl 4]: S6-S10

21. Wimo A, Karlsson G, Sandman PO, Corder L, Winblad B (1997) Costs of illness due to dementia in Sweden. Int J Geriatr Psychiatry 12: 857-861

Ethical Problems in Caring for Demented Patients

G.J. Agich

Caring for patients suffering from dementia involves some of the most difficult ethical problems in adult medicine because of the nature of dementia and the way that dominant ethical principles apply to the clinical features of dementia. In this paper I first discuss the features of dementia that complicate the care of demented patients, including the nature of diagnostic and prognostic information; cognitive, communicative, and other deficits associated with the disorder; and the dependence on others that dementia induces in patients. Second, I discuss the problem posed for the ethics of caring for demented patients by the ideal of respect for patient autonomy that dominates contemporary bioethics, especially as this ideal is expressed in the rights to informed consent, self-determination, and decision-making.

Features of Dementia

Dementia is clearly a medical concept, but it is not a diagnosis. It is a clinical syndrome that can be caused by more than 55 illnesses, some of which are non-progressive [1]. Unlike senility, all types of dementia are treatable, at least with psychosocial interventions, which makes accurate diagnosis essential for determining appropriate treatment, for providing information regarding prognosis and possible genetic risks, and for advising patient and family regarding healthcare options [2, p. 330]. Ethical questions are thus raised regarding the initial workup and subsequent interactions with the patient and family, including the use of mental status screening, laboratory evaluation, genetic testing, computed tomography (CT) or magnetic resonance imaging (MRI) of the brain, neuropsychological testing, electroencephalography (EEG), cerebral spinal fluid examination, as well as tests for biological markers or human immunodeficiency virus (HIV) [2–8]. The usefulness of these tests for diagnostic, prognostic, or treatment purposes is not only a matter of clinical judgment, but also a matter of ethical judgment.

Deciding to use these tests involves questions regarding the particular diagnostic or treatment usefulness of the results obtained, coupled with the cost and benefits of the test both for the patient and the patient's family. Because dementia is a disease that compromises basic human capacities, families and patients are apt to experience the symptoms with shame. The symptoms represent a loss about which it is difficult to speak in our society. Providing a rational and scientific way to understand the processes involved in dementia is one way to demythologize dementia, but the language of medicine and science can only do so much. Physicians should avoid the temptation to replace the patient's family's affect-laden understanding of the symptoms of dementia with a rational account, because patients and families need an opportunity to "make sense" of their experience *in their own terms*. Opportunity for the patient and family to discuss their interpretations should be provided. This process can begin with a candid discussion of the advantages and disadvantages of various tests.

Because the onset of dementia is often insidious, professional help is often sought after other adaptations have failed. Seeking professional help is for many patients and families an admission of their own failure. Thus, it is important to assist patients and families to understand that their responses are less failures than creative efforts to deal with what is clearly an extraordinary experience. Families and patients may feel the need to bargain about such matters, wanting to maintain control as long as possible. A spirit of negotiation is a far better way to understand this process than simply that of disclosure of information [9]. Even though bioethics has tended to downplay these everyday aspects of patient and family encounters, there is good reason to insist that they are central for understanding the ethics of caring for patients with dementia and for long-term care [10–12].

The first well-recognized ethical question that must be faced in the management of patients with dementia is whether the patient should be told the diagnosis [13]. For the patient at advanced stages, it is not clear that this question of disclosure is even meaningful. After all, disclosure of information is required because information is relevant for informed decision making. However, if the patient is not able to understand information or not able to make informed decisions, then the requirement that diagnosis be told may be irrelevant, *at least for the patient*. Even if the patient cannot be informed, the patient's family or other legal surrogate does need to be fully informed insofar as they will be making decisions on behalf of the patient [14]. At earlier stages, informing the patient is more ethically relevant. Because dementia is an end-state of a process of deterioration, the patient should be told about the full prognostic significance of the earliest symptoms and signs. Even though this information may not affect treatment decisions, it will significantly affect the patient's personal planning. Informing patients and family of the diagnosis along with the degree of diagnostic certainty is the important first step in helping the patient and family to begin the process of accommodating the relentlessly emerging symptoms of dementia.

An important clinical point needs to be stressed, namely, that early stages of Alzheimer's disease dementia are often accompanied by depression. The effect of disclosure of information on the patient's mental status should always be considered, but this effect does not provide a basis for withholding information. Nothing in the ethical or clinical literature on dementia justifies withholding information. Instead, full disclosure is imperative, but full disclosure should be accompanied by careful clinical attention to the patient's psychological state and communicative abilities. That means that disclosure warrants and, indeed, implies a more careful assessment of the patient's mental status and psychological and communicative capacities.

The obligation to inform patients of the probable diagnosis of dementia and its effect on cognitive functioning is confounded by the fact that dementia patients, even at early stages, do manifest problems with memory and communication that make meaningful interactions more difficult. However, these difficulties do not excuse the physician from informing and discussing the symptoms and their likely consequences. Disclosing the prognosis is important not just for purposes of planning healthcare, but for allowing the patient to adjust life goals. In informing the patient and family of the diagnosis, it is important, especially in the case of a diagnosis such as Alzheimer's disease, to stress the "possible" or "probable" character of the diagnosis since the diagnosis is made clinically and there is a degree of uncertainty that cannot be avoided. In communicating with a patient exhibiting the symptoms of dementia, a physician should adequately differentiate depression from other associated psychological manifestations and seek to treat reversible manifestations of the disease process.

Although communication with dementia patients is difficult, it is not impossible. Communication techniques have been suggested that might assist the dementia patient to participate more meaningfully and more fully in communication [15]. However, as I discuss in the next section, there is no compelling reason to think that respecting patient autonomy requires that the patient make all decisions for him-/herself and if he/she is unable, then a surrogate should decide. The problem with standard autonomy-based approaches is that they require full participation in decision making by the demented patient, which is often impossible, and they assume an unreasonably high level of functional capacity, so high, in fact, that few "normal" people actually exhibit such capacity in everyday life. As a result, an emphasis on actual, rather than ideal autonomy seems more ethically defensible [10, 11]. Such an approach, which is discussed in the next section, consists in respecting those beliefs, desires, preferences, wants, and values that the patient *actually* exhibits at the time of decision making and that reflect the *formed identity* of the patient. The values that define the individual patient's own unique identity should comprise the basis for decision making, not some ideal standard of free, rational choice.

Thus, it is imperative that the physician know who the patient is. Knowing who the patient is thus becomes more than an empty ideal, but a concrete as well as a practical clinical requirement that physicians caring for demented patients should attempt to gain an understanding of the patient's own life narrative. Understanding who the demented patient is can be augmented if the physician helps the patient to establish a values history during early stages of dementia and involves families, friends, and daily caregivers to modify this history in light of the patient's own experience and behaviors as the patient continues to live his/her life [16–18]. This last point is essential, because there is a tendency to believe that demented patients are "not there" or "not all there," meaning that the patient has no meaningful experience. However, anyone who spends time with demented people, especially those who are less than severely demented, recognizes that some patients are aware that their intellectual capacities are declining because of a progressive disease, that their behavior is abnormal, and that other people are being affected by their condition [19]. Of course, at the other extreme, there are patients who have no awareness that anything is wrong. There is clearly no uniform response and a wide variation exists.

Beyond a variation in the recognition of their illness, there is also a spectrum of emotional response [20]. If respecting patients means anything at all, it must mean more than dealing with the patients' cognitive deficiencies. It must also mean understanding what these patients are experiencing and feeling. As J. M. Foley [19, p. 42] has pointed out:

> There must be recognition of the variability, from patient to patient, and from time to time in the same patient. It is important to identify functions that are lost, but even more so to identify functions that are preserved. ...We must recognize that individual demented persons have their own unique attributes and that, despite metaphors loosely thrown around, they each remain a person, with their own gratifications and frustrations, their own unique background, and their own unique destiny.

It is thus important to recognize that the patient suffering from dementia can often still experience the world and still operate with a formed self-conception that should be elicited. Ethicists like to talk about the importance of identifying the patient's personal beliefs and values, and some clinicians think that some sort of arcane truth needs to be gleaned from the patient. This is an unfortunate misperception, because identifying a patient's beliefs and values is nothing more than understanding who the patient is, what the patient's life narrative is, and identifying the things that matter most to the patient. Nothing special beyond an open style of communication and an ability to listen is needed without which the physician is then forced to deal either with a surrogate decision maker or to rely on guesses. Allowing the patient's beliefs and values to guide clinical decisions is far sounder than any other approach. To do so, however, requires that the physician learn who his patient is. For some severely compromised patients, the life story and sense of patient values will have to come from family or friends, but the patient's own present experiences should also be considered and the day-to-day caregivers are the best source of such information. No matter how the patient's values are identified, the ethical management of a demented patient requires that they provide the guidance.

Even when a diagnosis can be made with high degrees of accuracy, prognostication for dementia is usually imprecise. The nature of behavioral, cognitive, and functional impairments that might occur in the future as well as their severity is quite variable [13, p. 948]. However, the obligation to respect a patient is not satisfied simply by conveying accurate diagnostic information, but rather consists in a respectful, communicative openness to the patient that involves educating the patient over time, responding to patient and family questions and concerns, as well as identifying and addressing the psychological and emotional states and needs of the patient and the patient's immediate caregivers. The importance of communicative openness in the care of patients with dementia has been stressed by many commentators [9, 11, 21, 22]. One study of functional communication found that requests for clarification occur more frequently in conversations with early-to-midstage Alzheimer's disease patients than with well, elderly speakers, suggesting that these patients have some insight into their communication problems and seek clarification to compensate for their communicative deficits [23].

Clearly, disclosing information about dementia and maintaining communicative openness with a demented patient makes sense only to the extent that the patient exhibits communicative capacity. There is, however, considerable confusion about communicative capacity, decisional capacity, and competence that encourages physicians to avoid what are understandably difficult and often frustrating efforts to inform the patient and educate the family caregivers. One pitfall that should be avoided is the tendency to dismiss the possibility of meaningful communication with dementia patients because of a "global" assessment of competency [24]. Recent studies suggest that mental status tests, commonly used in Alzheimer's disease assessment, are insufficient for determining the capacity to consent [25–29]. Competence cannot be assessed by such tests. The problem lies less with the tests than with the concept of competence itself, which is hopelessly muddled.

Competence is always an *instrumental* concept. Without the specification of a goal or purpose, the term is vacuous and its use can promote much mischief. It might be better if the use of the term *competence* were abandoned altogether. What is at stake are clinical assessments of specific patient capacities, for example, the capacities to care for oneself, to make informed decisions, or to communicate one's decisions.

Because assessing capacity for medical decision making varies with each clinical situation, it is important to remember that although patients may not have a functional ability to handle legal or money matters, they may still have an adequate capacity for making decisions about their own medical care [30, p. 878]. As a progressive degenerative condition, dementia will involve a series of everyday clinical decisions. These decisions may seem insignificant when each is separated from the others, but they accumulate to make future treatment decisions appear to be foreordained or habitual. For example, a mildly demented individual who was able to make her own healthcare decisions may have consented to treatment and, indeed, requested medical and other assistance in the past, yet her previous consent does not imply consent for treatment of serious or life-threatening illness in the present. In the absence of advance directives or explicit discussion with the patient, it is simply uncertain what the patient would want under present circumstances. This means that it is particularly imperative that physicians initiate discussions about end-of-life care and care for acute illnesses with their patients as early as possible [31].

Communication about an illness as complex and potentially devastating as dementia needs to include family members. Both patient and family will require time to adjust to the diagnosis and prognosis and only after a period of time has passed be able to face fully the difficult living, financial, and healthcare decisions that need to be made. A corollary question is whether family members should be told in cases of Alzheimer's disease about the advances in genetic testing, for example, apolipoprotein E genotyping [13, 14]. Though not directly necessary for the management of the patient with dementia, such testing will increasingly be of interest to family members. The physician should advise family about the availability of testing, educate them about the clinical and prognostic significance of test results, and offer or refer for counseling. Ample time and opportunity should be provided for family to raise questions not only about the patient's care, but about the meaning of the illness and/or test results for the family.

The physician's role in these discussions should be supportive of both family and patient needs, but patient needs and preferences should predominate. In these circumstances, the physician can truly be a patient advocate, and the physician's role should be one of respecting the patient's residual autonomy and prior decision making, rather than simply respecting autonomy in the abstract by succumbing to the decisions of the patient's surrogate. Physicians should also remember that patients who refuse a recommended treatment and choose a different course that threatens their well-being are not automatically decisionally incapacitated. It is also true that when a decisionally incapacitated patient "agrees" to treatment, this "agreement" does not constitute *informed consent* because the patient is incapable of consent. If patients are unable to consent to treatment, then they are unable to refuse treatment and vice versa. Too often, physicians accept a decisionally incapacitated patient's acceptance of treatment as a matter of consent, but turn to surrogates for consent whenever the patient refuses!

One problem that is especially difficult in patients with dementia is the issue of proxy or surrogate decision making [32]. In many instances, patients will simply not be able to make decisions for their own medical care and family members will need to be relied on as proxy decision makers. Decisions about in-home nursing care, nursing home placement, hospitalization for acute illnesses and end-of-life decision making place the family in an ethically conflicted situation. Even though most families may have the patient's best interest and the patient's own preferences and values in mind to guide decision making, it is natural for them also to consider their own stress, financial cost and gain, and the harsh burdens of caregiving. Care of patients with dementia has been called a "36-hour day" [33], and its effect on caregivers should never be underestimated.

Ethics of Caring for Demented Patients

Phenomenologically, the loss that is experienced by the demented patient is not just a matter of the loss of cognitive abilities, rationality or of self-determination, but a loss of dignity in the eyes of others. As Rick Moody has expressed it:

> We cannot grasp the dilemmas of dementia in a case study or a snapshot at a single moment of time. It is the whole history of the disease, of the patient, of rela-

tionships, which is crucial. In the slow deterioration of Alzheimer's disease the erosion of real autonomy takes place long before major decisions come into question [21, p. 87].

If we focus on decisions, they are likely to coalesce at a point well into the process of dementia, a point at which patients have lost mental capacity and so become a *problem* for medical or family decision makers. Recognizing that the care of the patient is a problem is often accompanied by a sense of frustration or confusion in reaction to a crisis that breaks through the systems of denial that enable family caregivers to cope with the daily demands that dementia patients present. Understandably, dementia brings a wide range of emotional complications that contribute to the ethical complexity of these cases. This complexity, however, is apt to be overlooked whenever the specific decisions are made the focus of attention. This focus misses a crucial fact about dementia, namely, that it is not a state so much as a process, a process that brings with it the emotional and psychological entanglements of relationships. In fact, much of the erosion of a demented individual's sense of dignity may have occurred long before ethical questions, such as, "Who will be a surrogate decision maker?" arise.

Understanding the actual experience of the demented individual is critical before adequate ethics for the care of demented persons can be developed. Although dementia is not a disease, it brings with it the explanatory models of modern medicine, an explanatory model that seeks to control and to treat disease. Dementia, however, is largely refractory to medicine's interventions. As a result, because patients, families, and physicians take for granted the power of modern medicine, expectations of cure or alleviation of symptoms are apt to be present, though frustrated with regularity. This frustration will often focus on the physician or healthcare institution as family members exhibit anger and make demands for diagnostic tests and interventions that may be clinically inappropriate. The physician at whom such feelings are directed might be tempted to comply with requests for interventions if only to assuage the family's emotional needs. Such a response, however, is unjustified. It leads to poor-quality medical care and makes the patient a pawn in a much larger emotional game. Rather, the physician should identify family frustrations and expectations and work to educate the family about the symptoms and their likely cause. A sympathetic understanding of the family's emotional reactions can considerably advance a spirit of cooperation, which is needed in order to achieve the best possible care for the demented patient.

Dementia raises particularly difficult ethical questions when viewed in light of contemporary bioethics. Bioethics has focused on a relatively restricted range of principles or ideals in terms of which to understand the ethics of patient care. Bioethics tends to conflate the ideals of dignity and autonomy. In so doing, it insists that respecting a patient's dignity involves respecting that person's decision making autonomy. Thus, the moment of decision becomes a focus of deliberation while background elements of the situation, including human relationships, are lost from sight (21, pp. 86–87). This has encouraged attention to situations in which conflicting choices have to be faced and clinical decisions have to be made urgently. Under these circumstances, bioethics has stressed respecting patient rights, including the right to informed consent and refusal of treatment. Recognizing that many patients do not exactly fit this paradigm has led to discussion of advance directives for treatment and research, as well as the use of surrogates and decisional standards such as substituted judgment and best interests [24].

Dementia patients are especially challenging when viewed in these terms, because the ideal of autonomy focuses on decision-making capacity and is a central concern precisely when choice and decision are critically at stake. These decisional nodes [12] compare situations in which difficult ethical choices need to be made, where the alternatives are relatively clear, and where an assessment of the risks and benefits of choices and their outcomes can be made. Dementia creates at least two important problems for this paradigm.

First, the demented patient poses critical decisional dilemmas at points in time *after* the dementia is either well established or recognized. Interactions with the patient before the diagnosis of dementia or its recognition by family or other caregivers occurs are not encompassed within this paradigm. This paradigm thus does not take into account the phenomenology of developing dementia in which both patient and family pretend that things are still normal, not simply by way of denial, but by way of maintaining hope and of sustaining respect for the patient. As the disorder of dementia develops, the patients increasingly become incapable of maintaining their relationships with others. Hence, it is no wonder that family members exhibit anger, denial, and guilt. The lived reality of the early development of dementia is hardly encompassed within bioethics' autonomy paradigm. Instead, the principle of autonomy would have us believe that family decision makers are disinterested and rational in ways that do not reflect the frustrations, guilt, helplessness, and shame that are natural concomitants of dementia. The autonomy paradigm does not help us understand the suffering of the family caregivers or the patient. In fact, the principle of autonomy would have us believe that the person's self is still present; yet how can this be whenever a demented parent can no longer recognize her own children? How can the patient be herself when she behaves in ways that contradict her own deep beliefs and values? Clearly, the phenomenology of dementia, particularly its early development, is beyond the scope of the autonomy paradigm.

The patient does not suddenly lose autonomy, but the autonomy is compromised over time in ways that are often so subtle that they go unnoticed. Though not noticed with full attention, both patient and family are aware of these changes. This horizonal awareness leads to all sorts of coping strategies in which the patient, in particular, exhibits autonomous and, sometimes, creative ways of adapting to the changes in her experience of self, others, and the world. Making sense of it all is one central function of autonomy, namely, of constructing a world within which we can live meaningfully. This aspect of autonomy is far more basic and pervasive than free choice, which has been the dominant focus of bioethics.

Second, the autonomy paradigm usually focuses attention on the issues of by whom, under what circumstances, and in terms of what standards can healthcare decisions be made. Although autonomy seems to be a *prima facie* relevant principle for these matters, it is ill equipped to deal with the details of the concrete situation. The demented person *as demented* is not able to make competent decisions for himself, so a standard device is to look to a surrogate decision maker, usually family members to make decisions for the patient. This move to surrogates is justified, because autonomy is itself regarded in extremely abstract terms, which means that if the patient cannot exercise his autonomy, then someone else can and must do so for the patient. This is paradoxical to say the least, if autonomy means *self*-determination. Many commentators insist that devices such as advanced directives or prior expression of wishes, beliefs, and values soften this paradox. By using advanced directives or a patient's informal expression of wishes regarding treatment as a guide, autonomy is presumably preserved. Critics, however, have insisted that this approach is without warrant. For example, Rebecca Dresser [34, pp. 72–73] has argued that this subjective approach, which relies on what has been called *precedent autonomy*, is exceedingly problematic. Because precedent autonomy is typically directed toward a future situation that the individual has never confronted, its expression may not be what the person would in reality choose. Even if precedent autonomy were a compelling consideration, autonomy is not the sole value relevant to treatment decision making. Even some of the staunchest defenders of precedent autonomy recognize that autonomy can be set aside or violated, for example, to avoid inhumane denials of life-sustaining treatment [35], to avoid the death of a mildly demented but pleasantly senile patient [36], or, more generally, to protect an incompetent patient's welfare [37]. Each of these examples recognizes the ethical salience of considerations of beneficence or welfare as a counterweight to autonomy.

Considerations of beneficence or promotion of patient well-being is a well-recognized and historically long-standing value in medical ethics [38, 39]. In simplest terms, promotion of patient well-being means that the physician has an ethical obligation primarily to the patient and not to the patient's family or others. As a result, consideration of patient well-being or beneficence requires that the physician protect the patient and seek to enhance the patient's capacity and minimize the patient's suffering. This obligation is clearest and easiest to discharge in situations in which other individuals or institutions seek outcomes that conflict with the welfare of the individual patient. In the case of the demented patient, these considerations require that the physician focus on the patient's well-being and advocate what is clinically, medically, and psychologically in the best interest of the patient above other considerations. This value reminds us that not all families are caring and loving. The physician should be vigilant for evidence of abuse or neglect of demented patients. Elevating the principle of beneficence above other considerations, however, creates ethical difficulties both generally and specifically for the care of the demented patient.

In general terms, an uncritical and irresponsible commitment to beneficence can sanction extravagant and ethically unjustified resource use. A slavish commitment to the pursuit of individual patient welfare is also problematic ethically, because beneficence has tended to support physician authority and power over that of patients' families or society [40, 41]. Hence, it is not surprising that beneficence has been touted as a primary principle for patient care in fee-for-service, third-party paid healthcare, an arrangement that had difficulty seeing that the welfare of any individual patient is inextricably bound up with the welfare of others, especially those most intimately related to the patient [42].

Because beneficence screens out considerations other than that of the individual patient's welfare, it is difficult to face rationally competing claims for resources. Beneficence thus errs in the direction of legitimating the *best* and most expensive care for an individual patient without regard for the opportunity cost of such a commitment. For example, many patients and families in early stages of dementia want to avoid institutionalization. Families are motivated not only by love and respect for the patient to keep the patient functioning at home, but also are motivated by very real financial considerations. Slavishly pursuing an ideal of patient welfare can too quickly dismiss the real and tragic choices that have to be faced in balancing the various beliefs, goods, and values that are actually at stake.

Despite these problems, considerations of patient welfare *are* ethically important in the care of a demented patient. For example, beneficence can be understood to be predicated not on the individual's isolated and abstract good, but on who the individual is in his concrete individuality. For the demented patient surrounded by caring family members, individual identity is inextricably interlocked with that of the family. The importance of sustaining relationships over time and of recognizing the need that family members have for participating in the care of patients is an important way of respecting the welfare of the patient as an individual with a concrete historical and social identity. Thus said, deciding the adequacy of patient care within the home becomes a more delicate matter involving not simply pursuing the patient's *medical* well-being, but of considering the patient's *personal* well-being within the concrete context of the patient's life within the family. As a result, the ethics of caring for an individual patient quickly becomes an issue of the ethics of responsibly respecting the interests, emotional needs, and capacities of family members. Hence, judgment will be required in order to sort through the wide range of incommensurable values that will be at issue. Because these values and goods are incommensurable, no algorithm or rule can be written that can save the physician from participating in this ethically difficult decision making.

Conclusion

Dementia is an important area for bioethical analysis, because its features challenge and extend the scope of bioethics' central principles. Ethical problems arising from the nature of dementia remind us that the importance of autonomy is not to be found in philosophical arguments, but in the practical difference that actual autonomy makes in patient care. This difference is almost entirely contained in the way that autonomy directs our attention to the actual experience of the demented patient and the patient's family. Caring for demented patients is thus not only a clinical and medical challenge, but a challenge for bioethics as well.

References

1. Mayeux R, Foster NL, Rossor M, Whitehouse PJ (1993) The clinical evaluation of patients with dementia. In: Whitehouse PJ (ed) Dementia (Contemporary neurology series, vol 40). Davis, Philadelphia, pp 92-129
2. Geldmacher DS, Whitehouse PJ (1996) Evaluation of dementia. N Engl J Med 335: 330-336
3. Corey Bloom J, Thal LJ, Galasko D, et al (1995) Diagnosis and evaluation of dementia. Neurology 45: 211-218
4. Katzman R (1986) Alzheimer's disease. N Engl J Med 314: 964-973
5. Post SG (1994) Genetics, ethics, and Alzheimer's disease. J Am Geriatr Soc 42: 782-786
6. Post SG, Whitehouse PJ (1995) Fairhill guidelines on ethics of the care of people with Alzheimer's disease: a clinical summary. J Am Geriatr Soc 43: 1423-1429
7. Roses AD (1995) Apolipoprotein e genotyping and the differential diagnosis, not prediction, of Alzheimer's disease. Ann Neurol 38: 6-14
8. Whitehouse PJ (ed) (1993) Dementia (Contemporary neurology series, vol 40). Davis, Philadelphia
9. Moody HR (1988) From informed consent to negotiated consent. Gerontologist 28 [Suppl]: 64-70
10. Agich GJ (1990) Reassessing autonomy in long-term care. Hastings Center Rep 20: 12-17
11. Agich GJ (1993) Autonomy and long term care. Oxford University Press, New York
12. Agich GJ (1995) Actual autonomy and long-term care decision making. In: McCullough LE, Wilson NL (eds) Long-term care decisions: ethical and conceptual dimensions. John Hopkins University Press, Baltimore, pp 113-136
13. Drickamer MA, Lachs MS (1992) Should patients with Alzheimer's disease be told their diagnosis? N Engl J Med 326: 947-951
14. Post SG, Foley JM (1992) Biological markers and truth-telling. Alzheimer Dis Assoc Disord 6: 201-204
15. Volans PJ (1989) Psychological approaches to assessment and management in dementia. In: Katona CLE (ed) Dementia disorders: advances and prospects. Chapman and Hall, London, pp 174-191
16. Gibson JM (1990) National values history project. Gerontologist 31: 447-456
17. Doukas DJ, McCullough LM (1991) The values history: the evaluation of the patient's values and advance directives. J Fam Pract 32: 145-153
18. Kane RA (1995) Decision making, care plans, and life plans in long-term care: can case managers take account of clients' values and preferences. In: McCullough LE, Wilson NL (eds) Long-term care decisions: ethical and conceptual dimensions. John Hopkins University Press, Baltimore, pp 87-109
19. Foley JM (1992) The experience of being demented. In: Binstock RH, Post SG, Whitehouse PJ (eds) Dementia and aging: ethics, values, and policy choices. John Hopkins University Press, Baltimore, pp 31-43
20. Sacks O (1990) The man who mistook his wife for a hat. Harper Perennial, New York
21. Moody HR (1992) A critical view of ethical dilemmas in dementia. In: Binstock RH, Post SG, Whitehouse PJ (eds) Dementia and aging. John Hopkins University Press, Baltimore, pp 86-100
22. Post SG, Ripich DN, Whitehouse PJ (1994) Discourse ethics: research, dementia, and communication. Alzheimer Dis Assoc Disord 8[Suppl 4]: 58-65

23. Ripich D, Vertes D, Whitehouse P, Fulton S, Ekelman B (1991) Turn-taking and speech act patterns in the discourse of senile dementia of the Alzheimer's type patient. Brain Lang 40: 330-343

24. Buchanan A, Brock DW (1986) Deciding for others. Milbank Q 64[Suppl 2]: 17-94

25. Fitten LJ, Lusky R, Hamann C (1990) Assessing treatment decision-making capacity in elderly nursing home residents. J Am Geriatr Soc 38: 1097-1104

26. Grisso T (1986) Evaluating competencies: forensic assessments and instruments. Plenum, New York

27. Marson DC, Schmitt FA, Ingram KK, Harrell LE (1994) Determining the competency of Alzheimer's patients to consent to treatment and research. Alzheimer Dis Assoc Disord 8[Suppl 4]: 5-18

28. McKinnon K, Cournos N, Stanley B (1989) Rivers in practice: clinicians' assessment of patients' decision-making capacity. Hosp Community Psychiatry 40: 1159-1162

29. Weiner BA, Wettstein RM (1993) Competency and guardianship. In: Weiner BA, Weltstein RM (eds) Legal issues in mental health care. Plenum, New York, pp 273-308

30. Caralis PV (1994) Ethical and legal issues in the care of Alzheimer's patients. Med Clin North Am 78: 877-893

31. Sachs GA, Cassel CK (1989) Ethical aspects of dementia. Neurol Clin 7: 845-858

32. Gutheil TG, Appelbaum PS (1982) Substituted judgment: best interests in disguise. Hastings Center Rep 13: 8-11

33. Mace NL, Rabins PV (1991) The 36-hour day, rev edn. John Hopkins University Press, Baltimore

34. Dresser RS (1992) Autonomy revisited: the limits of anticipatory choices. In: Binstock RH, Post SG, Whitehouse PJ (eds) Dementia and aging. John Hopkins University Press, Baltimore, pp 71-85

35. Dworkin R (1986) Autonomy and the demented self. Milbank Q 64[Suppl 2]: 4-16

36. Rhoden NK (1990) The limits of legal objectivity. North Carolina Law Rev 68: 845-865

37. Buchanan A (1988) Advance directives and the personal identity problem. Philos Publ Affairs 17: 277-302

38. Beauchamp TL, Childress JF (1994) Principles of biomedical ethics, 4th edn. Oxford University Press, New York

39. Pellegrino ED, Thomasma DC (1988) For the patient's good: restoration of beneficence in health care. Oxford University Press, New York

40. Agich GJ (1990) Medicine as business and profession. Theor Med 11: 311-324

41. Agich GJ (1990) Rationing and professional autonomy. Law Med Health Care 18: 77-84

42. Agich GJ (1987) Incentives and obligations under prospective payment. J Med Philos 11: 123-144